PATTERN OF EVOLUTION

ENCYCLOPAEDIA OF EVOLUTIONARY BIOLOGY SERIES

PATTERN OF EVOLUTION

By
Richa Arora

ANMOL PUBLICATIONS PVT. LTD.
NEW DELHI - 110 002 (INDIA)

ANMOL PUBLICATIONS PVT. LTD.
4374/4B, Ansari Road, Daryaganj
New Delhi - 110 002
Phones: 23261597, 23278000
Visit us at: www.anmolpublications.com

Pattern of Evolution

First Edition, 2004

ISBN 81-261-1526-2

PRINTED IN INDIA

Published by J.L. Kumar for Anmol Publications Pvt. Ltd., New Delhi and Printed at Mehra Offset Press, Delhi.

Contents

Preface

All living organisms share a master plan of structural and functional organization because they arose from a common ancestor long ago by the process of evolution. The tremendous work conducted by scientists in this field has increased our knowledge of evolution many folds. In the limited space, it is not possible to go into the details. Hence, a brief and comprehensive account has been produced in the light of recent work and latest information available from various sources.

The present title is an attempt to present an account of what is known or what is postulated about the life, and the subsequent evolution of the main animal groups. Certain basic scientific facts are included, it is true, but the majority of the contents consist of hypothesis which are never likely to be experimentally verified; and the reader should treat the book as a source of facts, ideas and theories on which thought and discussion can be based.

The present title claims to be a complete, thorough and authentic exposition of the topics dealt with. Emphasis has been laid on introducing basic as well as modern concepts. The language of the book is kept simple in lucid style. The subject matter has been interwoven for keeping up interest and curiosity of readers.

Though care has been taken to verify all passages quoted, but it is hardly likely that, in so large a mass of material, all errors shall have been avoided. The author and the publishers would welcome as a favour any suggestions or corrections submitted by interested readers.

Thanks are due to all friends and colleagues whose continuous inspiration have initiated me to bring out the present work.

There can be no claim to originality except in the manners of treatment and much of the information has been obtained from the books and scientific journals available in the different libraries.

—Author

Chapter—1

Introduction

Species were classified into groups by biologists before Darwin, it was not by the recency of their common ancestry: classification before Darwin was non-evolutionary. Whatever was possible before Darwin must still be possible now, and the advocates of transformed cladism have seized upon pre-Darwinian classification to defend their own school. Thus Nelson, in a remark we have already noticed, wrote: 'I think we are rediscovering preevolutionary systematics; or, if not rediscovering it, fleshing it out' (quoted in Patterson 1980 [1982]. Nelson and Platnick (1981), and Patterson (1983) have developed the theme. Thus, Patterson remarks. 'Today [phylogenetic] trees are usually presented in Haeckel's sense, as pedigrees or images of phylogeny, rather than in the sense of von Baer and pre-Darwinian systematists, as images of hierarchical organization, though cladist systematists have learned to distinguish phylogenetic trees from cladograms.' He then illustrates three models of classification and writes, 'the last two examples, and other pre-Darwinian examples...reproduced by Nelson and Platnick (1981) show that these models have no *necessary* phylogenetic implication'. And then, 'the equation of monophyletic groups with natural groups ([three refs]) is another sign of convergence between modern systematics and the pre-Darwinian systematists' attempt to map or mirror a discoverable natural hierarchy'.

That Linnaeus, Cuvier, Owen, MacLeay, von Baer, and Agassiz managed to classify without the theory of evolution certainly does show that it can be done. But the practical possibility is not in doubt. What is in doubt is whether cladistic classification can be justified without the theory of evolution. Practice alone is not justification: to justify transformed cladism it is not enough to show that classifications existed before Darwin; it must also be shown that those classifications were cladistic, because otherwise they could only justify non-evolutionary classification in general, not transformed cladism in particular, and that their practitioners had justified their techniques by a valid argument. The methods of Linnaeus and the other pre-Darwinian taxonomists must have been cladistic, their philosophy transformed cladistic. Their justification, moreover, must be valid not only for the biology of 1800, but also for 1980, because we are interested in the validity of transformed cladism now, not two centuries ago. This chapter, therefore, will not be strictly historical. We shall, to be sure, try to understand what earlier taxonomists thought, but only in order to carry their ideas forward into the 1980s, to see whether they are viable now; a sympathetic understanding of the thoughts of earlier ages is not the main aim. To borrow Ghiselin's (1976) distinction, this chapter is not history, but criticism.

It would be disproportionate to cover all the important pre-Darwinian classifications, and we do not need to. We only need to study a sufficient range of classifications to establish ideas that will stand for them all. We shall consider the system of Aristotle, the essentialist classifications of Linnaeus and Cuvier, and the ideal morphology of the first half of the nineteenth century. We shall ask, for each of them, what characters they used in their classifications, why they had chosen them, and whether they classified according to a unique natural hierarchy, or just one of many possible ones. Then we can discover whether they practised a coherent transformed cladism or some revealingly incoherent non-evolutionary method.

Aristotelian Classification

In an essentialist classification, species are grouped according to their shared essences, rather than any of their other characters (which are called accidental). Essentialism, therefore, implies character choice—choice of essential characters—but if it is to be practical, it must spell out what the essence of a species is, and how it may be recognized. We can distinguish two kinds of essentialism: the teleological essentialism of Aristotle and the theological essentialism of his later exponents. In the former, the essence of a species X is its *substance,* the *what-it-is-to-be* an X. The substantial characters of a species contrast with its accidental ones: without its accidental characters it would still be an X; but without its substantial ones it would not. In theological essentialism, a species substance is the respect in which it resembles (is analogous to) God. The two concepts are related. The theological concept descended from the teleological one: it was developed in those commentaries upon Aristotle, such as by Avicenna and Aquinas, in which the old pagan was adapted to the theologies of the prophet and the Roman Church. And it is easy to see how the development took place. One need only identify the purpose in life of a species with the reason why God invented it, then recollect that God creates species as emanations from himself, and the two concepts collapse into one.

But despite the similarity of the two meanings, we shall be concerned with only one of them here, the teleological version. It is more important to us. It exerted a stronger influence on taxonomists, as we shall see in the case of Linnaeus and Cuvier. They might perhaps have summoned the theological argument to justify the teleological version they practised, but direct thought on the nature of God did not enter into their choice of characters. Moreover, one of the proposed (but inappropriate) justifications of transformed cladism, as we have seen, appealed to the authority of Aristotle. Teleological essentialism therefore has a double interest lacking in the theological form. As it happens, we shall cover another form of theological justification later under the heading of idealism.

We have met the difficulties of teleological classification twice already; but when we turn to the system of Aristotle we meet another kind of difficulty, a historical one. Historians do not agree, at least in the amount of detail we require, on what his system was, and how (if at all) he operated it. If Balme (1961) is correct, Aristotle did discuss the principles of classification and the characters of animals that might satisfy his principles, but he never formally put his system to work (or no attempt has survived). We therefore cannot discuss his system practically, and are forced into abstraction.

Balme (1961) identified three main rules of classification in Aristotle. The first is the one we have met; that groups must be defined by their substances rather than their accidental characters, where the substance of a species is its purpose in life, the reason why it is what it is. It is difficult, but not meaningless. The characters that define groups are called *differentia,* and in *Historia animalium* Aristotle discusses possible *differetia* in four main classes: bodily parts, life history, activities, and psychological characters. Activities have priority over bodily parts because, in his teleological system, parts are adapted to activities and not vice versa. Once the substance of a group has been identified, the next step is to convert them into a classification. This is the business of Balme's second rule, according to which the successively smaller subgroups are defined by the successive division of the substance of the larger group. The division must be achieved by dividing the characters that define the higher groups. He makes this clearest in the *Metaphysics* (Z, 1038a).

> But it is also necessary that the division be by the differentia *of the differentia;* e.g. 'endowed with feet' is a differentia of 'animal'; again the differentia of 'animal endowed with feet' must be of it *qua* endowed with feet. Therefore we must not say, if we are to speak rightly, that of that which is endowed with feet one part has feathers and one is featherless (if we do this we do it through incapacity); we must divide it only into cloven-footed and not-cloven; for these are differentiate in the foot; cloven-footedness is a form of footedness.

And why must we? Because, if we do not, we may differentiate accidentally rather than substantially. Here is how Balme explains the matter (1961 [1975]); '[Aristotle's] reason is that you would thereby be dividing not the genus (footed animals) but another one (feathered animals ...); consequently you could not show that the feathers...belong essentially to the species in question, because you have not declared these attributes to belong to the genus in the first place.' If being feathered is not substantial in the genus, it cannot be substantial in a part of it.

Another advantage of successive division is its convenience in classification. Without it, a species can possess the differentia of more than one higher group. For instance, if we had divided, at a high level, into *feathered* and *featherless,* and then divided *feathered* non-successively into *tame* and *wild,* a member of the group 'featherless' might well also be wild (or tame). This is not possible with successive division: if we divided the differentia *feather* into different conditions of feathers, no member of the group 'featherless' can possess the differentia. If non-successive division is used, a species can be defined only by listing all its differentia 'in the way that one unifies speech by a connective', saying 'feathered and tame'; 'tame' by itself does not define the species. Without successive division, as many differentiate are needed as there are divisions; but with successive division, only one differentia, the last one, is needed, because it implies all the others. Aristotle clearly regards this as an advantage of classification by successive division (*Metaphyics Z.*, 1038a), 'the *last* differentia will be the substance of the thing and its definition, since it is not right to state the same things more than once in our definitions; for it is superfluous. And this does happen; for when we say "animal endowed with feet and two-footed" we have said nothing more than "animal with feet, having two feet."

Successive division has its own interest, but it also matters to us because it leads Aristotle to his criticism of definition by absent characters (which he calls privations), which we have seen being used by modern transformed cladists in self-justification. The relevant part of Aristotle appears in his critique of what he calls dichotomous classification. A dichotomous classification is a simple division into two by a presence and an absence (e.g. *footed* and *footless*). Aristotle disapproves of privations in dichotomous classifications because they cannot be divided: 'there is no differentia of a privation *qua* privation; for there cannot be species of what there is not, for example of *footlessness*'. Actually, this criticism is in itself excessive, because, as Balme (1972), comments, 'the argument is severe, if not captious...it is hard to see why Aristotle does not allow us to bypass the privative differentia...and to continue dividing the higher genus—the animals that remain after footed animals have been marked off. For footlessness in an animal implies other locomotive arrangements.' But however that may be, Aristotle only objects to privations in simple dichotomous classifications, which he objects to anyway: he does not object to them as such. This will become clear after we have seen Balme's third Aristotelian rule.

'His third rule is to divide the genus by a plurality of differentiae, not by one at a time' (Balme 1961 [1975]). No single differentia, even if it is positive character, will have enough divisions to distinguish all the species. 'This is why one should divide off the one kind straight away by many differentiate'. But if many differentiate are used the objection to privations disappears, for the group of differentiate (including privations) can be successively divided: 'In this way too the privations will make a differentiation, whereas in dichotomy they will not'. In the system Aristotle recommends, he admits privations; only in the system he rejects does he object to them.

So much for Aristotle. Despite the appeal of Nelson, Platnick, and Patterson, he did not justify transformed cladism. Privations, for a start, are not ancestral characters, and even if they were the Aristotelian argument would not help. The criticism of privations applies to classifications with single differentia; but Aristotle recommends the use of multiple differentiae. It also presupposes successive division. Transformed cladism, therefore, must find a justification of successive division, and of division by single differentia, before it can appeal to Aristotle. There may be a justification of successive division in Aristotle's metaphysical distinction of substance and accident; but some work would have to be done, for that distinction is not obviously acceptable in modern biology. As for a justification of simple dichotomies of single differentia, I can only suggest that one might be discoverable in the Platonic classifications Aristotle was criticizing; but I shall leave it to others to explore that mysterious territory.

The transformed cladistic appeal to Aristotle, therefore, does not work. It is not enough to confuse ancestral characters with privations, and then fetch an inappropriate justification, from a system that is at best obscure and is in fact probably no longer tenable.

Aristotelian Essentialism in Linnaeus and Cuvier

Linnaeus, if we concentrate on his botany, classified with the characters of the reproductive system. Why did he pick them? Because they were essential characters, of high importance in the life of the plant. They also met the requirement of Aristotle's third rule, for they allow many differentiae. Linnaeus's classifications were thus essentialist in the Aristotelian sense (Cain 1958). The same kind of character choice can be seen in the classifications of Baron Cuvier, which, also belong in part in the Aristotelian tradition. Cuvier summarized his philosophy of classification in the introductory essay to *La règne animal* (1829 edn). He clearly distinguishes technique and justification. The natural classification of animals into classes, orders, families, and so on, he, like his contemporaries, calls, *une méthode.* He discusses what the best method would be. It would be essentialist:

> When the method is good, it does more than teach names. If the subdivisions have not been established arbitrarily, but are based on true and fundamental relations, on the essential similarities [*rassemblances essentielles*] of beings, the method is the surest means of reducing the properties of beings to general rules, of expressing them in the fewest words, and of fixing them in the memory.

The classification described by Cuvier is natural in the usual sense of this book: the groups share other properties besides their defining character: the classification is 'the best means of covering the properties of beings by general laws'. Cuvier does not strictly offer this in justification of his method, although he clearly thinks it a desirable property. But in any case, we have already seen that simple phenetic 'naturalness' cannot support an objective classification. Cuvier says that a 'good method' gives natural classification, but what is the 'good method' of recognizing characters? Cuvier explained all correlations of characters by functional necessity. The characters that would be correlated with the greater number of other characters would be the functionally most important ones, which, accordingly, should be used to define groups.

> We assiduously compare beings, under the direction of the principle of *subordination of characters,* which is itself derived from that of the conditions of existence. The parts of a being must all be mutually adapted; some traits exclude

> others, others necessitate them: when, therefore, we recognize one such trait in a being, we can calculate in advance those that will coexist with it, and those that will not. The parts, properties, or traits that conform with the greatest number of others, or, in other words, that have the most marked influence on the whole of the being, are called the *important characters*, the *dominant characters*; the others are called the *subordinate characters*; and there are degrees of each.

And how are the dominant, important characters to be discovered? Cuvier gives two methods:

> This influence of the characters is sometimes determined rationally, by considering the nature of the organ. When this is impossible, we use simple observation, and a sure means of recognizing important characters, one derived from their own nature, is that they are the most constant, and that in a long series of diverse beings, arranged according to their similarity, these characters are the last to vary.

Cuvier, therefore, like Linnaeus, sought the essential characters of organisms to classify them. He wished his classifications to be natural in the sense that they would cover many characters; he believed that functional integration was the reason of correlations among characters; he therefore chose characters of the greatest functional importance. This method looks like the exact opposite of evolutionary taxonomy, but in fact the two schools are similar. They appear to differ because, although both aim to identify functionally important characters, Cuvier appears to do so in order to define groups by them, and evolutionary taxonomy in order to avoid doing so. The difference, however, is more one of interpretation than practice. Both schools classify with characters that show little variation; they therefore define similar groups. Cuvier did explain the invariance of characters by their functional necessity, in contrast with the evolutionary taxonomists's talk of broad adaptation, or even non-adaptation; but Cuvier's philosophy could probably be incorporated into evolutionary taxonomy. It would be easy if functionally important characters were likely to be broad adaptations. The philosophy of evolutionary taxonomy could also be imported into Cuvier's method: Cuvier sought characters that would be likely to indicate natural groups, which is just what the characters of evolutionary taxonomy are meant to do.

Cuvier's technique of character choice can be seen at work in his classification of vertebrates (Cuvier 1829). What are the important characters here? He recognized four groups of vertebrates, 'characterized by the kind or power of their motions, which themselves depend on the quantity of their respiration'. The differences in the method of locomotion was the most important character, and the circulatory system closely associated with it. They were therefore used to divide the vertebrates (Cuvier 1829).

1. Mammals. Double circulation. Aerial respiration in lung only. Strong walkers and runners.
2. Birds. Double circulation. Aerial respiration in lung and air sacs. Fly.

3. Reptiles. Single circulation. Aerial respiration in lung. Crawl.
4. Fish. Double circulation. Aquatic respiration in gills. Swim.

He put the same principle to work on each group. For instance, 'the variable characters that establish the essential differences among the mammals' (Cuvier 1829), are the organs of touch, because they determine dexterity and skill; and 'the organs of *manducation*, which determine the nature of the animal's food, and entail not only everything relating to digestion, but also a mass of other differences, even including intelligence'. These two classes of characters divide the mammals into nine orders. The first five have separate fingers on their limbs. Of these five, the first three have three sorts of teeth, grinders, canines, and incisors: Bimana (i.e. man, hands on two limbs); Quadrumana (hands on four limbs); Carnaria (non-opposable thumb). The second two are the Rodentia (which lack canines), and then 'those animals whose toes are cramped, and deeply sunk in large nails, which are generally curved; they have no incisors…', that is, the Edentata. Then there are two hoofed orders, Ruminantia and Pachydermata; and finally, Cetacea, which lack posterior extremities, and Marsupialia. The classification illustrates, by the way, that Cuvier was not a full Aristotelian: the successive subgroups are not defined by successive division. Vertebrates were defined by the circulatory system, mammals by their feeding mechanisms.

Cuvier and Linnaeus classified by characters of functional importance. They were essentialists in the teleological, Aristotelian sense. As such, their philosophy was neither phylogenetic nor phenetic; but whether such a philosophy could be formulated validly for the 1980s is (as we have seen) an open question. At all events, it could not justify the non-evolutionary application of cladistic techniques, for it does not lead to the cladistic technique of character choice. The characters chosen by the teleological essentialist differ from all three modern schools. In interpretation, if not practice, essentialist classification is the exact opposite of an evolutionary classification. It differs from cladism, because except by coincidence, functionally important characters need not be shared derived characters. Aristotelian classification differs from numerical phenetic classification because Aristotelian taxonomists choose their characters: they exclude characters judged accidental or subordinate or unimportant; the numerical taxonomist, by contrast, does not choose particular kinds of character at all. The technique of teleological essentialism is not cladistic, its philosophy is not transformed cladistic. The fact that it was practical, in pre-Darwinian times, should provide neither comfort nor justification for that school.

Morphological Idealism: Secular and Divine

Cuvier's morphology, according to which each part of an organism is well designed for life, would have been more familiar in Great Britain in the 'argument from design' of the school of thought for which Paley's name may stand. Its influence did persist through the century, but by 1850 it had already become old-fashioned. The years from Cuvier to Darwin, and beyond, saw its replacement by what may be called idealist morphology. Idealism, in this context, is the doctrine that nature is constructed from ideal forms; in morphology, it becomes the doctrine that the forms of animals are variants of a limited number of plans (or ideas). Species will then show similarities because they embody the same idea, rather than because they have a similar functional design. Morphologists thus came to emphasize the characters of organisms that (they believed) were non-functional but appeared to conform to the 'plan' of the group: male animals, for instance, possess useless nipples, which are part of the mammalian plan, and whales possess a set of useless bones planned on a hind limb. Vestigial organs are positively functionless, but the same theory would also point to similarities—such as between the skeletal designs of a horse and a man, which do not appear to be functionally necessary: if a horse and a human had been independently designed they need not have shown such detailed similarity of form.

Transformed cladists, as it happens, have been keener to find their ancestors among idealists such as Agassiz and von Baer than essentialists such as Linnaeus and Cuvier. The present section is therefore perhaps more relevant to an analysis of transformed cladism than was the previous one. As usual, we must proceed by asking whether the philosophy, idealism in this case, can justify an objective school of classification and, if it can, what kind of character choice (and what taxonomic techniques) it implies. If it is to justify transformed cladism, idealism must offer a valid non-evolutionary objective principle of classification by shared derived characters. Let us see if it does. Classification, according to idealist morphology, should represent the plan of nature. Is there any reason to suppose that nature has a unique, ideal plan? And if there is, by what techniques should it be discovered?

In answer to the first question, there were three reasons. Two of them are mentioned by Agassiz, in his *Essay on Classification* (1859). The one favoured by Agassiz (1859) himself is that 'the divisions of animals according to branch, class, order [etc.] ...have been instituted by the Divine Intelligence as the categories of his thinking. Perhaps Agassiz's idealism was more theological than that of other morphologists: but MacLeay (1825) said he would classify by 'the plan by which the Deity regulated the creation', Owen (1848) by the plan 'on which it has pleased the divine Architect to build up certain of his diversified living works', and, according to the historian Ospovat (1981), 'the supposition that the natural system was God's plan was widely shared'. It was perhaps less often expressed in the lands of *Naturphilosophie* or *anatomie transcendente* than of natural theology; they may have preferred the other reason for a natural system of types, which Agassiz cast down into a footnote:

> It must not be overlooked here that a system may be natural, that is, may agree in every respect with the facts in nature, and yet not be considered by its author as the manifestation of the thoughts of the Creator, but merely as the expression of a fact existing in nature—no matter how—which the human mind may trace and reproduce in a systematic form of its own invention.

In theory, this second philosophy could come in two forms, naively empirical or Kantian; that is, the natural classification might be said simply to exist in nature (naive empiricism) or we might be forced to discover it because our minds are built the way they are (Kantian). The difference, however, need not concern us. We can take the two empirical philosophies together, as versions of phenetic taxonomy. According to both, nature has a single hierarchical plan, defined by phenetic similarity alone.

We know how to deal with that philosophy. A biologist in the early nineteenth century might, quite reasonably, believe in a natural phenetic hierarchy. The theory, however, is no longer reasonable. It was finally exploded by the critics of numerical taxonomy. Nature does not have a unique phenetic plan. Agassiz, Owen, von Baer, could discover a plan of nature all right, but it would be one plan among many; the techniques by which they defined groups must have been chosen (from this broad perspective) subjectively. Their classifications can no longer be justified by an 'empirical' philosophy.

If the empirical philosophy must be abandoned, what about the divine one preferred by Agassiz? It differs from what I called theological essentialism. In both systems, the groups of nature are thoughts of God; but in Aquinas's analogical essentialism, the groups are also analogies of God: the plan of nature is not only a thought of God, but God is thinking about himself. Aquinas did prove that God can only think about himself (God is the only perfect being; a perfect being can only think perfect thoughts; therefore God can only think about himself), but his argument is theologically contestable, as God arguably could think about a natural plan somewhat independently of self-contemplation. The idealism of Agassiz says only that the plan of nature owes its existence to divine thought; it does not have to adopt the essentialist corollary that God is *ipso facto* thinking about himself. But that theological distinction is not important here. All we are concerned with is whether Agassiz's divine philosophy can justify his system of classification. To do so, it (and any modern transformed cladist who seeks comfort in the classifications of Agassiz) must show three things: first, that God exists; second, that at any time God thinks unique thoughts; and third, that God's thoughts are discoverable. The detailed nature of the divine thoughts would then lead to a technique of character choice in classification. If Agassiz is to justify transformed cladism as well, he must also have demonstrated a fourth proposition, that God thinks *cladistic* thoughts. Let me comment on each of the four questions. The first is difficult and uncertain, and too large to enter into here; but the existence of God does sound a strange premise for a system of classification for the 1980s. The second, however, has been settled. The very first Christian philosopher, Philo, a Jew of Alexandria and a contemporary of Christ, proved in his work *On the Unchangeableness of God*, that God can never change his mind; and later thinkers have agreed with him. The third question in the most important for us. Can we discover the divine thoughts, and thus find the proper principle of character choice for biological classification? Here, indeed, is a difficulty. A school of classification, such as idealism, will employ a principle of character choice: but it should also tell us why, if God thought up the classified plan, did he choose *those* particular characters? Neither Agassiz, nor anyone else has even tried to answer this question. Agassiz undoubtedly thought God caused the plan; but he did not try to show that God caused the plan Agassiz chose to classify. Agassiz merely assumed it, and rushed on to the consequence that pleased him so much, namely, that if God thinks of the plan of nature, then as the taxonomist himself discovers the plan, 'his mental operations ...approximate the workings of the Divine Reason'. The taxonomist becomes divine. He feared 'such a suggestion may appear irreverent', but concluded it was not; he was, after all, writing in Boston, not Rome. For all the argument Agassiz gives to justify his choice of characters, almost any mental operation in a taxonomist could be divine. His character choice, in other words, was subjective. He had not thought about the problem.

Perhaps it was not unreasonable of him not to. He could just assume, reasonably if naively, that the plan of nature he discovered was the divine one. He thought that, as a matter of fact, there was only one phenetic hierarchy, not an infinity of them. The question of whether his plan was the divine one would not have occurred to him. It is an anachronistic question; but nonetheless necessary for our purposes. If Agassiz is to be the prophet of any modern school, he must have answered the questions of modern critics: he must have revealed why God thinks only about particular kinds of characters; otherwise we cannot know whether any particular classification corresponds to the divine plan, and we can only use divine idealism to interpret classifications that have been discovered with techniques of secular inspiration. The interpretation is built on the classification; it does not justify it. This being so, the theology is quite superfluous. It has no use in classification. Darwin, indeed, used to complain about this at the time: 'but many naturalists think that something more is meant by the Natural System; they believe that it reveals the plan of Creator; but unless it be specified whether order in time or space, or what else is meant by the plan of the Creator, it seems to me that nothing is thus added to our knowledge' (Darwin 1859 [1969, p. 399]: Gillespie 1979 discusses the history). But, as Darwin was probably aware, in natural theology biology was supposed to add to our knowledge of God, not vice versa. Theology was not intended to justify a school of classification; if anything it was the other way round. No wonder it cannot justify transformed cladism.

What should we conclude? From the authority of Philo and his followers, we can say that, if God does think about the plan of nature, he will think unique and unambiguous thoughts. Agassiz's philosophy, therefore, could in theory justify an objective system of classification. But the conclusion is theoretical only. Ideal morphology is in the position that the critics of cladism attribute to that school it may have an objective principle, but it cannot be realized because it lacks satisfactory techniques. But whereas we do have some techniques to discover evolution, morphological idealism really is practically useless. We have no idea how to discover the thoughts of God. The justification works only if we can show that our techniques do discover the divine plan: without that argument, our techniques are unjustified, and our classification subjective. Because we do lack such an argument, the divine interpretation cannot justify morphological idealism.

Nor can it justify transformed cladism. To do so, it would have to show that God, in his biological faculty, thinks *cladistic* thoughts: that he contemplates shared derived characters only. That has not, and (surely) cannot, be shown. The fate of morphological idealism, in its empirical and divine forms, only proves still further that no phenetic or 'empirical' system of classification can be justified. When it is performed, it must be (from our modern viewpoint) subjective.

So much for the philosophy of the ideal morphologists. What about their techniques? They may not have been objective, but they did exist. If Owen, MacLeay, von Baer, and Agassiz, were ancestral transformed cladists, their techniques must at least have been cladistic. But were they? The quickest route to an answer is to search their classifications for non-cladistic groups. If they were cladists, they should not contain such non-cladistic groups as fish, or reptiles, or primates. The third chapter of Agassiz' *Essay*, 'Notice of the principal systems of zoology', is a synopsis of the classifications of the time. Let us turn first to the 'embryological system' of von Baer. Does it have reptiles and fish? Indeed it does: And what of the 'physiophilosophical system' of MacLeay? That too has Pisces and Reptilia. So does Owen, in his 'anatomical system'. And so on. Agassiz summarizes 18 systems in all, from Linnaeus to Vogt, and they *all* possess the tell-tale non-cladistic groups: not one of them, it appears, was cladistically defined. Not one of them, even though they were all pre-Darwinian, was a transformed cladist.

The classifications produced by von Baer and others, however, are only prima-facie evidence. They are not conclusive. Cladism is a method rather than a classification, and if von Baer (for example) had defined the reptiles by a shared derived character, then although we should now think him mistaken (for Reptilia cannot be so defined), he should be counted as a transformed cladist. He could have the method right, but bungled its application. We must therefore look behind the groups, to see how they were defined. Figure 1.1 reproduces von Baer's divisions of the Vertebrata, and reveals two groups of fish, the Urodela, the Anura, and the Reptilia. They are all defined by negative, absent characters. Absent characters can be derived, but all of them in this case are ancestral. He even partly defines the birds by an ancestral absence. The same general point is true of the other taxonomists summarized by Agassiz. Their groups were neither cladistic, nor defined cladistically.

If the morphological idealists were not forerunners of cladism, what were they? Of our three main schools, morphological idealism was closest to evolutionary taxonomy. Evolutionary taxonomy defines groups by homologies, and seeks to recognize and exclude analogies. Homologies were first distinguished from analogies by morphological idealists, and for a comparable purpose (Swainson 1835; Owen 1848; Agassiz 1859). They too classified by homologies, not analogies. Homology, of course, did not then have its modern evolutionary meaning. But it was still a principle of character choice. A principle must distinguish the true characters of organisms, which may be used to define groups, from the false ones. The homologies were the true characters, because they realized the ideal plan of nature; analogies were not part of

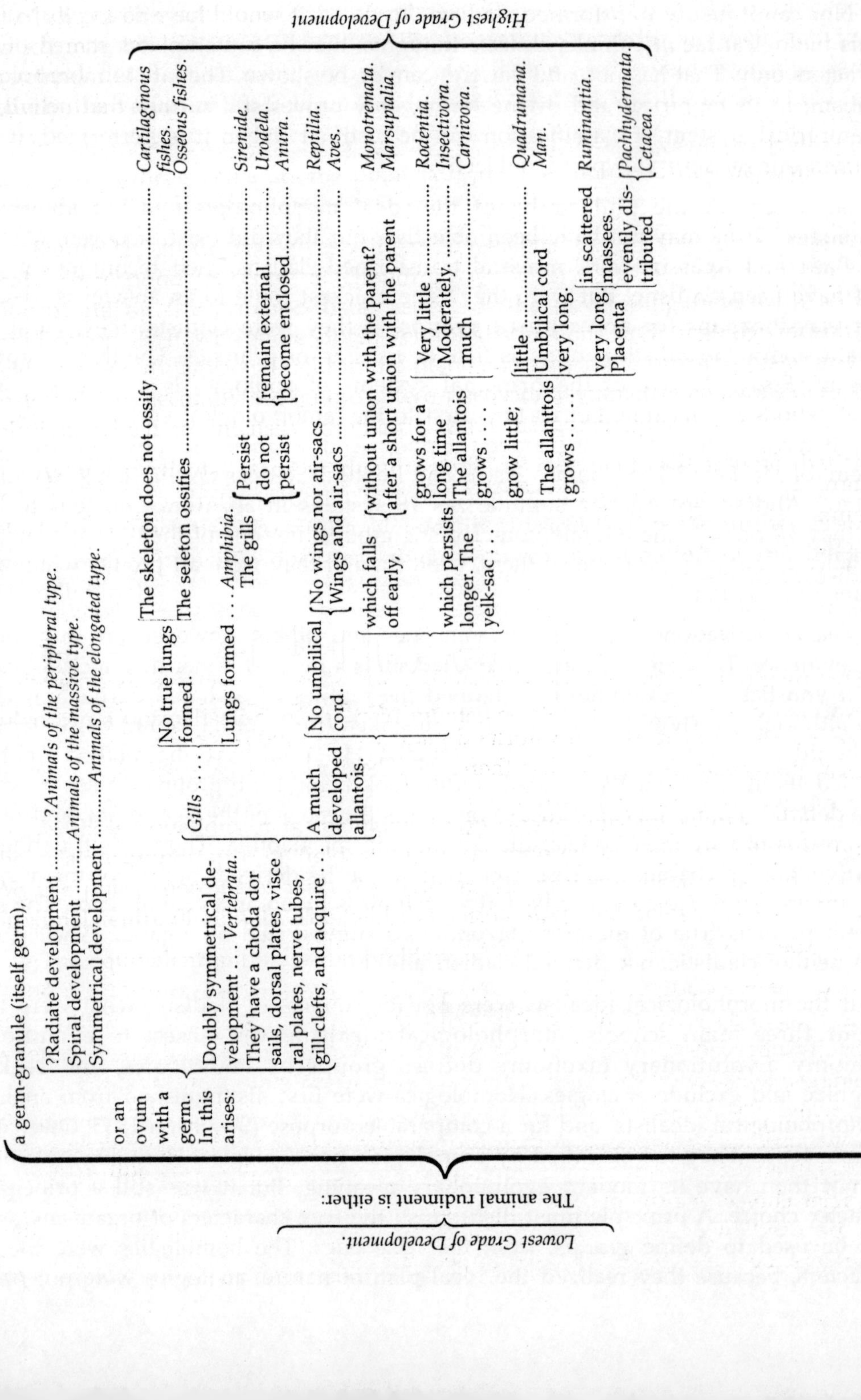

Fig. 1.1: Von Baer's divisions of the vertebrates (from von Baer 1828 [1853]).

the plan. Analogous similarities were imposed by functional demands. Homologous similarities were those not so imposed; they therefore must exist for some other reason: the plan of nature.

Owen's great achievement was to abstract all the homologies of the vertebrates to reveal the vertebrate idea, or 'archetype', in its true form. The archetype was an epitome of all the homologies—all the true morphological characters of vertebrates—which (as we have seen) Owen interpreted theologically. Different idealists did recognize homologies by different techniques. Agassiz preferred embryological characters (1859). Owen did not: he criticized embryology, and preferred to compare the anatomy of adult bones (Owen 1848). But they shared the same aim, which was to classify according to the ideal plan, which was recognized by homologous similarities. Although homologies were linked, in the idealist system, to an idealist philosophy, it identified much the same characters as homologies as an evolutionary taxonomist would now: Owen, for example, described the wings of birds and bats as analogous (1848) and tetrapod limbs as homologous. It was therefore possible for morphologists who lived through, and accepted, the Darwinian revolution, to continue business as usual (Coleman 1976). Evolutionary taxonomy has stripped the homology—analogy distinction from its idealist philosophy; but it still owes the distinction itself to that now discarded school of thought. If any school were to seek comfort in pre-Darwinian idealism, it should be a (non-existent) 'transformed' evolutionary taxonomy, not transformed cladism.

The purpose of this chapter was to discover whether the techniques and justifications of classification before Darwin could justify the system of transformed cladism. It has been successful in its purposes: it can conclude unhesitatingly in the negative. Pre-Darwinian classification does not justify transformed cladism. We have examined two pre-Darwinian systems of classification, essentialism and idealism. In order to justify transformed cladism, they would have to operate cladistic techniques, and provide a valid (in modern terms) justification for doing so. They do not do either. They are both utterly different from transformed cladism, their only similarity to it being a negative one, that they would not justify themselves by the theory of evolution. In all other respects the schools differ. Neither essentialist nor idealist classification uses cladistic techniques. Essentialism, in its teleological form (as practised by Linnaeus and Cuvier), groups species by shared purposes in life and thus defines groups by characters of high functional importance; idealists defined groups by homologies, which were thought to embody a transcendental plan of nature.

Nor can essentialism or idealism justify non-evolutionary classification even with the characters they do use. The essentialist justification is incomplete, and we cannot say whether it could be completed. Even if it could be, it would have to be an evolutionary principle, because natural selection gives organisms the only purposes they have. The status of idealism is easier to settle. Its naive empirical form is nakedly subjective: its divine form is impracticable: it can therefore be rejected in both its forms.

The favoured form was the divine one, according to which the ideal plan was supposed to exist because God was thinking it. Because the thoughts of God are undiscoverable, all an idealist could do was to pick characters subjectively, and then interpret the groups, after the fact, as the thoughts of God. As it happens, they chose characters very like our modern homologies. Transformed cladism, if it really were to 'rediscover pre-evolutionary systematics', could only converge on idealism by defining groups subjectively, with shared derived characters, and then declaring that these exist in the mind of God. No one will be able to prove them wrong. We do not know whether God thinks about the Reptilia: Agassiz and his contemporaries thought he did; the modern transformed cladist would maintain he does not. The conflict could never be resolved. The system is subjective. The chapter therefore confirms the existing conclusion. The only known objective, practical system of classification is Hennigian cladism. In it, the cladistic techniques are tied to the theory of evolution. If evolution is rejected, there is no reason to keep the cladistic techniques. Now we have looked at some pre-evolutionary methods, we have seen that they are, or boil down to, techniques that are subjective; we have also seen that they were not cladistic.

But perhaps this chapter has taken the advocates of transformed cladism too seriously. Perhaps the pre-Darwinian transformed cladists are meant only to be a comfortable myth, not a critical justification of the modern school. For most revolutionaries do invent historical myths about their ancestors. The agitators of the Great Rebellion encouraged themselves with the idea that ancient parliamentary liberties had been crushed beneath a 'Norman yoke', from which they would now, at last, liberate themselves (Skinner 1965). After the Calvinist revolutionaries of Scotland had forced their Queen to abdicate, their ideologists showed it was quite all right, for under the 'Ancient Scottish Constitution' the Scots were doing that kind of things all the time (Trevor-Roper 1966). And now revolutionary cladism would portray itself as the rescuer of a good old cause from the oppressions of a Darwinian yoke. In all these cases the history is equally mythical. The Anglo-Saxon Witan was scarcely a sovereign parliament; the wicked kings of Scotland never lived; and the ancient transformed cladists were neither 'transformed' in philosophy, nor cladistic in their method.

Our conclusion need not only be negative. The historical analysis has extended the argument in two respects. In the essentialist principle, we met again a possible alternative to the phylogenetic and phenetic principles. There may be other objective principles besides the phylogenetic one, and teleology may be one of them. The theological form of morphological idealism illustrates another point. An objective principle of classification must be practical. We may admit that the thoughts of God fall into unambiguous categories. But because we have no method of their discovery, the principle is practically useless. The phylogenetic principle is practical as well as objective. We have techniques by which it can, imperfectly, be discovered. But if they did not discover evolution, they would not be objective; they are tied by logic to the theory of evolution. And as this chapter has now shown, they are not only linked by logic, but by history too.

CLASSIFICATION AND THE TESTS OF EVOLUTION

The third proposed justification of transformed cladism is as a test of the theory of evolution. Evolution must be excluded from classification in order that the classifications may then be used to test the truth of evolution. The argument has not be discussed in any detail in the literature of transformed cladism; but it has been mentioned occasionally in print, and is often raised in spoken form. Here it is from the pen of Platnick (1979): 'if classifications (that is, our knowledge of patterns) are ever to provide an adequate test of theories of evolutionary processes, their construction must be independent of any particular theory of process'. The same argument can be found in Rieppel (1980) and in Janvier (1984); it is approved by Thomson (1982); and it fills the concluding sentences of one of Patterson's essays (1982a). He writes:

> As a science of pattern, cladistics holds out the possibility of a reconstruction of the history of life in space and time that does not depend on Darwinian or neodarwinian presuppositions. The interest of that reconstruction or cladogram is that theories of process—neodarwinian or any other—can be tested only against nature . . . But if we are taught, as we have been, to see that pattern through the spectacles of evolutionary theory, how could the pattern ever test the theory?

The essay ends there, with the clear implication that no answer can be supplied. These passages do suggest that transformed cladism has been justified as a test of evolution, but they should be used only tentatively: they are short comments, not formal arguments; and they were not actually presented as explicit justifications of transformed cladism, although Patterson does call it '*the* interest' of transformed cladism. The limited number of references hardly show that the argument is crucial to transformed cladism; but then it is difficult to determine just what are its crucial arguments, for, as Charig (1982) has also found, its position is 'constantly shifting'. I do not object to conceptual development; but, if transformed cladism is not a static body of doctrine, it does not much matter whether this chapter criticizes an idea believed by particular people. All that matters is that the ideas themselves should be important. And they are: they are not new to the controversies of classification; in one form or other, they have been with evolutionary taxonomy throughout its history (Hull 1967), as we have seen in the case of the numerical taxonomic critique. I do believe that many biologists think that transformed cladism can be justified as a test of evolution: I may be wrong; but even if no one believes it, even if this chapter contains nothing to frighten a transformed cladist, it remains a real discussion of a real question. And then, even if some readers do think this chapter unfair, others will probably agree that the argument should be considered.

The argument itself is clear enough. If asked why we think species evolved, rather than being separately created, we should probably include classification in our reply; any textbook certainly would. A transformed cladist may then reason that, if evolution was an assumption of the classification, surely the classification could not then test evolution: if we wish to test evolution, we must extirpate it from classification; and that (he would conclude) is why cladism must be transformed. It looks as if there may be a justification here. If a classification is built without the help of the theory of evolution, and then turns out to fit the theory, we should have clear evidence for evolution. Since the truth of evolution is an important question, non-evolutionary taxonomy would possess an important purpose.

That is the proposed justification. It only remains to discover whether it is valid. It would work with greatest possible power if two things are true: first, that classification is essential to test evolution; and second, that *cladistic* classification is essential to test evolution. We must therefore begin with three questions: Can classification test evolution at all? If it can, is classification necessary in order to test evolution? If it is, is it a *cladistic* classification that is necessary? Once we have answered them we can ask a fourth question, for none of these three directly challenges the proposed justification of transformed cladism. We must also ask whether it really is a circular argument to assume that cladism is phylogenetic, and also to use a cladistic classification to test evolution. We shall now take these four questions in turn. But, because the discussion will prove a little laborious, let me give, in advance, a four-word summary of the answers. If the justification of transformed cladism is valid, all four answers must be positive. The actual answers, however, taking the four questions in order, will be: Yes, No, No, No. Any one negative would destroy the justification, but three our of four, *cladis sunt tria partes*.

We must fix what I mean by a 'test' of evolution. A test can only operate between evolution and a specified list of alternatives. Before we can call something a test of evolution, we must specify what the alternative is. The most obvious alternative would be the biblical book of Genesis. But if taken literally, Genesis runs into trouble with some simple facts of geology. If I was asked why I do not accept Genesis, I should immediately seize on the geological evidence of the age of the Earth. That is exactly what the scientific party did at the recent trial at Arkansas, and it worked (Lewin 1982a, b). If we take some other alternative to evolution, a different test will probably be required. One other alternative is the theory of separate creation of the early editions of Lyell's *Principles of Geology* (e.g. 1830–3), according to which species originated separately, rather than by descent from a common ancestor. It need not be worried by the geological timescale. To refute it, we should need other arguments. As we consider how classification may test the theory of evolution, we must have some particular alternatives in mind. Let us here use these two, Genesis and the theory of separate creation.

Classification can argue for evolution against both of them. The hierarchical structure of taxonomy is often said to suggest evolution, and so it does, but the hierarchical structure alone is not sufficient proof. Any objects, whether or not they originated in an evolutionary process, can be classified hierarchically. Buildings, for example, do not physically reproduce; they have separate origins: but they can be classified hierarchically, by such criteria as whether they are large or small, ugly or beautiful, of brick or stone. Any particular set of buildings may not be divided hierarchically by these particular criteria; but an appropriate set of criteria should be discoverable for any case. This being so, the mere fact that biological classification is hierarchical does not prove that evolution has taken place. The true argument from classification is more subtle. It lies in the nature of the characters that define the classification. These characters have two important properties. One is that they are homologies, not analogies. The other is that they indicate natural, not artificial, groups. The two properties are closely related, for naturalness is use to recognize homologies, but it is convenient to take them separately. Homology, in a non-evolutionary context, refers to similarities between species that are not required by similarities in their modes of life: if fish and mammals had independent origins, the bones of the mammalian ear would not have to be the same bones as form part of the jaw in fish. In fact they are. If mammals had evolved from piscine ancestors the fact would be less surprising, because new parts must be formed by changes in preexisting parts. The homology, therefore, of ear bones in mammals and jaw bones in fish, is an argument for evolution. It argues against any alternative which postulates the separate origins of these two groups. Like ear bones, many other, homologies are used to define taxonomic groups, and in every case the same argument applies. If classifications were defined by any old kind of character, rather than homologies in particular, the hierarchical nature of the classification would not be an argument for evolution.

The second point was that biological classifications are natural, not artificial. The members of taxonomic groups resemble one another for many other characters in addition to those that define them (examples will come later). This suggests evolution, because evolution generates natural groups, but a hierarchical classification of non-evolutionary objects will usually be artificial: the members of the groups of ugly and beautiful buildings (for instance) will not resemble one another in other independent characteristic. The naturalness of biological classifications argues for evolution against any alternative which postulates an independent origin for the groups, at any level, in the classification. Naturalness and homology matter less because they are used in classification than because they suggest that the species are connected phylogenetically. The real test of evolution is the discovery of a phylogeny. It can, however, be called an argument from classification because the same evidence is used to classify species and to reveal that they are connected by a phylogeny. It does not matter whether the argument is called an argument from phylogeny or from classification; the argument itself is the same in either case.

Biological classifications, therefore, can test the theory of evolution. So far, so good for transformed cladism. Now let us turn to the stronger question of whether classification not only can test evolution, but is also essential to test it; of whether, in other words, classification is the only test of evolution. If it is, the justification of transformed cladism could (if the remaining steps in the argument succeed) turn out to apply in a strong form. If classification were the only test of evolution, evolutionary biologists would have to keep evolution out of classification; but if classification turns out to be only one among several arguments, then transformed cladism could only be justified in a weaker form. Evolutionary biologists would not have to exclude evolution from classification; and if we had any reason to include it, the justification of transformed cladism could be overruled.

The usual list of evidence for evolution includes several apparently non-taxonomic arguments. The crucial question is whether they genuinely are non-taxonomic, or whether they really reduce to the taxonomic argument we have just discussed. We can start with the geological argument, which Darwin discussed in two main forms. The first is the evidence of transitional stages between populations at different times. Darwin rejected this line of evidence, because of the incompleteness of the fossil record; and, despite the discovery of few cases of fairly complete records, Darwin's argument still applies today. The second is the 'geological succession of organic begins'. The grand pattern of the fossil record, for instance in vertebrates in which fish arise first, then amphibians, then reptiles, then mammals, and finally the subgroups of mammals, is exactly what an evolutionist would predict. Maynard Smith (repeating a remark of Haldane, he says) once told a televisual audience that his evolutionary views would be severely disturbed if a fossil rabbit were discovered in the Precambrian. The fact that one has not is evidence for evolution. And not only is the geological succession predicted by evolution, it is decisively not predicted by Genesis. Lyell (1830–3), in his pre-evolutionary phase, was well aware of the geological succession, and tried to explain it by changes in the environment. But even if we allow his rather forced explanation, the geological succession argues against the biblical alternative to evolution. The argument is non-taxonomic. It is no use a transformed cladist saying that the argument assumes the animals have been classified (into mammals, primates, etc.), and reduces to the taxonomic argument, because it does not. I only used those names for convenience. The actual argument does not require any classification. The argument is that complex evolutionary changes must take place in a series of small changes by the modification of pre-existing parts; to get from a fish-like form to a primate-like form requires transition through reptilian forms.

The geological succession provides one non-taxonomic argument for evolution. It is not the only one. There is also the evidence of the artificial selection of evolutionary change—which has often produced new biological species—and there is 'ecological genetic' evidence of natural evolution. A transformed cladist might, in desperation now, plead that these only demonstrate microevolution. But that would be a feeble thing to say. All that is needed to prove evolution is observed microevolution added to the philosophical doctrine of uniformitarianism, which (in the form that is needed here) underlies all science.

That is not all. The evolutionist can, and still without resort to classification, point next to what may be called universal homologies. These are properties possessed in common by all specie, but which would not have been if all species had been created separately. The genetic code is a universal homology. How do we know it is universal? Originally, because when Von Ehrenstein and Lipmann (1962) injected the mRNA for haemoglobin of rabbit into *Escherichia coli*, the bacterium then synthesized the protein; but this kind of experiment is now a biotechnological commonplace. Similarly, Weinstein and Schechter (1962; Weinstein 1963) found that a variety of different species all synthesized polyphenylalanine if injected with polyuridylic acid. So the code is universal; but how do we know it is homologous, and not convergent? The question (Crick 1966) 'is whether the code has a structural basis, or whether it mainly arose by chance. If the former were true there should be some stereochemical relationship between each amino acid and the triplets that code for it, or with the corresponding anticodons. So far this has not been demonstrated.' Nor has it yet. The code is therefore thought to be a 'frozen accident' (Crick 1968), descended from a unique origin of life. There are other biochemical universals too: all amino acids are laevorotatory; and all species use much the same molecular genetic apparatus. They independently support the same theory. They are evidence for evolution.

Although the argument from universal homologies has the same form as the argument from classification, it is independent of it. We need no more than the most rudimentary classificatory knowledge in order to demonstrate the universality of the genetic code. Von Ehrenstein and Lipmann's experiment did appear to presuppose that *E. coli* is a different species from a rabbit. However, even if this degree of formal taxonomic knowledge were necessary, it should not comfort transformed cladism. The evolutionary philosophy repudiated by that school is used to classify the higher categories, not to distinguish species. Moreover, the demonstration would have been just as powerful with less formal taxonomic knowledge: all that matters is that the two forms in the experiment should be very different from each other. The injection of haemoglobin mRNA into lots of randomly chosen and unclassified forms could equally well demonstrate the universality of the genetic code. And just as the demonstration of universality does not depend on classification, nor does classification depend on the existence of universal homologies. Universal homologies are not used in classification. Taxonomists classify living species, and no taxonomic act depends on, or can affect, the existence of universal homologies. Although the arguments are similar, from the standpoint of taxonomic practice that is a coincidence. No taxonomist need be professionally aware of the existence of universal homologies. They are independent of classification. That, however, has not prevented them from providing yet more evidence of evolution.

What might we conclude at this stage? Classification can test evolution, but it is only one test among many. That would seem to open up two possible uses of the theory of evolution in classification. We could either use classification to test evolution, or we could test it by other means and then (if we wished) assume the truth of evolution in classifying. If we had no reason to use the theory of evolution in classification, transformed cladism might be able to find a weak justification as a test (though an unnecessary one) of evolution. If we had reason to use it—which in fact we do—then we could dispense with the classificatory test, and with transformed cladism too. But that would only be the conclusion if the remaining two questions receive positive answers. Let us turn to them now.

We have demonstrated that classification can test evolution. Cladism therefore can as well (Cracraft 1983). That, however, does not add up to a justification of transformed cladism. It must be shown not only that cladism can test evolution, but also that the techniques of numerical and evolutionary taxonomy, if applied without presupposing evolution, could not. For the techniques of any school can be operated without assuming evolution. The technical acts of the taxonomist have no need of the assumptions that justify the techniques. Of course, it would be rather absurd to distinguish 'convergent' from 'ancestral' similarity in the evolutionary school if evolution were untrue, but it is exactly that kind of absurdity, in cladism, which is under analysis. We cannot ignore it. Transformed cladists make the comparable distinction between ancestral and derived characters without an evolutionary assumption, and evolutionary taxonomy could do likewise. The technical deed could be done. And because it could be done, if it is not going to be it must be for a reason, not because it is practically impossible. Any school that would justify itself as a test of evolution must show that it alone can provide such a test. If it does not, it is unjustified: we should be left at most with a justification of classification in general, and a subjective choice among particular schools. If transformed cladism is to be justified as a test of evolution, the taxonomic argument for evolution must work only with cladistic techniques. Let us see whether it does.

To state the taxonomic argument should be enough to settle the matter. It is (as we have seen) that the classificatory hierarchy is natural, not artificial, and is defined by homologies. Now, cladistic classifications are natural, but then so too are numerical and evolutionary classifications. Likewise, although cladistic classifications do use homologies, so too do evolutionary classifications. The need to test evolution could therefore at best only justify all taxonomic schools equally, and none of them in particular. It does not justify transformed cladism.

The same point can be made by example. E. B. Ford (1941, 1942, 1944a, b, 1947) pioneered the use of chemical characters in classification; but he also used them to test whether pre-existing classifications were natural. He worked on Lepidoptera. Pigment chemistry, of course, had not originally been used to classify the Lepidoptera. Taking specimens from the recognized families, Ford discovered whether the distribution of pigment chemicals matched the accepted classification. It turned out that they did. Such would have been the prediction of evolution, not of separate creation. But it did not require transformed cladism! Ford did not distinguish ancestral from derived states of pigment chemistry, nor classify by shared derived pigments. He supposed that similar pigments were homologous in the evolutionary taxonomist's sense. A similar kind of test has been performed more recently, although (now that the creationists and Popperians are among us) it was divulged differently. It also used a chemical character, protein sequences. Penny, Foulds, and Hendy (1982) reasoned that, if evolution is true, the pattern of sequence differences among species should be the same for different proteins. A taxonomy formed by any one of them, in other words, should be natural. They accordingly compared the tree of 11 species, formed from the sequences of five different proteins. The five trees were topologically identical; the prediction of evolution, not of separate creation was confirmed. But although Penny, Foulds, and Hendy thus offered evidence for evolution, and by a taxonomic argument, they (like Ford) did not use cladistic techniques. They did not distinguish ancestral from derived amino acid similarities. They supposed only that similar protein sequences were homologous in the evolutionary taxonomist's sense.

The end has now come for transformed cladism. It is not the only taxonomic school that can test evolution: any of the main schools can. The need to test evolution (which is real but which need not be met by classification) could justify any of the main schools. We could stop here, but I believe it is worth examining the fourth question too: not to turn the knife in the wound, but because it is of general interest. Let us admit therefore that classification is needed to test evolution (which it is not), and that the kind of classification needed must be constructed with cladistic techniques (which it need not). Then we can ask whether our philosophy of cladism must be Hennigian, which assumes evolution, or transformed, which does not. Is it circular to argue that classification tests evolution, if the truth of evolution was assumed in building the classification?

We have met this kind of question twice before. Numerical taxonomists criticized the evolutionary taxonomist's method of recognizing homologies because (they said) it was circular. A similar charge has been made against the cladistic technique of outgroup comparison. Actually, it is only a small step from the numerical taxonomic critique of how homologies are detected to the argument of transformed cladism. Numerical taxonomists, as it happened, fired their critiques against the methods by which homologies were recognized; but then, in the 1950s and 60s, you could scandalize Darwinian society more easily if you professed that the course of evolution could not be discovered, than that it had not taken place. But the step from one argument to the other is small, and it is not surprising that it was taken in one of those numerical taxonomic works:

> [Darwin] did not commit the earlier error of arguing in a circle. His principles of evolutionary importance were not derived from a pre-existing taxonomy, but from the results of artificial selection and from the study of heredity, variation and ecology. This is a point worth emphasizing. In many elementary text-books of biology, classification is treated as one of the lines of evidence for evolution. Darwin did not treat it thus; he discussed it quite late in the *'Origin'* as consonant with the theory of evolution, and explicable as a consequence of it. He never regarded it as primary evidence for evolution, and his caution was certainly justified. (Cain 1962.)

We have here an argument very like that of transformed cladism. And just as the arguments are very similar, so the same defence will serve. It would indeed be possible to formulate the argument as a circular one, and perhaps some authors have mistakenly done so. However, the circular formulation is the argument in its worst form, not its best, and we should be interested in the latter. The argument properly proceeds not in a circle, but by 'successive approximation'. The truth of a general proposition must first be independently confirmed, and then when this has been done, it can be used as an assumption in interpreting later observations; if the interpretation is successful, our confidence in the theory is proportionally increased.

First, an independent test must confirm the truth of evolution. This could be achieved by any of the non-classificatory arguments for evolution. Or we could classify a small group of animals without the assumption of evolution, and test the theory with classificatory evidence. We might, for instance test the theory of evolution on the birds. It would be important not to assume the truth of evolution while doing this, as it is an independent test. If the test was successful we could then, if we wished, accept the truth of evolution in future classifications. As each natural classification, and each homologous similarity, was discovered, the evidence for evolution would increase. We might, in other words, have needed to be transformed cladists for a short period last century, while classifying some small group; but the need has now passed. Now evolution has been confirmed, we can, without any circular reasoning, use the theory to build our classifications. In practice, the process of successive approximation is not as simple as that. In a logical reconstruction there must be an independent test, before subsequent assumption; but history need not proceed logically (Hull 1980). There need be no clear first stage in which the hypothesis is tested, and second stage during which it is assumed to be true. The real process is a continual feedback from the results of successive new tests, to our understanding of the facts in question.

The sequence—hypothesis test, further hypothesis (assuming the results of the previous test), further test—is not circular. Nor is it scientifically peculiar. What the numerical taxonomists, and now transformed cladists, really object to is not circular argument, but theoretical ideas. Theories in science, if they are interesting, are always taken beyond the evidence, and assumed in further tests. The machinery of molecular genetics has only actually been confirmed in a very few species, but biologists unhesitatingly agree that it is universal. They use this knowledge, obtained in *E. Coli*, in interpreting the results of experiments on all other species. We can test Newton's law of gravity with an apple in an orchard, and then apply it to the motions of the heavenly bodies; we can likewise test the theory of evolution in a genus of fruitflies, and apply it to all of life.

We therefore can both assume, and test, evolution in the process of classification. The only question left open by the argument of this chapter is whether we ought to. We have been through all that before. We examined the possible taxonomic systems, and argued that the best system is phylogenetic. The best known philosophy of classifications, therefore, depends on the theory of evolution.

Taxonomic controversialists seem fatally attracted to the discovery of circular arguments in the systems of their opponents. It is not only transformed cladists who have found circles in evolutionary systems; numerical taxonomists have found them in evolutionary and in cladistic taxonomy, evolutionary taxonomists have found them in phenetic taxonomy (Johnson 1970), and in cladism (Bock 1981); some cladistic factions have found them in the systems of other factions (Schaeffer, Hecht, and Eldredge 1972), and I do not suppose the list ends there. The remarkable point is that they have nearly all been wrong. It is all too easy to see circles where there are none; and, with this history, my suspicions are always raised when I hear the charge being made.

The conclusion of this chapter is that transformed cladism cannot be justified as a test of evolution. The justification contains three lethal errors. Although classification can provide a test of evolution, it is not a necessary test. Even if it were necessary, this would not justify transformed cladism, as opposed to any other technique of classification. Finally, there is nothing wrong, or circular, in both assuming evolution in order to classify, and using the classifications to prove evolution.

We have how examined all the proposed justifications of transformed cladism, and can draw a common conclusion. None of them work. Transformed cladism is either nakedly subjective, or can easily be stripped of its thinly deceptive justifications, and exposed in its truly subjective form. There is a good reason to use cladistic techniques, but it secures them by logic to the theory of evolution. If the theory of evolution is abandoned, so should be cladism.

THE FUNCTIONAL CRITERION

Although the rejection of evolution from classification is the most far-reaching ambition of transformed cladism, the same revolutionary movement has made several other controversial demands. Indeed, within taxonomy, the rejection of evolution has probably been discussed less than three other narrower practical issues. The main methods of classical 'evolutionary taxonomy', at least according to its practitioners, are functional analysis, to distinguish homologies from analogies, and palaeontology, to recognize ancestral forms: but many cladists have recently denied that either method works, and have called for their abolition. A third cladistic demand has also proved highly controversial, which is that classification should preferentially be dichotomous. In the next three chapters we shall consider these three controversies in turn.

The concepts of function, adaptation, and natural selection all go hand in hand. The 'function' of a part of an organism is the way in which it is adapted, to understand now an organ functions is to understand how it is adapted, and why natural selection brought it into existence. We have encountered two proposed classificatory uses for this set of concepts. The grandest was the possibility of a whole teleological principle of classification. In which sspecies would be grouped according to their principal adaptations; but I have no more to say about that. The other, which we met in evolutionary taxonomy, was to distinguish homologies from analogies. This we shall consider further in this chapter, together with a third possible use, to distinguish ancestral from derived character states.

For now we have decided in favour of the Hennigian school, the outstanding problem of classification is to recognize accurately ancestral and derived characters. We do possess some methods (as we have seen) but they are far from perfect, and we could well do with some more. Moreover, because our existing methods often mistake shared derived characters, we also need methods of reconciling apparent contradictions among characters.

EVOLUTION AND CLASSIFICATION

What is the correct relation of classification and evolution, of (in the terms of transformed cladism) 'pattern' and 'process'? The central doctrine of transformed cladism (as we have seen) is that the classification of organic diversity must be kept separate from the theories that may explain it, which means in particular that the theory of evolution must be kept out of the procedures of classification. Transformed cladists have expressed the point briefly by saying pattern and process must be studied separately, pattern before process. Another quotation, from Cracraft (1983.) this time, will re-explain the terms: 'it is becoming increasingly apparent that systematics is the area of biology that defines the *pattern* of organic change through space and time and, consequently, specifies that body of knowledge that theories of evolutionary *process* must be capable of explaining.' So confident are many transformed cladists of their distinction of pattern and process that they even scold their critics with it. Thus, the 'so many . . . commentators who criticize transformed cladistics' learn from Humphries (1983.) that 'they confuse the cladistic activity of ordering characters to their level of universality with a causal explanation of the hierarchy'. If it were an error to suppose the two are connected, this would be a good reply. I do believe classification and evolution are strongly connected; my aim has not been to confuse them unwittingly but to fuse them together with argument. We have examined the doctrine of transformed cladism in its many applications, and seen how evolution and classification are connected in particular places: I now wish to conclude by stating the argument both in its abstract form, and in the terms of transformed cladism itself.

The double distinction of pattern and process has a closely related triple form, the distinction of 'cladograms', 'trees', and 'scenarios' (Eldredge and Tattersall 1977; Cracraft 1979; Eldredge 1979; Eldredge and Cracraft 1980). A cladogram is a branching diagram of species, clustered by their shared derived characters; the lines among the species indicate only the sharing of derived characters. A tree is a phylogeny, with ancestral species placed at the branching nodes, rather than at the tip of a branch, as in a cladogram; the lines among species represent evolutionary lineages. A 'scenario' is an account of the evolutionary process that causes the tree and its cladogram. The distinction is real, but it has been put to a transformed cladistic use which we must oppose—the doctrine that, as a matter of method, the cladogram must be constructed first, and only then can one go on (if one so chooses) to construct a tree, and even try to explain it. The flow of information would then be strictly one-way: a cladogram would be needed before a tree could be suggested, a tree before a 'scenario' (references above, and those cited in Hull 1980, see also Fisher 1981 and Platnick. Because the cladogram is the 'pattern', and the scenario the 'process', in the triple distinction, as in the double, the theory of evolution is separated from classification. Because the cladogram-tree-scenario distinction is so closely related to the pattern-process one, we can take them together. In both, transformed cladism proposes to expel the theory of evolution from cladistic classification.

Previous chapters have accumulated four sources of opposition. The first, and one of overwhelming importance, is that evolution is needed to justify the whole system. The remaining three matter less, and are, secondly, that natural selection can provide a cladistic technique and thirdly, that both trees and cladograms are built from the same common fund of knowledge the fourth argument matters least, because the philosophy of cladism does not demand dichotomous classifications, but it is that, if a dichotomous preference is to be maintained, it must be justified by a theory of speciation. A fifth criticism may be taken from Hull (1979, 1980), and added to the list: even if the distinctions of pattern from process, and of cladogram, tree, and scenario, were true in philosophy, they would not necessarily, for that reason alone, be true in practice. Let us examine these points, taking Hull's first.

Let us pretend, for sake of argument, that the distinctions are philosophically valid. Let us suppose that classification is epistemologically separate from, and prior to, evolutionary explanation. Transformed cladism would then deduce the practical conclusion that we must study pattern before process, and build cladograms before trees, and trees before scenarios. Hull (1980), however, remarks 'of course, no taxonomist has ever proceeded in this fashion, and according to the philosophy of science explicitly espoused by the cladists, no necessary or even important connection exists between either logical or epistemological priority and the actual temporal order in which scientists perform their task (Gaffney [1979]). In other words, even were classificatory pattern philosophically prior to evolutionary process, the practical conclusion would not necessarily follow. But is the philosophy itself valid? How, in practice, should the theory of evolution be used in classification?

Naturam expelles furca, tamen usque recurrat. The branching hierarchy of phylogeny, as Hennig realized, is the only unique, unambiguous hierarchy in nature that can be the aim of classification. It is the only known objective system. When phenetic classification discards evolution, it becomes subjective; when 'evolutionary' classification overrules the phylogenetic principle, it becomes subjective; and when transformed cladism discards evolution, it too becomes subjective. The way forward for all three schools, as each is well aware, is towards objective classification, towards unambiguous natural hierarchies. That means embracing, embracing more keenly, or re-embracing, the theory of evolution and the branching hierarchy it implies. They must become cladistic, in the evolutionary, untransformed and reformed, sense. Evolution, in Henning's argument, is needed to justify cladism. Even if that theory of process can be kept out of the methods, as they are actually operated to detect patterns, it cannot be kept out of any justification of those methods. Cladists may not think about evolution while they apply such methods as the embryological criterion, or outgroup comparison; but they need to as soon as they try to explain why they are using those techniques. The theory of evolution has a further use. It is needed to improve, or to modify, the cladistic techniques. Whenever the application of a cladistic technique to a character is not perfectly straightforward, the cladist should return to the principle from which the method was developed. It is from there that the technical improvements must come, and only there that they can be justified. For example, an understanding of the circumstances that favour neoteny would help to sort out the paradoxes of the embryological criterion.

Philosophical justification, and technical improvement, thus flow from the theory of evolution into classification, menacing the transformed cladist's distinction on the way. The transformed cladist might still plead that the methods, at least in their most mundane, operational work do not contain a theory of process. But this too would be a mistake. All the cladistic methods stem from the theory of evolution. Outgroup comparison makes an assumption about the frequency of convergence; the embryological criterion assumes that new evolutionary stages are always added on at the end of the adult stage; and there would be little reason to suppose that ancestral stages appear earlier in the fossil record if they did not evolve before derived character states. With the functional criterion, the dependence on a theory of process, natural selection in this case, is particularly clear. Evolutionary theory is put to particularly active use in this method. Each character has to be referred back to the theory of natural selection, which acts as a continual reminder of how classification uses the theory of evolution. Information here will obviously be flowing from the theory of evolution into the classification: from 'scenario' to 'cladogram'. The 'scenario' will have to be understood in order to construct the cladogram (Fisher 1981). Cracraft (1981) believed that the functional criterion 'confounded process analysis with pattern recognition', and was therefore in error. His premise is correct, but the conclusion upside-down. The functional criterion does indeed put together pattern and process, but it is the pattern-process distinction, not the functional criterion, which comes out the worse.

It the functional criterion so clearly damages the cladistic distinction of cladograms and scenarios, what about the middle part of the triple distinction? What about trees? A tree differs from a cladogram in that ancestors are tentatively recognized. A tree is consistent with several cladograms—with several hundred in a typical case (because ancestors may appear in many places in a cladogram, but in only one place in a tree)—and is in that sense a more exact theory. It might therefore be thought that, to convert a cladogram into a tree, we should need extra knowledge, and that cladograms must be discovered before trees. Transformed cladism argues exactly that. I would not, however. The 'extra' knowledge mainly concerns which character states are ancestral, because we need to know whether particular species in the cladogram possess ancestral or uniquely derived character states. If they have ancestral states, they may be ancestors; if derived, they cannot be. That is the 'extra knowledge'. It is not 'extra' at all. It is obtained during the initial cladistic analysis of character states. In discovering which character states are derived, cladism has to discover which are ancestral. It does then throw that knowledge away before constructing its classification, but that is its own business. Cladograms and trees draw on the same common pool of knowledge, and could be built at the same time. Actually, whether characters are ancestral or derived is only the most important knowledge needed to convert a cladogram into a tree. Palaeontology can also be useful. A species is more likely to be an ancestor if it preceded its hypothetical descendants in time. But that knowledge comes out of the cladistic analysis too! The palaeontological criterion for distinguishing ancestral from derived character states will have discovered all the relevant facts about temporal priority. Cladogram and tree can be built at the same time, from the same knowledge. I therefore conclude that the cladistic distinction of pattern from process (or of cladogram from tree, from scenario) is in error. Cladograms have neither technical, nor temporal priority to trees. They can be built at the same time, from the same cladistic knowledge. 'Scenarios' sometimes may be needed before cladograms and trees, and sometimes they may not. (Hendy and Penny 1984 offer another reason against distinguishing cladograms from trees.)

Notwithstanding all that, the triple distinction of cladogram, tree, and scenario does contain some truth. Its legitimate meaning is that as we move from cladogram, to tree, to scenario, we move to hypotheses of increasing precision. A cladogram is consistent with several trees, and a tree is consistent with several scenarios. In this sense, 'more knowledge' is needed at each successive step. In the case of the step from cladogram to tree, that 'more knowledge' is disingenuously so called because the knowledge was acquired in the cladistic analysis, only to be thrown away before building the cladogram. No such general remark can be made for the next step, to the 'scenario'. The form of the distinction in terms of the greater precision of the theories at each stage is valid; but it does not justify either the philosophy, or practical conclusion, of transformed cladism.

We have often observed how transformed cladism finds the theory of evolution unnecessary. Unnecessary, however, does not necessarily mean undesirable in all cases, and Platnick (1979) finishes his paper by asking whats the proper relation of evolution and classification might be in transformed cladism. Here is his answer: 'presumably, its role is exactly as early evolutionists perceived it, namely, as an explanation for the existence of a natural hierarchical system'. For Platnick, this evolutionary rationale is just a surface gloss which the taxonomist, after his work is done, may affect, take up or put down, according to his wish; it is an 'unnecessary' assumption. There is another similarity with numerical taxonomists here; they too used to declare that phenetic relations must be worked out first, and then, when that is done, the groups may be interpreted, by evolution if you wish, as a later affectation (Hull 1979, gives references).

But for the Hennigian cladist, evolution is much more crucial than that. Cladism needs evolution in its techniques, although not actually to operate them in the day-today practical sense. It needs evolution in its philosophy. The theory of evolution is necessary—not just desirable or decorative—in the justification of cladism. The theories of evolution and of natural selection are part of the technical machinery of the system, for they are the source of new methods, and the court of appeal for the old; but the main contribution of the theory of evolution is to underwrite the philosophy of the whole system. The investigation of classificatory patterns without the theory of evolution is doomed to be, in technique stagnant and impoverished, in practice hypocritical, in philosophy incoherent and subjective.

Chapter—2
Evolution and Classification

Evolutionary taxonomy has been developed by the best philosophers of classification and has dominated the activities of the museum work-bench for over a century. It possesses both a reasoned philosophy and practical techniques; it has accumulated rich and broad, tolerant and flexible set of concepts, and a huge literature. We need concentrate, however, only on its leading ideas, which can be found in such standard works as *The Methods and Principles of Systematic Zoology*, by Mayr, Linsley, and Usinger (1953, new edition Mayr 1969), and *Principles of Animal Taxonomy* by Simpson (1961).

The justification of evolutionary taxonomy is simply expressed. Evolution has produced the natural groups of life (Sprague 1940); classification should seek to be natural rather than artificial; classification should therefore represent evolution. Evolution results in natural groups because all evolutionary change takes place in the same phylogenetic tree: all changes in all characters must have taken place within the same pattern of lineages. The distribution of character states will define a pattern of groups according to where in the phylogeny they arose. If a character defines a monophyletic group (that is, one with a common ancestor that also possessed the character), other characters, which changed at the same time, should also fall into the same grouping. Other characters may be distributed more narrowly within it (if they changed later) or more widely (if they changed earlier), but, except for convergence, they should not contradict it: they should not be found in some members of the group, and in some members of another group. If, however, different characters were not forced to change within the same phylogenetic tree, taxonomic groups would not be natural. If, in some scarcely conceivable non-evolutionary system, different characters changed within different patterns of lineages (joining the same set of species in the end), they would not end up distributed in the same groups. The case of convergence illustrates the point. Convergent characters do not define monophyletic groups; they define polyphyletic groups, whose common ancestor did not possess the character. There is no reason why other characters should (except for functional correlation) fall in the same groups: groups defined by convergent characters are liable to be artificial.

If the groups of the classification are to be evolutionary, the course of evolution must be known. Its discovery is the technical problem of evolutionary taxonomy. The material available consists of such characters of specie as may be observed in the museum, or (exceptionally) in the field. What can the evolutionary taxonomy do with these? It might just assume that shared characters between species indicate shared common ancestry; and the greater the number of shared characters, the more recent the common ancestor. But that simple rule is not enough. Because some characters are convergent, different characters often give conflicting indications. If convergence is ignored, polyphyletic groups may be defined, which, as we have seen, are unacceptable in evolutionary classification. Much of the practical work of evolutionary taxonomists has been to remove polyphyletic groups from earlier classifications. A typical example of convergence is that between the marsupials in South America and Australia, and the placentals elsewhere. In this case, an Argentinian *autodidáctico* in 1906 had exaggerated the age, and overlooked the convergence, of his native fossil fauna, and concluded that almost all the mammalian groups had originated in his homeland (Simpson 1980). That is no longer believed; but it took several evolutionary taxonomists several years of work to remove the convergent groups from mammalian classification.

Once it is accepted that convergence must be kept out of classification, the whole enterprise of evolutionary classification becomes less easy. Taxonomists cannot use any characters to define groups; they may use only non-convergent characters. And if convergent characters are to be spurned they must first be recognized, for convergence is too common a phenomenon to ignore. 'The first step, then, towards the achievement of a phylogenetic classification is an analysis of the taxonomic characters to determine which of them are derived from common ancestors (*homologies*) and which are spurious similarities (*analogies*), usually convergent adaptations correlated with similar habits' (Mayr, Linsley, & Usinger 1953).

The principle of evolutionary taxonomy may be to exclude convergent characters (and polyphyletic groups); its practical realization is another matter. In their published accounts, the methods are not rigorous; they are often vaguely formulated, and without any pretence to the contrary. But even if they are vague, they do exist. We shall discuss them only incompletely in this chapter because the techniques used in evolutionary taxonomy have been improved by numerical taxonomy and cladism, and we shall elaborate the most modern techniques later. We do not need to know their primitive forms, but should give evolutionary taxonomy some credit for the techniques to be discussed. For now, let us consider two kinds of technique to reveal convergent characters, by theory and by observation.

By theoretical analysis, I mean the application of the theory of evolution to estimate the probability that each character is homologous or analogous. Two kinds of criteria have been used. Both were discussed by Darwin in the *Origin* (1859 [1969 ed]). According to the first criterion, non-adaptive characters are more likely to be homologies than are adaptive characters; and according to the second, characters uniformly adapted to a broad range of environments are more likely to be homologies than are characters that fickly change, among a limited numbers of states, with every small change in the environment.

Let us take the first criterion first. Convergence is caused by the natural selection of adaptations; the obvious method of avoiding convergent characters, therefore, is to avoid adaptations. Groups should be defined by non-adaptive or, more precisely, selectively neutral characters. Whatever its merits, the idea has certainly been influential. Nearly every enthusiasm that has swept through evolutionary taxonomy has been inspired by the doctrine of selective neutrality. Developmental characters and molecular sequences are two outstanding instances.

In the late nineteenth century, a biologist would have classified his material by one (or both) of two main kinds of character—adult characters and developmental characters. Different biologists held different opinions about which of the two should be preferred, and the preferences changed with time. The golden age of the embryological criterion began in about 1870, and ran for more than two decades. Embryologists made hay in that long summer undisturbed by any murmuring sceptics. The doubters did come, of course, but they do not matter to us (See Gould 1977a; Ridley 1986). What we should ask is why biologists in 1870 held embryological evidence in such high esteem. Let us turn to the principal paper of its principal British advocate: the paper on the 'genealogical classification of animals', by Edwin Ray Lankester (1873). As Lankester observed the classifications of his time, he saw, in an updated form, the methods and the results of Cuvier; species were classified according to the comparative anatomy of adults. Lankester disapproved of what he saw. He preferred the method of embryology, and

> It would not be surprising if the facts of development were to lead to another primary grouping of the animal kingdom than that indicated in the four Cuvierian types or the six or seven types now generally adopted . . . . They are confessedly groupings based upon the anatomy of the adult organism; and therefore necessarily there has been a tendency in forming them to attach great importance to distinct plans of structure due to a secondary adaptation, whilst the fundamental community of organization has been ignored with something like intention.

According to the 'recapitulation hypothesis', however, developmental characters are often ancestral, 'this tendency to recapitulate' being 'the fullest expression of the phenomenon termed heredity'. Recapitulationists such as Lankester were quite sophisticated when they came to discuss how natural selection controlled development, but we need not go into that here (see Ridley 1986). They preferred embryological evidence in classification because they thought adult characters unreliable, because they were adaptive and liable to convergence; developmental characters, however, were sometimes ancestral. No modern biologist would agree exactly with Lankester, but that is not the point: the nature, not the validity, of his argument is what matters.

Developmental characters were to be trusted because they were, under certain circumstances, unaltered by natural selection.

In modern molecular taxonomy, the character is different, but the inspiration is the same. If the proteins, or DNA sequences of specie were thought to be adaptations then they would provide 'just another character' for use in classification. As such, they have the practical disadvantage, relative to other characters, that they are difficult to see, but the advantage that they are easy to compare. It is difficult to compare the differences between two limbs with the differences between two skulls; but it is easy to compare the differences of cytochromes with those of haemoglobins. The large number of molecules is also an advantage. These advantages and disadvantages, however, have little to do with the popularity of the technique. That is due to the 'molecular clock'. The molecular clock is a theory concerning the rate of protein evolution, according to which proteins evolve at an approximately constant rate. The difference between the proteins of species, therefore, should be exactly proportional to the recency of their common ancestry: species with more similar proteins should have a more recent common ancestor. I do not wish to discuss here whether the theory is true; the quantity of evidence is too great. Let us concentrate on the justification. Why should molecules evolve at a constant rate? The answer, of course, is that molecular evolution is selectively neutral (Kimura 1968, 1983). Molecules are good classificatory characters because their evolution is non-adaptive. (I am not denying that selective explanations of the molecular clock have been devised; but I would deny that they have inspired molecular systematics.)

With the wisdom of hindsight, we know that Lankester was wrong (we shall return to this later as well). So too was anyone else who thought that embryological characters were non-adaptive, and, for that reason, of especial classificatory value. It is too early to judge molecules. We cannot yet say whether classification by molecular evidence will be superior than had similar efforts been made with other kinds of evidence. However that may be, the argument itself has run a similar course, one century on, to the biogenetic law. If Fritz Müller's *Für Darwin* was published in 1863, and Ernst Haeckel's *Generelle Morphologie* in 1866, Kimura's important paper was in 1968; if the great age of biogenetic enthusiasm was in the 1870s and 1880s, we know the 1970s and 1980s; if the support for the biogenetic law was crumbling in the 1890s, well . . . . As molecular taxonomy progresses, it will move on to another stage. We cannot predict what it will be, or when it will come; but if any molecular taxonomists dislike my historical analogy, I would offer them some comfort in the prospect of a long period of power. The theory of recapitulation was still influential in the 1920s, and even in the 1930s biologists could still win their spurs in combat, if a little uneven by now, with the shade of Haeckel.

Developmental and molecular characters are not the only characters whose taxonomic merits have been argued in this way. Classification by animal signals is another instance. According to a classical theory, animal signals are so distorted during their evolution that the final relation of the form and meaning of the signal is quite arbitrary. Therefore, as Lorenz (1941 [1971)] argued in his great work on the classification of the Anatidae by their displays, signals (and 'releasers' in particular) should be good taxonomic characters:

> There is a second factor which makes releasers particularly useful in phylogenetic considerations: because the special form of a releaser is not directly derivable from its function (in contrast to mechanically operative motor patterns) and is not influenced by the latter, the *possibility of convergence* can be fairly confidently excluded when there is a correspondence between the elicitatory ceremonies of two related species.

Lorenz's idea, as it happens, never really caught on; but the nature of his argument is revealing. He knew how he had to justify a proposed new class of classificatory evidence. He had to show that natural selection would not cause it to converge.

The taxonomic interest of developmental, molecular, and behavioural characters followed, in all three cases, from their supposedly non-adaptive nature. That is historical fact. But the argument itself is not convincing. Its premise can always be challenged. It is scientifically difficult, perhaps even impossible, to prove that a character is selectively neutral, because neutrality is a negative or 'null' hypothesis. Null hypotheses can in principle be refuted—they are good hypotheses in the Popperian sense—but that is little help to the scientist.[1] Neutrality can be confirmed with respect to a particular environmental variable, or even a series of environmental variables; but the danger will always remain that someone will demonstrate that some other environmental variable is selectively influencing the character. The hypothesis of neutrality will then be refuted. This is as true of embryological and ethological evidence as of molecular. In the wake of the theory of recapitulation came doubts as to whether embryonic characters were any less adaptive than adult ones (Garstang 1922, 1928). Many students of animal signals would doubt whether the relation of form and meaning in signals is arbitrary (e.g. Morton 1977; Hamilton and Zuk 1982). It is at present controversial whether molecular evolution is controlled by selection (Goodman 1981). Many characters that have been thought to be non-adaptive have turned out, on investigation, to be adaptive; and judgements of non-adaptiveness are precarious generally (Cain 1964). Modern evolutionary taxonomy may prefer not to justify its techniques with such a shaky philosophy. It may prefer to rely instead on the second reason why some characters might be more conservative in evolution than others.

It is that some characters are adapted to a wider range of environments than are others. Broad and generally adapted characters will not change as much in evolution as will narrower, more specialized characters. The more slowly a character evolves, the more it reveals its ancestry, and the better a taxonomic character it provides. This kind of evolutionary constancy does not result from non-adaptation: it depends instead, as Darwin (1859 [1969 edn]), said, 'on such organs having generally been subjected to less change in the adaptation of the species to their conditions of life'.

Broad adaptation is a better justification of the taxonomic value of a character than is selective neutrality. It is theoretically more plausible. Many modern biologists are aware this is so, and, indeed, probably often intend it when they describe a character as non-adaptive: the translation is easy (Simpson 1961). But biologists do often treat ancestry and adaptation as alternatives. We could just let this pass; but it may be worthwhile to point out one of the reasons for the mistake. Most evidence, and especially crude evidence, of adaptation is provided by the comparative method. If one character repeatedly evolves in one particular environment, we properly infer that the character is an adaptation to that environment. We necessarily lack this kind of evidence for characters that do not often change in evolution: and these are the ancestral characters. We therefore usually cannot infer, by the comparative method, how ancestral characters are adapted. This mistake is to confuse the inapplicability of the method with the absence of the subject it investigates: one kind of evidence of adaptation is absent for constant characters; but it does not mean they are not adaptive. The gnawing apparatus of the Rodentia illustrates the point (Cain 1964). It is constant in all rodents, but it is certainly adaptive; it is an adaptation for feeding. Methods other than comparison, such as functional morphology or experiment, can reveal how evolutionarily constant characters are adapted to their environments.

Although relative evolutionary conservatism is a theoretically sound method by which to distinguish taxonomically good characters from bad ones, it is difficult to apply in practice. It can be applied only rather loosely. No one doubts, for instance, that a backbone is a conservative character, and that plumage colours are not. But it is one thing to prove that the distinction is meaningful and theoretically valid; and quite another to provide a method by which all, or even many, particular characters can be quantitatively valued. The general direction of the rule can be stated—backbones are good characters, plumage colour bad—but that is not as complete a taxonomic technique as is desired.

So much for the theoretical method. Let us turn now to what I called the 'observational' one. It, too, has two main forms. The justification of both is that if a character has been proved to be reliable in one case, it will probably be good in another. The method assumes that reliability is a property of a character; therefore, if a character is reliable in one group, it will be in others too. This could be justified by the arguments about broad and narrow adaptations that we have just considered. If some characters do indeed fit a broad range of environments, and change less often, than others, then they are the reliable characters that evolutionary taxonomy seeks to recognize. It is, as we have seen, difficult to recognize them by theoretical analysis. But if a character is a broad adaptation it will remain constant in form, which should be detectable by simple observation. The argument of what I am called the observational method is as follows. The constancy of a character among a set of species suggests it is a broad adaptation. If the breadth of an adaptation is more a property of a character than of a taxonomic group, then what is a broad adaptation in one group will probably also be one in other group. If a character is particularly constant in existing groups (and is therefore likely to be homologous), it is also likely to be homologous in other groups. Such a character can accordingly be relied on in the classification of many groups. Of course, the group may matter: a character might be a broad adaptation in one group, but not in another. The method supposes this to be relatively unimportant.

The observational method, therefore, is to discover which characters are recommended by the study of other specimens, and to rely on them. If the characters are found in fossils, the method could theoretically be used even in the absence of any previous classification. Let us suppose that we have scored our specimens for their numbers of tail vertebrae, numbers of neck (cervical) vertebrae, tooth shape, and dental formula. All these characters are preserved in fossils. If some are more constant than others, if the number of neck vertebrae, for instance, was more constant than the number in the tail, that would be evidence they would be more reliable characters in other specimens. In practice the fossils will have been classified already, which assists the test of constancy because the test is strongest for a known sequence of ancestor—descendant populations (that is, the samples are classified in one taxon). The same method is possible without fossils. Instead of fossils, it can use the evidence of other, previously existing taxonomic groups. The fit of a new character to established taxonomic groups measures the character's reliability. If the character fits a group, being constant across the range of environments occupied by the species of that group, then it is probably reliable; if it changes within the group as if it had been repeatedly evolved and lost, it is probably unreliable. The strength of the test depends on how well established the existing groups are. If the groups are fairly certainly evolutionary, rather than convergent, and a new character fits them, that is strong evidence; if the group is newly erected, tentatively defined by only a few characters, and a new character fits it, that is less strong evidence. In practice, increasingly natural groups are developed, which provide increasingly strong tests of the reliability of new characters. The test simply checks whether a new character has the same taxonomic distribution as other characters, for the groups that the new characters are tested against are themselves defined by characters.

The coelom may provide an example. In the nineteenth century, the coelom was found to define a natural group of animals. The Polyzoa, however, in their original extension, comprised two groups, the Entoprocta and Ectoprocta. The former possess a true coelom but the latter do not; the Entoprocta are pseudocoelomate. For this reason above all, Hyman (1951) insisted on their separation: 'the insuperable difficulty [for their inclusion in one phylum] is the pseudocoelomate nature of the body cavity of the Entoprocta whereas the Ectoprocta are typical coelomate animals'. The coelom was known to be a good character from other animals and therefore could be supposed to be here too. The separation favoured by Hyman, however, is not universally accepted (Nielsen 1971, 1977); and the coelom 'may be polyphyletic' (Clark 1964).

The method is not valid unless the character in question has been tested, whether on pre-existing groups or fossils. It must have been shown to correlate with other characters, to define a natural group. If it was used to define the pre-existing group, other characters must be shown to have the same taxonomic distribution; if it was not originally used to define the group, it must subsequently have been found to fit the group well. If a character has been so tested, and has passed the test, we have evidence that it is homologous. In the manner already described, we can then reason, provisionally and in proportion to the case, that it is probably homologous in other, unclassified specimens.

I say *provisionally*. Evolutionary classifications are always subject to improvement. Each new piece of evidence can be used to test the existing groups, to test the current ideas about which characters are homologies and which analogies. If evidence piles up against a classification, it can be changed; and the revised classification will be superior to the one it replaces. Of course, the taxonomist's initial crude ideas of the classification of a group will influence the assessment of whether the next character under study is taxonomically good or bad: errors may be perpetuated, 'but the effect of the early classification *decreases* as re-classification takes place in the light of new evidence' (Hull 1967). The whole process is one of 'successive approximation'. 'Classification and re-classification . . . goes on all the time in evolutionary taxonomy in the light of the discovery of previously unknown species and additional evidence' (Hull 1967).

Critics of evolutionary classification—and especially the numerical taxonomists we shall meet in the next chapter—have often mistaken this procedure of successive approximation, or of 'reciprocal illumination' (Hennig 1966), for argument in a circle (see especially Hull 1967). The process clearly is not circular, but a normal scientific sequence of an initially crude hypothesis, a test, a hypothetical extrapolation, a further test, and so on. It is a process of theory-building. Its critics were led to their conclusion by a different understanding of evolutionary classification. Sokal and Sneath (1963), for instance, remark that,

> circular reasoning arises from the fact that new characters, instead of being evaluated on their own merits, are inevitably prejudiced by the prior erection of [another] taxon A based on other characters (X). Such a prejudgment ignores the fact that the existence of A as a natural (or 'monophyletic') group defined by character complex X has been *assumed* but *not demonstrated*.

In other words, according to Sokal and Sneath, characters are not tested before being used to assess new specimens. Although it sounds unlikely, untested characters may sometimes have been assumed to be homologous. But the misuse of a method is not a reasonable objection to it because any method, be it ever so good, can be misused. The passage just quoted 'is not a characterization of the best in evolutionary taxonomy but a parody of the worst' (Hull 1967). Circular reasoning is only a possibility in evolutionary taxonomy, not an inevitability. The system should be philosophically judged in its best form. Characters can be tested against other characters, and even against unclassified evidence, such as from fossils. Once it has been shown to be homologous, that demonstration can be put to work. Any proposition in science, once proved, can be assumed in further investigations. Sokal and Sneath (1963), could not see the difference between 'the principle of reciprocal illumination' and 'the much-condemned vertical construction of hypothesis upon hypothesis'. The difference is that between stacking a hypothesis on a successfully tested hypothesis and stacking a hypothesis on an untested hypothesis.

Opinions differ on the proper parts of the theoretical and observational techniques within evolutionary classification as a whole. I think of the observational method as testing the reasoning of the theoretical one. I therefore think they should be used together. Some authors, however, think the reasoning used in the theoretical method so imprecise as to be contemptible, and prefer to recognize homologies as those characters that are observed to have the same taxonomic distribution as a majority of other characters. They would first cluster the species according to all the characters, to give the most 'parsimonious' groups; only then, after the species are classified, would they recognize the homologies as the characters that fit the groups and the convergent characters as those that do not. This is a step towards the numerical taxonomy of the next chapter. In the extreme, all the characters could be fed in without prior analysis; but there is a gradation of methods from this extreme through to the individual testing of characters by the theoretical and observation techniques I have described. Other authors distrust the unanalysed statistical evidence of many characters, and prefer to rely on fewer, thoroughly thought-out characters. According to Hecht and Edwards (1977, and Gaffney 1979, agreed), for example, 'only one character is needed...From the viewpoint of lineage detection, it is more important to use a few well-analysed morphoclines than many poorly ... analysed ones'. Philosophers differ, but taxonomic practice is fairly uniform. Faced with a conflict of characters, evolutionary taxonomists almost invariably try to understand which characters are more reliable. Parsons and Williams (1963), for instance, in an important paper on the relationships of the modern Amphibia, discuss a long list of characters, to try to understand whether the similarities among groups are convergent or ancestral. It would be laborious, and probably unnecessary, to prove that Parson and Williams are not exceptional; they illustrate a rule. Biologists trying to discover homologies think about characters; they do not just multiply them.

After the homologies have been recognized, the next stage is to classify the species into groups according to their homologies. Although this can be done easily enough, and often without ambiguity, some groups pose a conceptual paradox. The classification is intended to represent evolution: but evolution has more than one aspect, and its different aspects do not always imply the same classification. An evolutionary tree cannot be simply converted into a classification. As Darwin said of Charles Naudin, 'his simile of tree and classification is like mine (and others), but he cannot, I think, have reflected much on the subject, otherwise he would see that genealogy by itself does not give classification' (Darwin 1887, vol. 2, see also Hull 1970). Darwin's remark is correct for a number of reasons of which only one matters here. Evolution has two main aspects: the order of splitting and the rate of divergence. As we saw in Chapter 1 (Figures 1.1–1.3), they can either agree or disagree. If splitting and divergence both take place at approximately constant rates, there is no difficulty. The paradox arises if divergence and splitting do not take place at constant rates (Figure 1.3). Bird and crocodile share a more recent common ancestor than do crocodile and lizard. Classification by order of splitting puts a crocodile with a bird; classification by degree of divergence separates the bird from the group of crocodile and lizard. When the different aspects of evolution disagree, evolutionary taxonomists have to decide which to represent. They have indeed decided. They have consistently chosen to classify groups according to their degree of divergence, not order of splitting. But how has the apparently ambiguous aim to represent evolution led to a consistent practice of representing one aspect of evolution rather than another?

At this stage, we shall merely set out the reasons given by evolutionary taxonomists. We shall not judge them after we have examined the competing arguments of phenetic taxonomy and cladism. Within evolutionary taxonomy itself, we should distinguish the justifications offered before and after the rise of cladism. They differ not in their logic, but in their decisiveness. Originally, evolutionary taxonomists recognized a difficulty; and, after some inconclusive reasoning, in which they pointed out that phenetic divergence is a real component of evolution, they came down in favour of the classification based on phenetic divergence (Figure 1.3 b) (Mayr, Linsley, and Usinger 1953; Simpson 1961 and elsewhere in that chapter). If the differential rates of evolution are large enough, the evolutionary taxonomist would allow it to override the order of branching of the group. Birds are raised to high taxonomic rank, equal with that of all the reptiles, because of their rapid evolution into an aerial niche. The difficulty with this argument is that it applies as strongly to convergence as it does to differential divergence: for convergence is a real part of evolution too.

The aim can be re-expressed by means of two crucial terms: homology and monophyly. In these terms, evolutionary taxonomy aims to classify monophyletic groups, defined by homologies. There are, however, homologies and homologies, monophyly and monophyly. Some homologies are evolutionarily ancestral, and others derived. Within the group of birds, crocodiles, and lizards, feathers are homologous within birds and scales are homologous within crocodiles and lizards, but scales are the ancestral homology of the whole group, feathers are a derived homology within birds. The evolutionary taxonomist classifies according to homologies in general, using ancestral and derived ones indiscriminately. A case like that of the birds and reptiles arises when one group evolves many unique, derived homologies of its own, while the group from which it arose retains the homologies of its ancestors. The birds have evolved many unique homologies while the reptilian groups have retained many ancestral homologies. The reptiles accordingly possess many homologies, and so too do the birds. They are both good groups in evolutionary taxonomy, which takes no notice of the fact that they are defined by different kinds of homologies. I shall use the term homology like evolutionary taxonomists, to refer to both ancestral and derived characters. Some cladists (e.g. Patterson 1982b) would restrict the term 'homology' to derived homologous similarities; but I shall not follow them in this work because the broader meaning is coherent in theory and, I believe, more widely comprehended. There is not much to be gained from, and little hope of effecting, the change.

The evolutionary taxonomist defines groups by homologies in order to ensure that the groups of the classification are monophyletic. This term is also ambiguous. Fortunately we do not need to discuss all its proposed meanings (to which Holmes 1980 gives many references.) For our purposes we need only distinguish two important meanings, although I shall mention two others as well. One unimportant meaning can be swiftly dismissed. It will not do to 'employ the term monophyletic as meaning the descendants of a single group of populations, descendants of a single species' (Mayr 1942), because in that sense any two groups could be combined in a 'monophyletic' group by going back far enough to their common ancestor (Hennig 1966). It is, however, easy to add to Mayr's definition the requirement that the ancestral species itself must be a member of the group. Then a monophyletic group is simply one whose common ancestor would be classified as a member of that group, to be contrasted with a polyphletic group, whose common ancestor would not be classified as a member of the same group. Such is the first important meaning of the word. Reptiles and birds are probably monophyletic in this sense. Some evolutionary taxonomists then worried that their aim to classify monophyletic groups was violated by groups such as mammals, which (it was thought) were polyphyletic by the definition (Simpson 1959a). Simpson (1961) duly redefined monophyly to save the phenomenon. 'Monophyly is the derivation of a taxon through one or more lineages from one immediately ancestral taxon of the same or lower rank' (Simpson 1961). By changing 'ancestor' to 'ancestral taxon', mammals could be made into a 'monophyletic' class even though they are (it was then thought) derived from several reptilian groups, because the relevant reptiles came from a taxon of lower than class rank. But the definition only thinly disguises its philosophical error, which is to allow artificial, convergent groups past the definition of monophyly, and into evolutioanary classifications. Mayr (1969) accepted Simpson's definition, and duly altered the glossary of his book from the vague formula of Mayr, Linsley, and Usinger (1953). But I wish only to mention Simpson's definition. The other important definition of monophyly, besides that labelled above the 'first important meaning', states that a monophyletic group contains all the descendants of a common ancestor. It is Hennig's (1966) definition. The difference between the two important definitions is that the evolutionary taxonomist's admists groups like reptiles, which are excluded by Hennig's (Figure 2.1). Hennig first published his definition in 1950, and it has since been approved by Sokal and Sneath (1963), most of Hennig's followers, and, I think, an increasing number of other taxonomists. But it is not approved of by Mayr (1969), who believes it is 'completely contradicted by common sense' and 'in opposition to the phenomenon of evolutioanary divergence'. Evolutionary taxonomy is certainly tied to the former, broader meaning. It does not define groups that are monophyletic in Hennig's sense. With the broader meaning a group can be called monophyletic even if it lacks some of the descendants of its common ancestor: the Reptilia can be defined and termed monophyletic. It is defined, moreover, by ancestral homologies. Ancestral characters reveal whether a group shared a common ancestor: it is hardly surprising that evolutionary taxonomists put ancestral homologies to use.

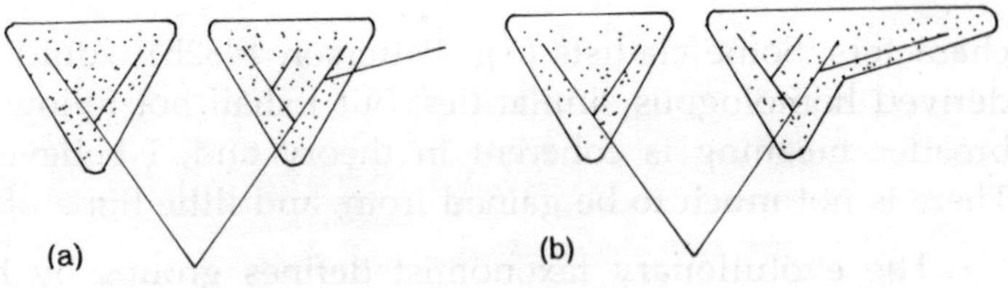

Fig. 2.1: Two different meanings of monophyly. (a) The evolutionary taxonomist's meaning: a group is monophyletic if it shares a common ancestor that would be classified in that group. (b) The cladistic meaning: a group is monophyletic if it contains all the descendants of a common ancestor. Not all the descendants of the common ancestor need be included in a group for an evolutionary taxonomist to call it monophyletic.

Evolutionary taxonomy has practically interpreted its philosophy of representing evolution as the aim of classify monophyletic groups, recognized by homologies. Homologies define monophyletic groups in the sense that they share a common ancestor that possessed the homologies; but if some of the descendants of that ancestor then lost the homologies, they will not be included in the group. The interpretation leads to a compromise among aspects of evolution, but a principled compromise, of a particular kind: phenetic resemblance can override the pattern of splitting; but only for phenetic resemblance in ancestral homologies. The kind of phenetic information that the evolutionary taxonomist will not admit (if we ignore Simpson's definition of monophyly) is convergent similarity. Convergent characters do not reveal common ancestry: they therefore must be excluded.

But in the original treatises that I cited, the conflict between rates of divergence and order of splitting was not taken to be a crucial issue. It was noticed as a difficulty, but not as one of great importance. They were content with some uncertainty as to whether splitting or phenetic divergence should be preferred in particular cases. Since then, however, the cladistic school has appeared. This has forced evolutionary taxonomists to state their preferences more boldly, and both Mayr (1974, 1981) and Simpson (1975, also 1978), as well as others (e.g. Johnson 1970, Bock 1974; Gould 1977b; Hecht and Edwards 1977; Szalay 1977; Halstead, White, and MacIntyre 1979; Ashlock 1980; Martin 1981; Charig 1981, 1982) have done so. They have interpreted their philosophy of monophyly and homology decidedly in favour of representing rapid phenetic divergence in classification.

They would not, however, always call it rapid phenetic divergence. According to Mayr (1974), birds are not simply classified by their phenetic divergence from reptiles. The evolution of birds, he stresses, was by means of an adaptive innovation: the change was not simply phenetic, it was adaptive; and the evolution of a new adapted type should be classified as a new 'grade' (Huxley 1958). Thus: 'Rensch, [etc.]. . . have particularly emphasized the importance of these levels of adaptation, designated by

Huxley as *grades*. All members of a grade are characterized by a well-integrated adaptive complex' (Mayr 1974 [1976]). And 'In the history of the vertebrates we know many such case of the formulation of successful new grades, such as the sharks, the bony fishes, the amphibians, reptiles, birds, and mammals. Each of these is characterized by a certain type of adaptation to the environment' (ibid.). Other revolutionary taxonomists agree (Simpson 1953, 1959b, 1961; Bock 1965).

The distinction of adaptive from phenotypic change matters because, as the next chapter will show, phenotypic similarity is an ambiguous concept. The status of the concept of adaptive similarity is less certain. It may be ambiguous as well, or it may be capable of unambiguous definition; I would only remark that the matter has yet to be usefully considered. Simpson and Mayr may say that such groups as birds and mammals are characterized by a well-integrated adaptive complex; but before the concept can be discussed it must first be defined. It has not been, at least in satisfactory form. Discussion cannot really even begin, let alone conclusions be drawn. Classification by shared adaptations is a separate principle from the two-phenetic and phylogenetic—we have been concerned with. For reasons that will become clear, I call it teleological classification. If we were to group species by their grades, we should need techniques to recognize shared adaptations, which yield unambiguous and hierarchical relations between species according to the similarity of their adaptations. We scarcely know whether species possess unambiguous adaptations, and we certainly lack any formal techniques to classify them. Teleological classification is a problem for the future. But that is not an end to the difficulties. Evolutionary taxonomy does not propose to group species purely teleologically by their grades. It would classify by evolution in some cases, and grades in others. The two concepts are quite distinct, and may even contradict each other: according to Simpson (1959a, b), the mammalian grade is polypheletic; and even if this case would now be denied (Crompton and Jenkins 1979; Kemp 1982a), a grade could still theoretically be polyphyletic. Evolutionary taxonomists have yet to explain why their philosophy demands a phenetic (or teleological) classification in the case of rapid phenetic divergence (or adaptive breakthrough), but not in the case of convergence. Even if the technical problem of recognizing shared adaptations is solved, the problem of when to classify into grades and when phylogenetically will remain. Major groups may arise by adaptive innovations, but so too do species; the difference is a matter of degree (Simpson 1953). Evolutionary taxonomy must specify where on the continuum of degrees of adaptive change it would abandon phylogenetic classification, and classify by grades. Moreover, the specification must be principled, or the whole system will collapse into subjectivity. It would, I think, be very difficult to find any such principle; unless, perhaps, macroevolution and microevolution are different in kind (Goldschmidt 1940; Gould 1980). Both the teleological principle and evolutionary taxonomy's use of it are too informal. Shared adaptations are practically recognized mainly by phenotypic similarity, and if evolutionary taxonomy recognizes separate groups characterized by 'adaptive breakthroughs', it does little more than classify according to phenetic divergence.

The original justification (as we saw) for doing so was simply that phenetic divergence is a real part of evolution, which should be represented in an evolutionary classification. For now, I am only pointing out that the justification is not consistent, for convergence is a real part of evolution too; but stronger criticism will come later. We have also to deal with two more justifications, which have been advanced more recently: humanity and information content. It is inhuman, some say, to dismantle such groups as reptiles and fish, which are 'more in tune with straightforward human perception' than groups defined only by the order of splitting (Martin 1981), and so obviously real (Halstead 1978; Halstead, White, and MacIntyre 1979). I have no more to say about this, because it is too far removed from my theoretical concerns. The other argument is that a classification incorporating two kinds of criteria, phenetic and cladistic, is more informative than one with only one of them. We shall come back to that.

In summary, the evolutionary taxonomist tries to classify species according to their evolution. Convergent similarities of appearance are vigorously excluded: their exclusion is the 'first step' in the system. The main method is the comparison of characters, to distinguish taxonomically good characters, which are those likely to be homologous, from taxonomically bad ones, which are those more likely to be analogous. Good characters are weighted highly in the classification, bad ones lowly: the classification represents monophyletic groups, defined by homologies. The system seems to have two main difficulties. Its philosophical aim to represent evolution is ambiguous: although it has been interpreted unambiguously, it remains to be seen whether the interpretation is valid. And secondly, its methods of distinguishing homologous from analogous characters, although theoretically coherent , are practically imprecise and permit subjectivity. The two more recent schools of taxonomy share a dislike of the subjectivity of the methods of the evolutionary school. But whereas the phenetic taxonomist's reaction is to remove them, the cladist's is to reform them. Phenetic taxonomists remove the whole difficulty by ceasing to try to represent evolution in classification; once that is accepted it is possible to develop more reliable techniques with conceptually more modest aims. The phenetic taxonomist would remove from classification all cladistic information. The cladist's decision is the exact opposite. Cladists would remove all the non-cladistic, phenetic information; their task is to improve the techniques of detecting evolution.

THE TECHNIQUES AND JUSTIFICATION OF PHENETIC TAXONOMY

Phenetic taxonomy, which I shall here identify with the numerical taxonomy of Sokal and Sneath (1963), can be understood as a reaction to the character weighting of evolutionary taxonomy. We have seen how character weighting is fundamental to evolutioanary taxonomy, but is only imperfectly formalized. When the phenetic taxonomist observes the vague weighting criteria of evolutionary taxonomy, he does not like what he sees. He has a precise mind, a taste for measurement and computers, and finds little pleasure in what evolutionary taxonomists so unblushingly call an art.

But he has a solution. All the artistry is to be swept aside. If, he says, there are no objective techniques of weighting, then there shall be no weighting; and if the only point of weighting was to represent evolution in classification, then, he says, something else will be found for classifications to represent. In numerical phenetic classification, all the characters are measured, equally weighted, and aggregated in a quantitative measure of the similarity of the taxa under investigation. The classification then represents not evolution, which the phenetic taxonomist thinks in any case in undetectable, but what it has always really represented, if in an imperfect form: phenotypic similarity.

The justification of phenetic taxonomy follows directly from its reaction to the character weighting of evolutionary taxonomy. Classification could in theory be either phylogenetic or phenetic, but because Sneath and Sokal (1973), and other pheneticists (Cain 1982) judge the methods by which phylogenetic relations are determined to be too uncertain for practical use, or even circular, practical classification, they say, ought to be phenetic. Phylogenetic divergence took place in the past; and phylogenetic relations cannot be known with the certainty of simple observation. Therefore, phenetic classification, which its advocates sometimes call 'empirical' taxonomy, would rely on quantitative, operational, untheoretical measurements of characters: leg length, arm length, chaetal number. There would be no need for subjective assessments. All the characters, once measured, would be weighed equally in the classification: all characters would be equally good. The whole procedure (they believed) would be perfectly objective and repeatable; from the same starting point, any taxonomist would inevitably arrive, in the end, at the same classification. This happy result could be contrasted with the endless disagreements of evolutionary taxonomy, which were only to be expected given the subjectivity of its techniques. Thus, if we turn to the section of Sneath and Sokal (1973), entitled 'the advantages of numerical taxonomy', we find two main, related words—repeatability and objectivity.

So, repeatability and objectivity were the aim of numerical taxonomists. And they devised statistical techniques to realize it. I may remark that the development of statistical techniques to compare phenetic resemblance is uncontroversially an important contribution of the phenetic school. I am not, however, going to spend long on the techniques or go into detail (which is abundantly supplied in their book). We are concerned with the controversial part of phenetic school, the assertion that statistical comparisons of phenetic resemblance not only should be made, but (because of their objectivity) should be used to classify. We shall only need to examine the techniques in sufficient detail to discover whether they truly are objective: further review would be superfluous here.

How, then, is a numerical phenetic classification produced? It is what Sneath and Sokal call 'the fundamental position of numerical taxonomy' (1973) that 'classifications are based on phenetic similarity' and 'a priori, every character is of equal weight in creating natural taxonomy'. The procedure of classification begins with plenty of phenetic measurements, of the more characters the better (leg length, nose length, hair length, immunoglobulin concentration, and so on), in each of the taxa under investigation. These taxa they term, in the rude idiom of a barbarous philosophy, 'operational taxonomic units', but that is too much to swallow: we shall term, for sake of example, species; they could, of course, in practice be any Linnaean level, or just organisms.

The techniques operate on numbers called distances, which can be calculated, for any pair of species with respect to any one character, by the subtraction of the measured value of the character in one species from that in the other. If the measurements for two characters are plotted on a two-dimensional graph, there will again be a distance (the Euclidean distance) between the two points for any species. Likewise there is a distance between two species in the 102-dimensioned plot of 102 characters; but 102 dimensions are difficult to draw. We shall confine ourselves to the two-dimensional case: exactly the same points apply in higher dimensions. Thus, if we have measured two variables, say immunog lobulin concentration and body length, in each of five species, an average point can be plotted for each species (Figure 2.2). Phenetic taxonomy then converts the measurements into a hierarchical classification, by means of a cluster statistic. We shall use, for example, a 'single-linkage' (or 'nearest-neighbour') cluster statistic. The method is to move out from each point, and as more and more distant species are encountered, steps are formed in the hierarchy. The nearest pairs are 1 and 2, and 3 and 4, each separated by a distance of one unit. These are the first steps. At 1.8 units the cluster of 3 and 4 captures 5 (whose nearest neighbour is species 4, 1.8 units away). The two clusters are not jointed until we arrive at 5 units, to give the final classification (Figure 2.3).

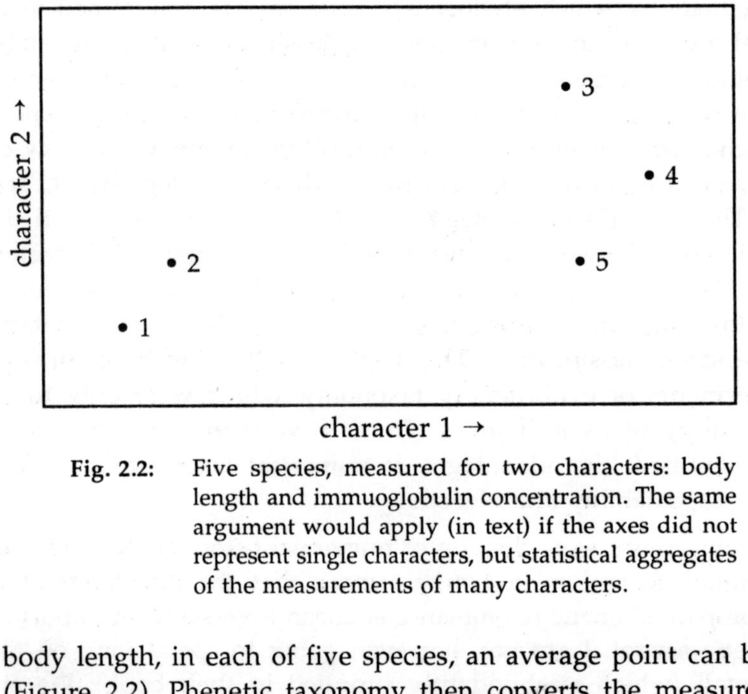

Fig. 2.2: Five species, measured for two characters: body length and immuoglobulin concentration. The same argument would apply (in text) if the axes did not represent single characters, but statistical aggregates of the measurements of many characters.

We have now seen the method, and (in a simple example) the resultant classification. Two points should be made before we move on. The first is that the classification is purely phenetic. It does not attempt to represent phylogeny. Evolutionary convergence is quantitatively represented in the classification as taxonomic similarity. It is no use objecting that convergence should not be allowed to obscure the relationships of species in the classification, because numerical taxonomists completely deny that it is an obscurity. They are not trying to represent, or discover, phylogeny: they are trying to represent something which is to them more easily known than phylogeny: simple phenetic similarity. It is not true that 'equal weighting and random selection of characters are implicitly based on the assumption that during evolution the genotype as a whole changes harmoniously and that all components of it change at approximately equal rates' (Mayr 1969). That would be true if the technique was intended for evolutionary taxonomy. However, it is not; it does not make any assumptions about evolution. It is intended only to discover phenetic similarity. Likewise, although sibling species should properly be distinguished in evolutionary taxonomy, if they really are phenetically indistinguishable they should not be in a phenetic classification. It is therefore not a criticism to say that 'sibling species are not perceived by numerical taxonomy, whereas some race differences may loom unduly large' (Dobzhansky 1970, also Mayr 1969). That is what the method aims to do. It is not a criticism of a school to point out that it successfully fulfils its declared intentions.

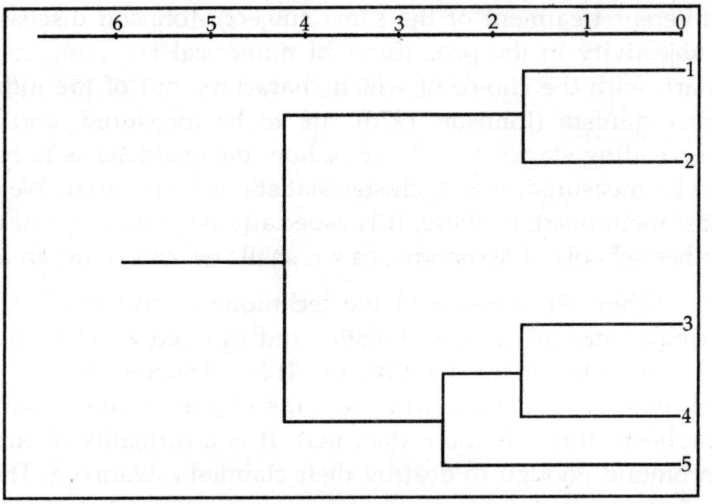

Fig. 2.3: Numerical phenetic classification of the five species of Figure 2.2.

The second point is that the procedure is, as the numerical taxonomist claims, perfectly repeatable. The whole procedure is exactly specified, and any two taxonomists who started from the same point would finish with the same classification. If they started with the same measurements they would arrive at exactly the same hierarchy. If each were left to choose which characters they were going to measure, the exact hierarchy might differ, but not by much if the number of characters was large. Although a classification made with measurements of only immunoglobulin concentration and body length would probably differ greatly from one made with measurements of body temperature and the number of neurones in the brain, if instead hundreds of characters had been used in each case, then even if there were little overlap between the to sets of hundreds of characters, the final hierarchies would probably be very similar. In the sense the techniques of numerical taxonomy are repeatable.

But we must now look less sympathetically at the 'objectivity' of numerical taxonomy. We must turn to the arguments in suspect acknowledged importance by L.A.S. Johnson (1970, first published in 1968; Sokal 1973, for instance, described it as '*the* outstanding critique of numerical taxonomy ever written'; Wiley 1981, gives a different treatment of the same subject). Johnson discussed several kinds of inherent subjectivity in the procedures of numerical taxonomy. Subjectivity enters right at the start, with the choice of which characters, out of the infinity of possible characters of an organism (Johnson 1970), are to be measured; and it can then be found at all succeeding stages, the choice of how the character is to be described, how 'distance' is to be measured, which cluster statistic is to be used. We shall concentrate on one, the last-mentioned, of these; it is especially important because it is a ground on which the other schools of taxonomy, as we shall see, can claim an advantage.

When we considered the technique of numerical taxonomy. In simply picked a 'single-linkage' cluster statistic, and applied it as it there were no difficulty in the choice. In fact there is a difficulty. It is a difficulty which is little discussed by numerical phenetic taxonomists, who treat the choice of cluster statistic as, at worst, a practical problem. But it is more than that. It is a difficulty of fundamental philosophy, and is profound enough to destroy their claimed advantage. The difficulty is that there is no natural aggregate measure of similarity among species.[2] Of course, techniques do exist, and the single-linkage, or nearest-neighbour, technique is one of them; the difficulty is that there is more than one, but no criterion (within the phenetic philosophy) by which one may be preferred to another. The point can be demonstrated by working through a particular example. Suppose that we have measured, in seven species, two different characters, and that the measurements look like those in Figure 2.4. (Again, the point is the same in 102 as in 2 dimensions, but the latter is easier to draw. In fact, this particular difficulty of phenetic classification deepens as more dimensions are added.) If we again follow through the single-linkage statistic with

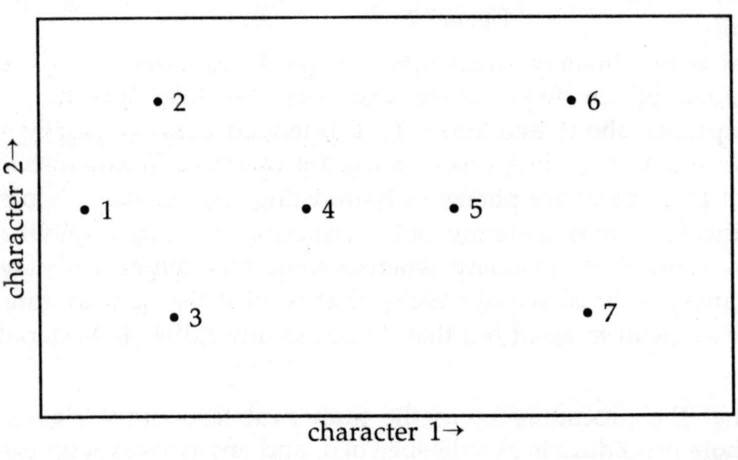

Fig. 2.4: Seven species, plotted in a two-dimensional taxonomic space. The axes could represent measurements for single characters, or the statistical aggregates of many characters.

which we are familiar, species 1, 2, and 3 soon form one cluster, and species 5, 6, and 7 another. Species 4 will first be captured by the cluster of 5, 6, and 7 because species 5 is its nearest neighbour. The resultant classification is Figure 2.6b.

The 'single-linkage' statistic, however, is just one of many cluster techniques. We could instead have used a 'furthest-neighbour technique in which a species is joined to the cluster to which it has the nearest furthest neighbour than the nearest, nearest neighbour. Or, instead of taking that cluster with the nearest or the furthest neighbour, we could calculate the average distance from a species to a cluster, and join the species to the cluster which was the shortest average distance away. In Figure 2.6 I have marked the average and nearest linkages from species 5 to the two clusters that are competing for it. Cluster B has the nearer nearest linkage, but cluster A has the nearer average linkage. If we use the average-linkage statistic we obtain one classification (Figure 2.6a); if we use the nearest single linkage we obtain a different classification (Figure 2.6b).

The hierarchy, in short, depends on the cluster statistic. This point is general. It does not apply only to single-linkage and average-linkage techniques. There are many more cluster statistics, and they all produce their own kinds of hierarchies. Sneath and Sokal (1973), have discussed dozens of different cluster statistics —principal-components analysis, factor-analysis, discriminant-function analysis, single-, complete, and average-linkage statistics, and so on—and an infinity can be imagined. The reader can turn the pages of *Numerical Taxonomy* and observe, in the illust-rations conveniently provided, the different hierarchies produced by each different statistic from the same raw set of measurements. And the point is not only theoretical. Johnson (1970) provides an illustrative list of references to numerical taxonomic papers in which various authors have obtained different classifications of real organisms from the same data.

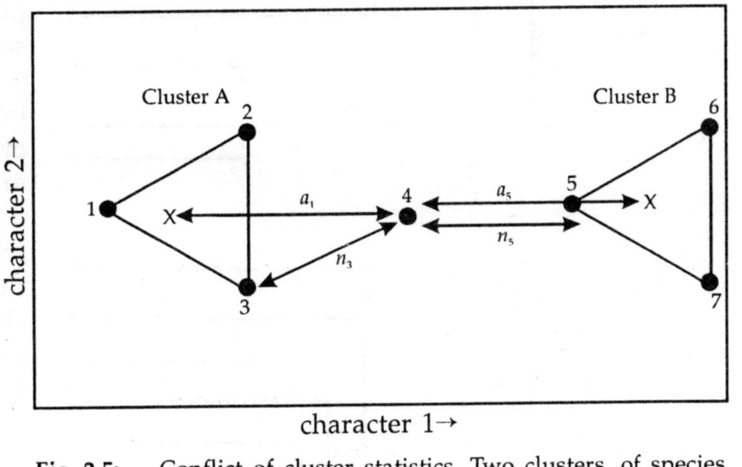

Fig. 2.5: Conflict of cluster statistics. Two clusters, of species 1, 2, and 3, and of species 5, 6, and 7, have formed: which cluster should species 4 join? Its nearest neighbour is species 5 ($n_5 < n_3$); but its average distance to the cluster (A) of 1, 2, and 3 (a_1) is less than that to the cluster (B) of 5, 6, and 7 (a_5). Thus a nearest-neighbour cluster statistic species 4 to one cluster; an average-neighbour statistic to the other: the resulting classifications differ.

The choice of cluster statistic is not the only problem. Johnson discussed several others, of which we may mention one more: the measurement of distance. We have so far been clustering according to the Euclidean distance between species. In two dimensions, for two species, the Euclidean distance is the hypotenuse (calculated by Pythagoras's theorem) of the triangle of which the other two sides are the distances apart of the two species in each dimension. But this is not the only measure of distance. One of the early works of numerical taxonomy, by Cain and Harrison (1958), suggested that the distance between species should be measured by 'mean character distance' (MCD). The mean character distance is the average of the distances of each dimension (rather than the sum of their squares). As with different cluster statistics, so with different measures of distance, the hierarchical classifications may differ according to which is used.

Although the different techniques have been classified and their different properties discussed—some produce short hierarchies with all the species similarly related, others longer hierarchies with more steps—no one has discussed the real relative merits of the techniques from fundamental principles. We do have discussion of computational time, of algorithmic consistency; quantitative methods have been developed to compare different statistics (Sneath and Sokal 1973, Rohlf 1974; Rohlf and Sokal 1981). But the question of principle has not been faced. There are to reasons. One is that the phenetic taxonomist feels no strong need to: if phylogenetic classification is impossible, classification can only be phenetic; and Sneath and Sokal (1973) are accordingly untroubled by Johnson's demonstration that there is no optimal phenetic classification. The other is that they have not, because they cannot. There are absolutely no criteria to support such a discussion. But although there is no fundamental reason within phenetic taxonomy for preferring one statistic to another, the practical phenetic taxonomist has to use one of them. He will

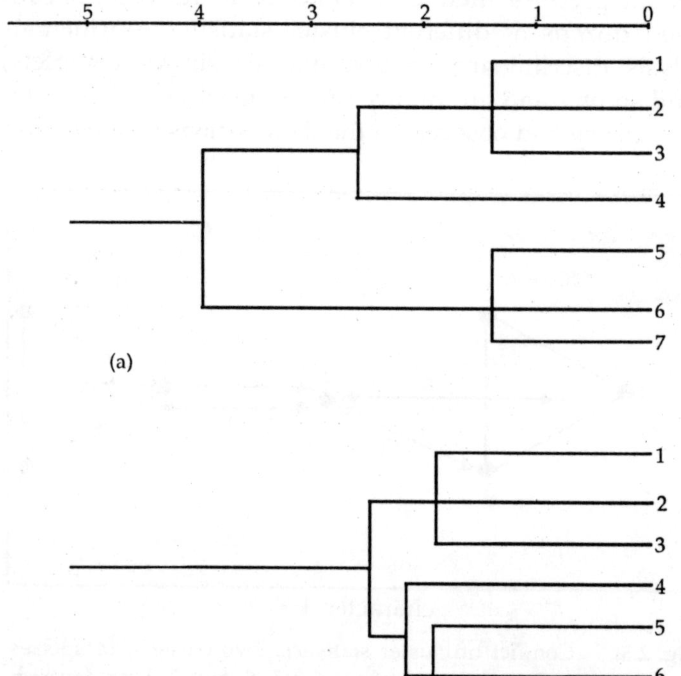

Fig. 2.6: Numerical classification of the seven species of Figure 2.4: (a) by average-neighbour statistic; (b) by nearest-neighbour statistic. They differ topologically: species 4 is classified differently by the two statistics.

have to make a choice. It is of course perfectly possible to do so, but the choice, when it is made, will be . . . no, surely not?

It will be completely subjective. The reason is easy to see. The only objective method of deciding between cluster statistics would be by reference to some higher criterion, which the statistics are actually meant to be discovering. But no such criterion exists. If there were, in nature, a true hierarchy of phenetic similarity among species, then it would be easy to test the different statistics against the hierarchy, find out which one was best at discovering it, and then preferentially use that statistic. In fact, the only measures of phenetic similarity are provided by our statistics; there is no independent, external criterion against which we can test them. For this reason, Johnson (1970) stated that there is an inherent circularity in the philosophy of phenetic taxonomy. The cluster statistic is meant to measure aggregate morphological similarity, but that does not exist independently of the cluster statistic: there are as many concepts of aggregate morphological similarity as there are cluster' statistics. When the choice is made, even if it is concealed by the circumlocutions of information technology, it will have to be made on practical considerations, or may even be quite arbitrary. In the late 1950s the numerical taxonomists came in exploding petards beneath the subjective old taxonomists they were going to eliminate, only to be hoist, in their turn, and within a decade, on their own favourite device.

The subjectivity of phenetic taxonomy has two broader consequences. The first concerns the evolutionary school, which is damaged by the same criticism as phenetic classification. Evolutionary taxonomy does contain phenetic element, and (as we have seen) insists that it is real and merits classification. Evolutionary classification is not as purely phenetic as numerical taxonomy, because it excludes convergence; but it does include differential rates of divergence. How can differential divergence be measured? Any attempt to measure it as simple phenetic change is bound to run into Johnson's difficulty, as Johnson (1970), himself realized. If it is to be thought of not as phenetic, but as adaptive change, we need techniques to measure adaptive similarity, and techniques that avoid the ambiguity of purely phenetic techniques. No such technique has ever been suggested, although one may be possible. The teleological principle might then supply the need for some theory by which to choose among cluster statistics. Ecological adaptation might be the property of the real world we could check our statistics against. Evolutionary taxonomy, however, has more pressing problems of subjectivity, and discussion does not customarily reach this far. But the evolutionary taxonomist' troubles with weighting are in principle soluble; one day they may be solved. If they ever are, another philosophical difficulty will be expecting the attentions of that school.

The second consequence concerns the distinction of natural from artificial classifications. Biologists disagree over the formulation of this distinction; but, according to one popular form, in a natural classification the members of a group agree in other characters besides the one (or more) used to define it, whereas in an artificial classification they do not. Because the distinction mentions only resemblance of characters, it is phenetic. So, just as phenetic similarity cannot be objectively measured, nor can a single 'most' natural classification be identified. Is the classification of Figure 2.6a more natural than that of Figure 2.6b? The question is unanswerable. The degree of naturalness of a classification cannot be used to judge its objectivity.

This conclusion is strictly circumscribed. It does not mean it is nonsense to discuss the naturalness of a classification, nor that our ideas of phenetic similarity are an illusion. They are both concepts with a meaning. The members of a group, such as the Reptilia, which is phenetically defined, do appear more similar to each other than to other species. In most respects a lizard and a crocodile look more like each other than does either to a bird. It has not been the purpose of this chapter to argue they do not. Phenetic measurements can still be made, as indeed they must be for any evolutionary study of particular characters; the naturalness of classifications can still, at least informally, be compared. We shall continue to use both concepts when their subjectivity and informality do not matter. What the conclusion doe mean is that there is no *single* meaning of phenetic similarity or naturalness that could underwrite an objective system of classification. Phenetic similarity alone cannot supply a consistent, unambiguous, objective, hierarchical classification of all life. Lizards and crocodiles may resemble each other more closely than do either with birds: but the implicit measurements and weightings would not necessarily give a hierarchical classification of all life; nor would they necessarily be the same in the case of the gross phenetic similarity of some other animals, such as millipedes, centipedes, and insects; nor would the implicit cluster statistics in these different cases necessarily be consistent. And even if our intuitive ideas of phenetic similarity did supply a unique, consistent, hierarchical cluster statistic, that statistic would still be (in the sense of this book) subjective: it would owe its power to mass agreement, not natural principle. We certainly can measure the phenetic similarity of species with respect to characters; we can aggregate the measurements in multivariate comparisons, if we have some reason to do so; but that does not make phenetic classification objective. If it is objective, one particular measurable cluster statistic must represent a unique natural hierarchy. The argument we have been through suggests that, within the phenetic system, that is impossible.

I have one final point, which will take us on to the next chapter. Phenetic taxonomists, as we have seen, oppose the weighting of characters. They dislike the subjectivity of the methods of character weighting in evolutionary taxonomy. But in fact, if an explicit, principled, quantitative case can be made for weighting, there is no real reason (within its philosophy) why numerical taxonomy should not admit cluster statistics that systematically weight some kinds of characters heavily, and others lightly. When I quoted earlier from Sneath and Sokal, they said numerical taxonomy was only opposed to weighting a priori: it could be introduced if a case is made; and they discuss statistics that do so. If only evolutionary taxonomy could provide a quantitative rule to govern its technique, it could then be incorporated into phenetic taxonomy. Phenetic taxonomy could become, in the technical sphere, a kind of evolutionary taxonomy, or vice versa. If only there could be found within evolutionary taxonomy, what cannot be found in the original form of numerical phenetic taxonomy—a real hierarchy for the cluster statistic to aim at—then evolutionary taxonomy could solve the problem of the cluster statistic. In fact, there is in evolutionary taxonomy a natural hierarchy that can serve this purpose—it is the hierarchy of phylogenetic branching. For this reason, numerical taxonomy may be led, by its own logic (Farris 1977, 1979; Mickevich 1978), to develop itself into the school of classification that provides the subject of our next chapter. It is led to become cladism.

REFERENCES

1. Which has not stopped some Popperians from arguing that neutralism is unscientific (Ayala *et. al.* 1974); Maynard Smith (1978, p. 37) points out that 'the neutral theory is a good "Popperian" one'.
2. Like Charig (1982, p. 371), but for a different reason. I prefer the term 'aggregate similarity' to 'overall similarity'. Like him, I shall use the term we prefer. 'Overall similarity' is the usual term.

Chapter—3

Ontogenetic Evolution

'The eye to this day gives me a cold shudder.' (Darwin, 1887)

Our understanding of embryology is today in a state of quite remarkable confusion, which, as I shall attempt to show, is not the result of any scarcity of relevant information but must almost certainly be explained in terms of the vagaries of the scientific process: the presumptions contained in the survey which follows may perhaps be excused in the light of this confusion and justified by the importance of the implications for kindred subjects such as evolution. Embryology finds itself in a situation in many ways comparable to that of evolution immediately before Darwin, the requirement being not yet further data but a unifying conception of the appropriate generality to rationalize vast tracts of detailed but unconnected information. Therefore, if, as may well be the case, such a conceptual framework awaits us in embryology, we might anticipate that it will cut across traditional, entrenched subject barriers. It may be surprisingly simple; it will turn out to have been amply evident in the facts for some time and will probably have been unsuspectingly alluded to by many authors in the past; it will not require, and probably will not be amenable to, any proof in terms of any single, new experiments; indeed, the underlying molecular mechanisms may remain uncertain for some time to come or may even be quite incorrectly conceived, though with no serious effects on the particular formulation of the general issues, as with Darwin's misconceptions regarding mechanisms of inheritance.

I. EMBRYOLOGY

Current Theoretical Issues and Models

Mosaic development and the nature of preformed cytoplasmic factors in the egg

Preformationist views of embryology might well be thought to be of no more than historic interest but, to quote one recent influential textbook: 'Localization as used here is the specification of cell fate according to the section of egg cytoplasm inherited by an embryonic cell lineage. . . . The localization phenomenon is particularly interesting to contemporary students of development because it suggests that specific programmes of development are sequestered in the egg cytoplasm' (Davidson, 1976). The argument for cytoplasmic formative factors is based on the independent ('mosaic') differentiation of separated egg or early embryo fragments in accordance with their normal expected developmental fates. While the cells of all vertebrate embryos retain totipotentiality until the stage where the embryo consists of many hundreds or thousands of cells (as proved by the formation ('regulation') of whole embryos from embryo fragments (Fig. 3.1; Spemann, 1938), many invertebrate embryos begin differentiation at a time when the total number of cells is much fewer and correspondingly their eggs are small. At the relevant pre-differentiative stages, tests of cellular potentialities by fragmentation and transplantation experiments comparable in detail to those done in vertebrates are difficult to perform. Yet the question of the degree of prior commitment of embryo fragments can only be judged by adequate testing. Possible initial multipotency can only be excluded if the earliest stages have been so tested. Mosaic behaviour as such may be explicable in different terms; isolated cells cleave in a predictable and characteristic way that may itself be a sufficient precondition for their future characteristic patterns of independent cellular specialization. In some cases (Conklin, 1905), experiments involved killing of selected blastomeres rather than separation; it is known that, even in regulative organisms such as amphibians, this can result in non-regulative development of the surviving blastomeres (Fig. 3.2), suggesting that the killed cells are still exercising a normal role (presumably here only a mechanical one) of restricting the behaviour of other cells.

However, the most damaging evidence regarding preformed development is the fact that in the species which classically exemplify the phenomenon, such as ascidians or *Amphioxus* (Conklin, 1905, 1933), an entirely different, regulative, result can be obtained (see, for example, Wilson, 1893; Reverberi, 1971). Regeneration of complete organisms from adult components, which is a common potentiality even in species showing mosaic development (Berrill, 1961), is incompatible with the idea that cell fates are fixed by their descent. Furthermore, there is positive evidence in mosaic species that, far from being fully committed as a result of

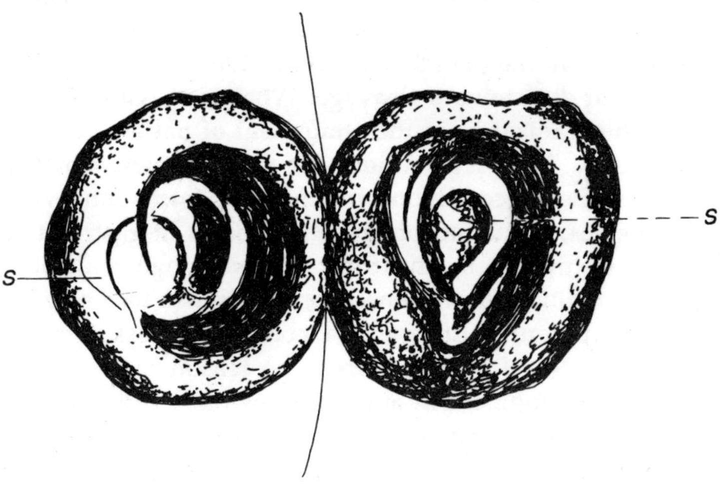

Fig. 3.1: 'Regulation' in the amphibian embryo: two normal larvae (*S*) derived from two initial blastomeres separated by a thread. (From Herlitzka, 1897).

their pattern of derivation from specific regions of the egg, the development of some embryonic components can be made to vary according to the nature of neighbouring tissues (Reverberi, 1971). A good example is shown in Fig. 3.3: the blastopore lip was transplanted to the abdominal region in an *Amphioxus* embryo; the resulting induction of neural tissue was closely comparable to the results of the same experiments in amphibian or avian embryos.

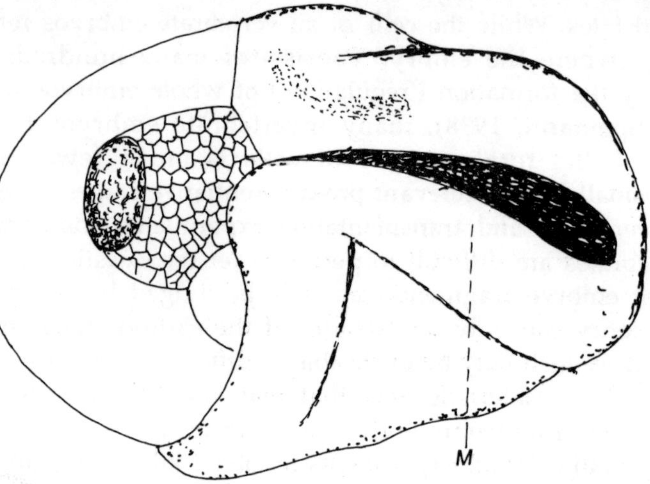

Fig. 3.2: 'Mosaic' development in the amphibian: after killing of one of the two initial blastomeres, the remaining cell goes on to form a rostral half-embryo (*M*). (From Roux, 1888).

I am therefore arguing that the time has come to lay the ghost of preformation. Many of the most cherished examples such as 'germ plasm', muscle or pigment cell precursors in ascidians and the amphibian 'cortical organizer' are now open to alternative interpretations (Nieuwkoop & Sutasurya, 1979; Gerhart *et. al.*, 1981; Horder, 1981). There is no denying the theoretical possibility of intracellular spatial organization in egg cells; this reaches some complexity in protozoan cells white protozoan inheritance of 'cortical organization' (Beisson & Sonneborn, 1965) may be a mechanism available in controlling metazoan determinate cleavage. What is at issue is the question of the *nature* and *role* of such organization in the attainment of patterns of cell differentiation in multicellular organisms. The preconditions for development which must be programmed into the egg are the preconditions for the first formation event in embryogenesis, which, in vertebrates, is the morphogenetic process of gastrulation, and it happens that there is direct evidence, in the form of pseudogastrulation (Smith & Ecker, 1970), for corresponding morphogenetic preconditions in the egg, based presumably on the well-attested potential of eggs for complex cytoskeletal organization (Rappaport, 1970; Harris, 1979). Moreover, there is no reason to regard the evidence in species showing mosaic development in any fundamentally different way. The mosaic development of isolated early embryonic fragments may reflect no more than the continuance of a predisposition to certain patterns of cleavage, which, as the determinate cleavage of such species indicates, must be tightly controlled; it is not a test of the state of commitment of cells to their prospective patterns of differentiation at the time of isolation. It should be pointed out that, at best, such techniques reveal only low degrees of elaboration of preformed pattern: they cannot demonstrate a greater number of preformed features than the number of blastomeres or fragments that can be isolated and shown to behave differently.

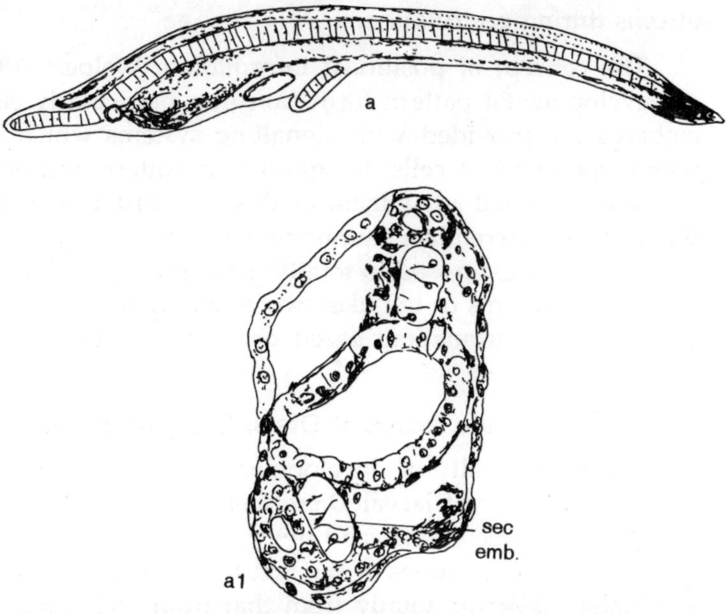

Fig. 3.3: Primary induction in *Amphioxus*: the blastopore lip of one early gastrula was transplanted to the blastocoele of another: *Top*: larva with transplant lying ventrally. *Bottom*: section showing notochord (sec. emb.) derived from transplant and neural tube induced by it from host abdominal ectoderm (Tung *et. al.*, 1962).

Embryonic Regulation; Gradients; Positional Information

Having put preformation in perspective, we have to face up squarely to the central problem of development: on what basis can a pattern of discrete, differentiated, anatomical parts emerge from the starting conditions provided in the egg and early embryo, starting conditions which the phenomenon of regulation shows must be diffuse, totipotent and self-organizing? The idea of a gradient has an immediate intuitive appeal and it has surfaced on numerous occasions and in varying guises: e.g. 'fields', 'equipotential harmonious systems', 'polarity'. The existence of gradients has been posited in a wide but somewhat ill-assorted range of biological situations. Surprisingly, the most frequently quoted examples are not in embryonic development but in regeneration in invertebrates (Child, 1941) and the usual test of graded organization is 'susceptibility' of the formed organism to toxic agents. The choice of species is relevant because in organisms such as *Hydra* or planarians, with their relatively simple head to tail organization, the anatomical structure may well engender graded effects under certain experimental tests; whether this has any meaning as regards actual developmental controls is another matter. Segmented organisms such as annelids, on the other hand, may regenerate merely by repetition of their sequential polarized pattern of development. Slime mould aggregation is often put forward as a cast-iron case of a gradient controlling a pattern, but we are perhaps entitled to question the universality of mechanisms seen in highly specialized organism facing quite unique 'developmental' problems. The widespread effect of a transplanted dorsal blastopore lip on a recipient amphibian embryo (Spemann, 1938) may be transmitted, not by a diffusing chemical, but by movement of cells during gastrulation.

The concept of positional information (Wolpert, 1969) encapsulates the problem of developmental pattern formation in a particularly clear-cut form. It is argued that embryos are provided with signalling systems which inform cells of their positions within the array of cells; integration of pattern and embryonic regulation reflect the operation and self-adjustment of this long-distance communication channel; and cells themselves differentiate autonomously on the basis of this information. The model was originally expressed in totally general terms without necessarily being linked with diffusion gradients or the like: two recent, quite independent and explicit formulations of positional information, based on particular classes of experimental evidence, will be considered here.

The Evidence from Imaginal Discs; Compartments; Homoeotic Mutations

The subspecialty of developmental biology concerned with pattern formation in the imaginal disc of larval *Drosophila* is a particular interest because it has offered a direct link between genes and development of complex patterns but it is important to examine the implications of the fact that it is based not only on a new phylum but on technology differing totally from that from which most embryological concepts were derived.

Genetics has been exploited to investigate clonal patterns of origin of cells in imaginal discs. Somatic crossing-over can be induced in single cells at chosen stages of development, and resulting expression of markers used to visualize the pattern of the clone of derived cells. Clones formed early are large and irregularly shaped, but later formed clones show ordered features such as elongation in the long axis of the formed wing (Bryant, 1970), most easily interpreted as the result of controlled orientation of mitoses (Bryant & Schneiderman, 1969; Vijverberg, 1974) together with stretching out of an initially small patch of cells during evagination of the disc. Garcia-Bellido *et al.* (1973) have proposed that certain regions where clones never overlap may represent a hidden constraint to development (a 'compartment' boundary) actively involved in pattern control. In some situations where similar 'compartment' behaviour occurs, there are detectable anatomical boundary structures which could be associated with mitotic restriction. For example, clones early become restricted to the confines of single discs in conformity with the anatomical partitioning characteristics of these discrete epithelial structures: Lawrence & Green (1975) describe indications of a boundary between abdominal segments based on changing pigmentation patterns and cell shape parallel to epithelial folds; dorso-ventral clonal restriction in the wing corresponds precisely with the anterior and posterior margins of the wing (see Brower *et al.* (1982)), who also report indications of cell alignment (including peripodial membrane cells) parallel to the antero-posterior 'compartment boundary'). In such cases clonal restrictions correspond to eventual anatomical features because they share a common denominator (i.e. mitoses are orientated by disc folds which themselves anticipate adult morphology). In principle, there is no reason why mitoses should not be restricted (Simpson, 1976), perhaps by patterns of tension in the disc epithelium, without there being any corresponding visible structural feature. The causal connection with the common denominator may not be one to one: antero-posterior wing boundary and vein IV may partially share an underlying morphological determinant even though their anatomical correspondence is not exact. This boundary is not a unique feature in the wing: clones elsewhere show a tendency not to cross veins (Bryant, 1970). There are situations in development (Lawrence & Morata, 1976). and in regeneration (Bryant, in Ashburner & Wright, 1978) where clonal boundaries shift *but* morphological patterning is not correspondingly distorted, thus indicating against a causal role of clonal restriction in pattern formation.

A second component of the compartment model is the correspondence between clonal restriction patterns and the spatial distributions of the effects of homoeotic mutations, which characteristically transform the pattern of differentiation of one disc into that of another. This raises the vexed question of the timing and early forms of developmental commitment of disc cells. It is frequently assumed that, because isolated fragments of discs fulfil their prospective fates accurately, all cells have their fates fixed but, as was the case in studies of embryos showing mosaic development, such a test is open to serious criticism. Specific patterns of development finally displayed by fragments may be the indirect result of varying specific patterns of growth, adhesiveness and shaping existing at the time of isolation (Garcia-Bellido, 1967) or of later induction by patterns of underlying mesoderm (El Shatoury, 1955). It is not justifiable to infer that within each fragment (which is rarely smaller than one-tenth of the whole disc) single cells are individually committed. A second source of evidence for disc differentiation is that based on apparent selective reaggregation of randomly mixed disaggregated cells derived from marked disc or disc fragments, selective reaggregation being identified by the formation of 'mosaic' patterns, i.e. composite structural patterns of cuticular organelles showing integration across cells of two separate origins (Tobler, 1966). However, if cells are not yet developmentally committed, mosaic patterns could be the result of 'inductive' interactions across randomly associating clones. Poodry *et al.* (1971) were able to show that reaggregation was extremely limited: mosaics, which reached only limited anatomical complexity, were made up of clones of cells derived largely by mitosis from between one and six original cells. Therefore the final pattern of differentiation was not the result of ordering of already committed cells: each founding cell was the ancestor of a *number* of final differentiated structures. The situation is further complicated by the fact that surviving clones are probably derived from small undissociated cell clumps rather than single cells (Bryant, in Ashburner & Wright, 1978): preconditions for patterned development therefore exist regardless of reaggregation. Ultimately, however, the most telling evidence pointing to the labile state of commitment of imaginal discs is transdetermination (Hadorn, in Ashburner & Wright, 1978) and the heteromorphosis of discs induced by teratogens (Waddington, in Counce & Waddington, 1973; Bryant, in Ashburner & Wright, 1978).

The most positive evidence regarding the nature of pattern control comes from examples (reviewed by Waddington, in Counce & Waddington, 1973) in which the formation of one structure can be shown to depend on the presence of another: e.g. bracts (Tobler, 1966), veins (Garcia-Bellido, 1977), siting of bristles (Stern, 1968), movement of sex-comb (Horder, 1976), growth stimulation by dominant zones (Lawrence & Morata, 1976), mitotic competition (Simpson, 1976), orientation of cuticular ridges (Lawrence, in Counce & Waddington, 1973). Induction is considered to be involved widely in insect development (Counce, in Counce & Waddington, 1973). It might be thought that there is counter-evidence in the form of the supposed position specificity of 'autonomously' developing clones of homoeotic cells (Postlethwait & Schneiderman, 1971) which fail to interact with surrounding wild-type tissue. However, there are difficulties in judging the appropriateness of the position of a given homoeotic structure in the context of a non-corresponding foreign structural background; in some cases (Postlethwait & Schneiderman, 1971) individual cuticular organelles are clearly abnormal, so perfect autonomy cannot apply. In studies such as these, clones may be large enough so that the relative normality of pattern within them may be accounted for by internal interactions. There is the further possibility (Horder, 1976) that, since discs are homologous, they can provide equivalent pattern-defining cues for homoeotic clones sufficient to allow fairly normal development.

The compartment model seeks to link the apparent decision-making homoeotic genes with the control of integrated pattern formation within specific developmental fields; but the available data suggest pattern is built up gradually by multiple sequential interactions between differentiated structures. The model is difficult to assess given the existing uncertainty as to the meaning and significance of homoeotic mutations. Even *engrailed,* the prime example of transformation of posterior wing into anterior, shows unexplained anatomical abnormalities over and above those expected of a straight-forward anterior mirror duplication across the antero-posterior compartment boundary, and her it is known that pattern depends on local direct interactions between cells which can actually distort pattern in the face of a largely intact posterior compartment (Lawrence & Morata, 1976). Several considerations point to the interpretation that homoeotic mutations owe many of their effects to the organization of insect development rather than any special feature of the wild-type genes as such. It is notable that mutations with comparable effects are not prominent among vertebrates, whereas it is an obvious characteristic that insects are not only organized segmentally but that the segments may be homologous and therefore share many aspects of their development (Horder, 1976). Mutations blocking one divergent developmental pathway may readily lead to reversion to another pattern of homologous development. Heteromorphoses and teratological phenocopies suggest the ease with which alternative pathways may be selected. The restriction of the effects of homoeotic mutations within the bounds of single discs may reflect the physical separation and independence of disc structure.

The Polar Coordinate Model; Intercalary Regeneration: Distal Transformation

This model (French *et al.*, 1976) is closely related to studies of regeneration in *Drosophila* imaginal discs. After division of discs the larger fragment regenerates a complete organ while the smaller tends to form a set of duplicate parts (Bryant, in Ashburner & Wright, 1978). This pattern applies regardless of the orientation of the dividing line. The seeming equivalence of all points around the circumference of the imaginal disc led to the proposal that organization is based on a continuous circumferential position-defining mechanism. The 'shortest intercalation rule' would explain complementary regeneration/duplication; for a large fragment, regenerative intercalation of the shortest sequence of circumferential values between its cut ends would result in a complete disc, but for a small fragment replacement of intervening positions by the shortest sequence will produce duplication. Complimentary regeneration-duplication patterns of regrowth of amphibian limbs are always explicable as 'distal transformation' (Bryant, 1975) which itself is a replay of the rule underlying the polarized pattern of normal development, namely that distal parts are sequentially laid down on previously formed proximal parts. In discs, there is no correspondingly marked epimorphic sequence in development and duplication does not fit an equivalent explanation because, depending on surrounding tissue context, one cut edge can duplicate *or* regenerate. For discs one alternative model might run along the following lines. Regeneration of a complete disc, which is associated with fragments which are large, may reflect the ability of that fragment to impose normal and integrated overall pattern on new tissues: duplication, associated with small fragments, may reflect absence of overall integration and domination of regionalities specific to the fragment. The polar coordinate model has difficulties explaining substitution of regeneration in place of duplication in long-term disc cultures (Schneiderman, 1979) and instances of spontaneous duplication (Wildermuth, 1968), as well as the evidence for direct interdependence of disc pattern elements already referred to.

The model has at the same time been applied to regeneration in adult amphibian and cockroach limbs, although certain differences are unexplained. Whereas the cockroach lib shows true intercalary regeneration when proximal limb parts are transplanted onto distal without rotation (Bohn, 1976), this does not occur (French *et al.*, 1976) in amphibians. This different pattern of behaviour may find an explanation in the different nature of the tissue involved; unlike the amphibian, the cockroach undergoes continual structural modification in the form of moults involving stretching out of an epithelium which gives rise to the intercalated tissue. Intercalation in the longitudinal axis leading to the formation of supernumerary limbs is subsumed by introducing a second rule (the 'complete circle rule'), whereby distal transformation in the proximo-distal axis of the limb can only occur if a complete set of circumferential values exist at the initiating proximo-distal level. Again there are paradoxical features to some of the data; if, as Bohn (1976) shows, segments of the cockroach limb are homologous (longitudinal intercalary regeneration occurs as if to complete each proximodistally mismatched segment, regardless of whether mismatched tissue is from the same or different segments), then, after limb rotation, why does longitudinal supernumerary intercalary regeneration continue to include all distal segments? But it is undoubtedly with the vertebrate limb that the model becomes most obviously inappropriate and specific predictions of the model have not been borne out (Tank & Holder, 1981; Wallace, 1981) with the result that the model has been revised (Bryant *et al.*, 1981). It is difficult to see how such a model would ever explain findings such as those in which different components of limbs, such as skeleton, muscle and skin appendages, are independently patterned (Saunders *et al.*, 1976; Maden, 1980). Circumscribed wounding of the limb (involving a small fraction of the circumference) or flank (which has no circumference (Fig. 3.5) is sufficient to initiate outgrowth of a supernumerary limb (Tank & Holder, 1981; Wallace, 1981). We know enough about the simplicity of the organization involved in initiation of a vertebrate limb (Figs. 3.5–3.7, and see later) for there to be much easier explanations for such a result than the strained and arbitrary modifications that are required of the polar coordinate model (Bryant *et al.*, 1981).

Experimentally Established Foundations of the Analysis of Development

A complete alternative to mechanisms based on the clonal history of a cell or on long-distance communication exists in the findings of classical experimental embryology which provide overwhelming, and undisputed, evidence for the decisive role, in the control of cell differentiation, of purely *local* and *progressively arising* circumstances.

Embryonic Induction

Lens induction by the optic cup or neural plate induction by archenteron roof are facts beyond serious contention, but these stand for a considerable number of examples, so that we have effectively reached the point where there is evidence, scattered in the literature, to show that each tissue and organ in the vertebrate is dependent for its development on interaction with an appropriate inductor. In each case, the common feature is that an already specified tissue, such as the eye cup specified by previous events, including neural tube induction, comes to lie immediately adjacent to unspecified embryonic tissue: induction of a new specific cell type such as lens then results. This interaction is *necessary* (removal of the inductor prevents the appearance of the induced organ (Fig. 3.4a) and *sufficient* (transplanted inductors can change the fate of competent tissue throughout the embryo (Fig. 3.4b)) for the formation of the induced structure.

To interpret inductive signals as form of 'positional information' is a semantic device which undermines the principal claim of the positional information model (its integrating long-distance channel of communication), and leaves unexplained why the inductor acquires the properties of a source of positional information (this presumably requires introduction of new and arbitrary parameters into the originally simple model) and also why tissue whose fate is changed by an inductor transplanted to a foreign site does not respond to the more general positional information still available in the rest of the surrounding embryo.

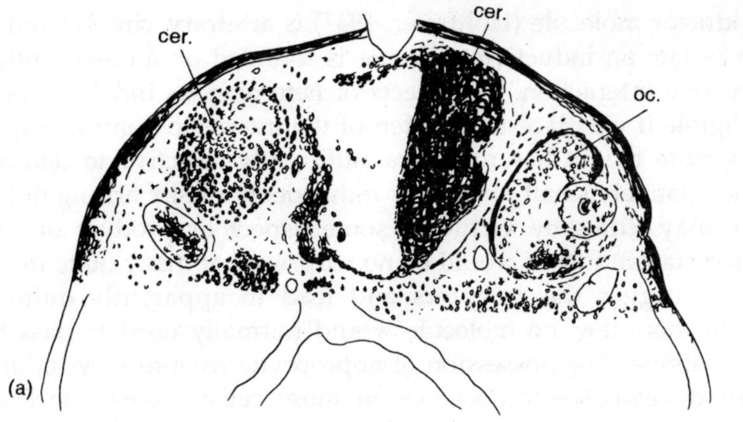

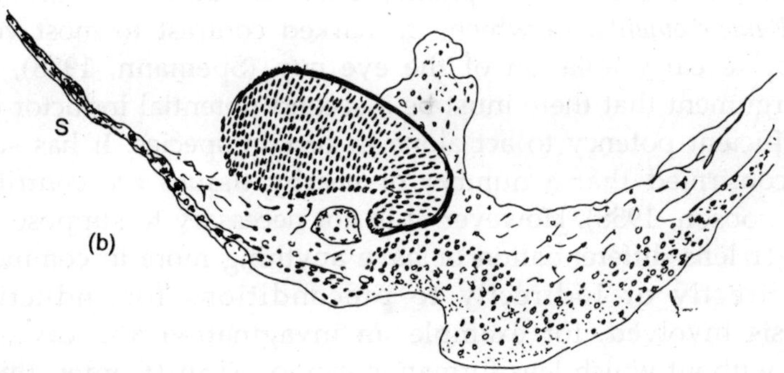

Fig. 3.4: Lens induction. (*a*) Removal of one eye primordium at neurula stage in *Rana fusca*: lens fails to form from prospective lens ectoderm on left side. *l*, normal lens; *cer*, brain; *oc*, normal eye. (From Spemann, 1901).

(*b*) Transplanted optic vesicle of *R. palustris* can change developmental fate of trunk ectoderm of *R. sylvatica* to form lens (marked S): an example from the experiments of Lewis (1904).

Why induction has not found its rightful place as the very bedrock of all theories of development is not easy to explain, although a number of past contentious issues have presumably served to cloud the subject generally.

(*a*) It may be thought that the fact that, despite exhaustive investigations, no single inducing factor has been definitively isolated or chemically identified speaks against the reality of the phenomenon. Worse, in the cases of a number of inductions (Saxen & Toivonen, 1962), a large number of heterologous agents, with no possible relation to biological reality, have been found to act in a fashion identical to that of normal inductors. For example, boiled salamander heart can induce a lens in urodele ectoderm (Holtfrter, 1934). To introduce concepts such as that of 'sublethal cytolysis' (the idea that induction of ectoderm into neural plate under tissue culture conditions at appropriate pH must be the result of release by the tissue itself of suitable, normally masked, inductor molecule (Holtfreter, 1947) is arbitrary, circular and unnecessary. The attempt to isolate an inductive molecule is founded on an assumption that induction is an instructive interaction. The effects of heterologous inductors become very much more intelligible if we place the burden of the interaction on the responding cell; if we were to suppose that it has receptors sufficiently selective to detect and distinguish between the relatively small number of inductors normally arising in the embryo (which themselves may differ by virtue of some aspect or product of their own already differentiated states), then it would be no surprise that many quite incidental extraneous agents could trigger the receptors and lead to apparently quite normal induced responses. In this view, no molecule would normally need to pass between inductor and induced tissue. The possession of appropriate receptors by induced cells does not imply prior developmental decisions in those cells; a complete battery of receptors may be an intrinsic property of all cells from the egg onwards.

(*b*) Confusion surrounds the phenomenon of 'double assurance'. The famous anomaly of *Rana esculenta,* in which, in marked contrast to most vertebrates, a lens develops despite early ablation of the eye cup (Spemann, 1938), gave rise to the convoluted argument that there must be a second potential inductor of the lens which is only of sufficient potency to act alone in certain species. It has subsequently been abundantly confirmed that a number of nearby tissues are contributory to typical inductions (Jacobson, 1958). However, it is not necessary to suppose that these tissues actually trigger lens differentiation or have anything more in common than that they contribute directly or indirectly to preconditions for induction such as the morphogenesis involved, for example, in invagination and cavitation of the lens primordium, without which lens formation cannot occur (Horder, 1981).

(*c*) One renowned and beautiful experiment (Spemann, 1938)—transplantation of urodele abdominal ectoderm onto anuran head with resulting induction of a correctly located balancer—poses intriguing problems. How could the anuran provide an inductor for an organ it itself cannot produce? At first sight this is strong evidence for 'universality' (with respect to species) of continuously distributed 'positional information', but not necessarily. The result may illustrate the retention through evolution of homologous structures and inductive mechanisms: the anuran has as an homologous organ, the structure which the urodele ectoderm normally regards as its inductors (Horder, 1976).

(*d*) It could be argued that the concept of embryonic induction provides no solution to the problem of development, because it does not explain regulation and it implies an infinite regress of pattern back into the egg: each inductor must be preceded by its own correspondingly positioned inductor indefinitely. Such an argument omits the essential involvement of morphogenetic cell movements!

Morphogenesis

If induction mediates the switching on of differentiated cells, it becomes essential to account for the control of the formation and spatial arrangement of inductors. An especially characteristic feature of embryos is morphogenetic cell movement, of which gastrulation is the earliest, largest-scale and most consistent example among species. In vertebrates, the massive cellular rearrangements of gastrulation lead to the formation of the archenteron, which induces the neural plate, which itself, by say of additional further morphogenesis, becomes further subdivided to produce a further generation of more localized inductors such as the optic vesicle. Starting from simple preconditions, morphogenetic rearrangements are therefore the mediators of step-wise localization and increasing diversification of inductors. Spemann (1938) showed that the entire anatomy is conditional on the dorsal lip of the blastopore, which mediates gastrulation, and there is no difficulty in envisaging the necessary morphogenetic preconditions for this event within the egg (Smith & Ecker, 1970; Freeman, 1981). Regulation, after halving an embryo or as shown in the spreading influence of a transplanted blastopore on the blastula, can be explained as the result of the infolding and redistribution of the invaginating arhenteron, according to the extent of blastula tissue available.

Morphogenetic variables provide a basis for long-distance coordination of pattern formation in the embryo compatible with short-distance inductive interactions as the immediate determinants of specific differentiation in individual cells. Due consideration of morphogenetic forces alone can lead to ready explanations for the diversity with which often small numbers of apparently identical specialized cell types can become arranged into characteristic organ systems. By way of illustration the development of the limb will be discussed as a specific example.

At the heart of pattern formation in the limb is the condensation of initially homogeneous mesoderm (already 'induced' in the lateral plate) into precartilagenous rudiments. Thereafter, limb development follows a relatively predictable and autonomous (Chalmers & Ray, 1962), though complex, growth and patterning by means of processes such as joint formation, ossification, longitudinal epiphyseal growth and functional modelling. Precartilagenous condensation is a morphogenetic event which probably serves to space out rudiments competitively (Hinchliffe & Johnson, 1980) within the existing mesodermal field; their size and number is not absolutely fixed in any one species but appears to vary according to availability of mesoderm (Ede, 1971). Thus, it can be argued that the problem of control of skeletal pattern will reduce to the question of the control of the shape and dimensions of the early limb bud which in turn determines positions of

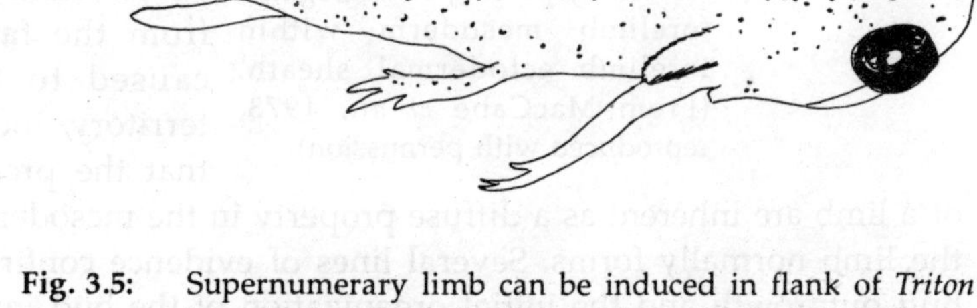

Fig. 3.5: Supernumerary limb can be induced in flank of *Triton* larva by implant of celloidin. (From Balinsky, 1927).

condensations. Indeed, Slack (1977) describes how, after inducing enlarged limb outgrowths, the number of cartilage rudiments is proportional to limb bud size.

There is considerable evidence to show that the dimensions of limb buds and regenerating blastemas are actively controlled in each species: deliberate attempts to double or halve the bud (Harrison, 1921) or the blastema (Goss, 1961) usually (although results are complicated by regression or fusions of supernumerary elements) result in early adjustment to normal dimensions; early in the formation of supernumerary limbs (which originate from tissue originally destined to form a single limb) both buds are equal and normal in size (Harrison, 1921); and in transplantations of limbs to hosts of the same species but varying size, or to other species whose limbs are smaller or larger (Harrison, 1969), the transplant may catch up but never exceeds a growth rate characteristic of its own normal capacities. Because of the distortion of the blastema by the stump and the substantial contribution of covering epidermis, it is difficult to assess the dimensions of regeneration blastemas in relation to those of limb buds, but they may be comparable (Tank *et al.*, 1977).

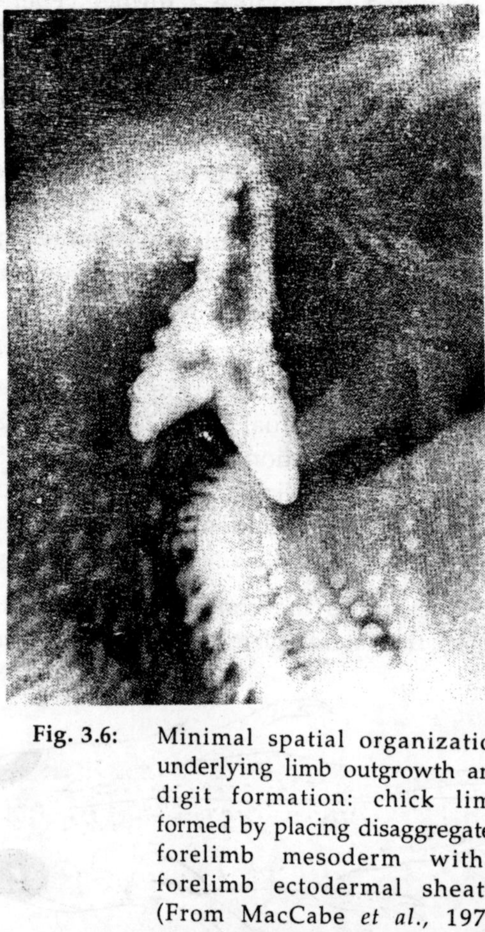

Fig. 3.6: Minimal spatial organization underlying limb outgrowth and digit formation: chick limb formed by placing disaggregated forelimb mesoderm within forelimb ectodermal sheath. (From MacCabe *et al.*, 1973, reproduced with permission).

How is the specificity of limb bud shape controlled? It must be inferred from the fact that new limbs can be caused to form throughout a wide territory, including the flank (Fig. 3.5), that the prerequisites for the formation of a limb are inherent as a diffuse property in the mesoderm and ectoderm from which the limb normally forms. Several lines of evidence confirm that the initial impulse to bud outgrowth and the initial organization of the bud involve minimal complexity in spatial terms (Fig. 3.6). Thus supernumerary limb outgrowth can be triggered by a wide variety of extraneous agents and manipulations (Wallace, 1981; Tank & Holder, 1981; Fig. 3.5) whose lowest common denominator appears to be nothing more than non-specific wounding with delayed healing. Formation of duplicate limb parts in the developing chick limb can be induced by a diffusely organized zone of mesoderm (the 'zone of polarizing activity', ZPA) (Fig. 3.7) whose low level of complexity is indicated by the fact that it is not species or sage or limb type-specific and that it can be mimicked by various tissues from outside the limb (Saunders, 1977) and by vitamin A (Tickle *et al.*, 1982). It is hard to resist the conclusion that, just as a transplanted ZPA can induce new limb parts by interacting with non-ZPA limb tissue, normal limb outgrowth itself may be triggered in the first instance by interaction of the earliest ZPA rudiment with surrounding flank tissue. The same interaction explains limb duplication following amphibian limb bud rotation (Harrison, 1921). Although the initial trigger is therefore probably mesodermal, continued outgrowth is known to be the result of the interdependence of mesoderm and ectoderm: limb mesoderm induces the formation of the 'apical ectodermal ridge' (AER) and, after ZPA transplantation, delays its spontaneous regression (Saunders, 1977); if the AER is ablated outgrowth ceases, while transplantation of the AER shows that it mediates the shaping of the mesoderm (Saunders *et al.*, 1976). The AER is particularly prominent during the formation of the foot or hand plate and looks as thought it is involved in the process of flattening out of the terminal bud, thereby leading to the characteristic shaping associated with the pattern of terminal skeletal parts.

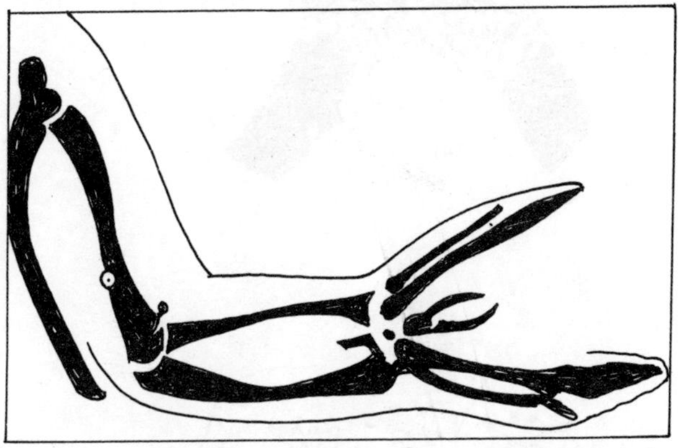

Fig. 3.7: Minimal spatial organization needed for control of digit number and type: chick forelimb duplicated distally by inserting disaggregated ZPA (zone of polarizing activity) cells into anterior limb bud. (From Tickle, 1981: reprinted by permission from *Nature* **289**, 295-8. © 1981 Macmillan Journals Ltd.).

At the present time it would be unwarranted to be more specific regarding mechanisms. The main points are, first, that the preconditions for outgrowth and eventual autonomous formation of a limb are diffuse and simple and, secondly, that all the important tissue components involved in continued outgrowth (AER, ZPA, apical mesoderm (Kieny, 1964)) retain the same potentialities at all stages. Therefore, from simple starting conditions, poximo-distal elaboration of pattern must arise in multiplicative fashion, each successively laid down distal limb region acquiring more complex feature as the result of forces, such as AER shape changes, imposed by already formed more proximal tissues. There is no doubt that the distal limb tip can be influenced by proximal structures, e.g. by transplanted supernumerary skeletal elements (Goss, 1961), and this is reflected in the regulation of tips transplanted to older or younger limb stumps (Kieny, 1964).

With regard to the skeleton as a whole, we might expect, given the uniformity of cell types, frequent correlations in structure among regions of the adult skeleton of

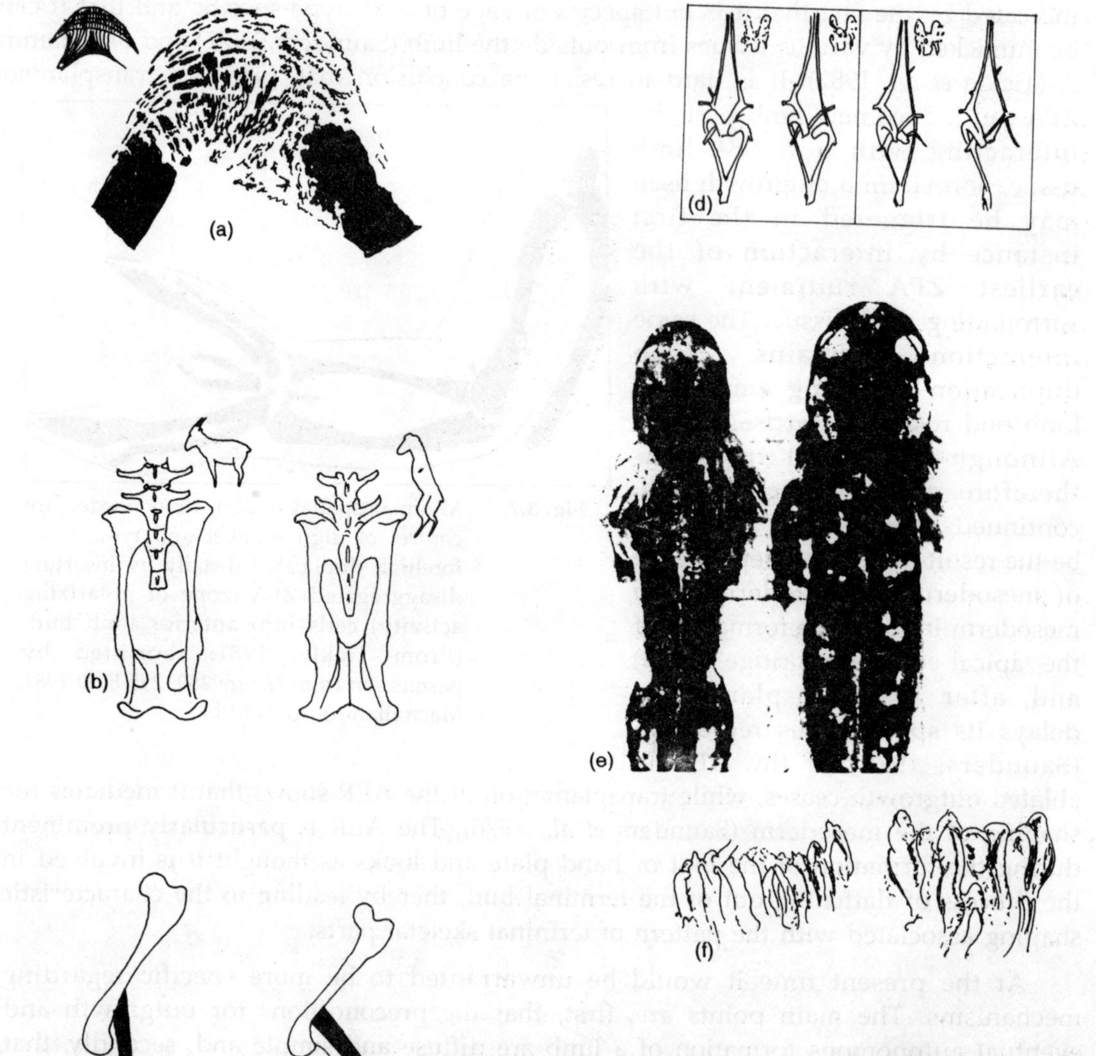

Fig. 3.8: (*a*) Ankylosed human knee showing formation of new trabeculae in lines of applied stress (inset *b*). (From Weiss, 1934.) (*b*) Transformation of pelvis towards form typical of bipedal species in congenitally bipedal goat: normal on left. (From Slijper, 1942.)

the kind described, for example, between fore- and hind-limbs, trunk and tail in lizards (Kramer, 1951). However, skeletal pattern achieves considerable regional diversity because it is arrived at by a variety of embryological routes, e.g. ossification may be membranous or cartilagenous and both forms of ossification are themselves conditional on a wide variety of prior morphogenetic permutations, such as those determining migration of neural crest and sclerotome. As the limb demonstrates, slight variations in starting conditions are capable of amplified and complex effects on later pattern. A major source of diversity and of detailed adaptive patterning and growth results directly from the functional demands to which skeletal elements are subjected (Fig. 3.8*a*, *b*, *c*).

Summing Up

Induction must once and for all be acknowledged as a central mechanism in embryogenesis. This mechanism is massively documented throughout the embryology of both vertebrates and invertebrates. The failure of present technology to substantiate induction in molecular terms is not important since the phenomenon can be unambiguously and fully defined operationally as described above while, for the purposes of explaining pattern formation, this level of description is both appropriate and adequate. Further, the concept of a battery of receptors lends itself readily to extension to include the role of other known developmental variables such as the effect of hormones or morphogenetic, mitotic and functional stimuli: any realistic definition of induction must embrace the integration of such multiple factors. The much neglected evidence for the role of 'function' in morphogenesis (Fig.3.8) indicates the responsiveness of specific cells to particular features of their circumstances which, conceivably, could be the basis of their response to their inductor, of later maintenance of their differentiated state, and of growth or, equally, atrophy and degeneration in the case of organs no longer involved in 'functional demand' (Goss, 1964).

Morphogenetic mechanisms for the deployment of cell types are inextricably linked with induction. It is obvious from the few examples already mentioned that spatial integration and timing of pattern formation depend crucially on the scale, rate and form of morphogenetic events. For an explanation of how these embryonic phenomena are themselves acquired and inherited in various species, we turn to a consideration of evolution.

(*c*) Adaptation of muscular attachments following experimental creation (on right) of a tibial pseudarthrosis in a dog. (From Wermel, 1935.) (*d*) Modelling of pattern of major vessels by blood flow: aortic arches of 4-day chick embryo were blocked with metal clips in various patterns shown above. (From Rychter, 1962.) (*e*) Functional control of gill growth: salamanders raised under 11% (*left*), 100% oxygen (*right*). (From Drastich, 1925.) (*f*) Functional control of growth of intestinal villi: 54–70% removal of the small intestine in a dog leg (*right*) to 30–50% taller cells and 4–5 x dilated crypts of Lieberkuhn. *Left*: control. (From Goss, 1964, after Flint, 1910, reproduced with permission.) (*g*) Control of melanocyte function by temperature (°C): threshold for pigmentation on extremities in Himalayan rabbit. (From Schmalhausen, 1948).

II. EVOLUTION

It is ironic that it is Darwin's influence by way of Haeckel that has in one way or another dominated our thinking regarding the relation of evolution and embryology. De Beer (1958) demonstrated the way in which Haeckel's explanatory use of 'recapitulation' had overinterpreted the relationship, but, in his determination to refute that view, he came to concentrate on particular aspects of development in which the time relations in evolution had been changed ('heterochrony'), neoteny and the problem of larval forms being the classic examples. A closely related concept is allometry (Thompson, 1942; Huxley, 1972), by which evolutionary trends are analysed directly in terms of differential tissue growth rates controlled perhaps by 'rate genes' (Goldschmidt, 1940); again the emphasis is on temporal differentials in development. Such approaches establish the important point that evolutionary change could very plausibly be explained in terms of what are, in principle, simple embryological trends or transformations. However, they are essentially descriptive and beg the question of what controls expression of the hypothetical rate genes. As indicated by the multitude of factors affecting limb development, it is surely unrealistic to think in terms of single, organ-specific controllers of growth: the attainment of a predictable species-specific rate of growth (Harrison, 1969) is explicable rather as the result of the balancing of combinations of multiple positive and negative forces. One of the few laws of evolution of any generality that have emerged, 'Cope's law' (which state that there is a tendency for all established species to evolve into larger forms), has recently been shown to have explanations other than in terms of growth control (Stanley, 1973).

Approaches such as these share with Haeckel's a form of teleological thinking; conclusions were drawn about embryological processes by inference from the adult form, an approach which betrays a hidden assumption that the embryo is simply a miniature adult, whereas the embryological processes discussed above show the attainment of the adult to be confusingly indirect. Moreover, the starting point of these approaches is the adult form of ancestral species, since new species are considered as heterochronic or allometric variants of the original; as de Beer realized, this entirely leaves out of account any true evolutionary 'novelties'. In so far as the adults of ancestral species are the products of their own embryologies, then it follows that gradual evolutionary modification must entail modification at embryological stages. While it must be true that all processes in embryogenesis are potentially open to modification through natural selection, it must also be true that the selective advantages of changing certain processes will be very different from others. It is inconceivable that modification of vertebrate embryos could include any fundamental revision of the basic processes of DNA replication and translation because so much depends on them. For the same reason, the earlier the embryological event the more dependent will be a greater number of subsequent embryological processes and so the greater the likelihood that it will be conserved in the evolution of new forms of embryos. It follows that there will be a tendency for early embryological features (such as early stages of limb development) to be retained and shared in later-evolved species, with characteristically increasingly complex, sequential embryologies superimposed. It is therefore inevitable that a form of 'recapitulation' should be seen in embryology. Von Baer's Laws, dating from well before Darwin, express this fact free of any interpretation, evolutionary or otherwise.

Embryology and evolution are therefore fundamentally interdependent. Assuming the embryological mechanisms described above, what potentialities and constraints do they impose for evolutionary change? How have these mechanisms themselves evolved and been perpetuated through generations. These matters will be considered in the context of a specific example.

Speculations on the Evolution of Lens Induction

In many respects, the development of the vertebrate eye has been successfully charted out: may of the component tissue contribute in patterns which parallel their deployment elsewhere throughout the embryo, while shaping and growth are the end result of many, often morphogenetic, interactions involving fluid pressures, scleral skeletal elements, bony orbit and so on (Coulombre, 1965; Horder, 1981). But the unique and characteristic early event of lens induction represents an obvious test case for any account of the evolution of embryological mechanisms.

There is no doubt as to the strength of the evolutionary pressures in favour of visual sensitivity. Salvini-Palwen & Mayr (1977) estimate that photoreceptors have evolved independently in various phylogenetic lines at least forty times. We can trace the vertebrate eye reasonably securely from *Amphioxus* in which light receptors occur diffusely throughout the cavity of the central nervous system; as is true throughout evolution (Eakin, 1973; Salvini-Plawen & Mayr, 1977), photoreceptivity is associated with cilia, which are features of the ventricular cells.

Lateral and pineal eyes could then have evolved as gradual evaginations from a photoreceptor-rich region of the neural tube, which itself evolved into the diencephalon. Consistent with this scheme is the retention among vertebrates of light sensitivity in the diencephalon and varying degrees of anatomical overlap between diencephalon and the pineal (Hulsemann, 1967; Horder, 1981). This proposal satisfies an essential requirement which should be met by any hypothetical evolutionary sequence; a continuous sequence of morphogenetic events in an embryo is a repetition of a continuous sequence of morphological steps built up through the preceding evolving series of embryos, each stage of which must have been functionally advantageous in the transitional organism. This will be referred to as the continuity principle. In the case of the eyes the morphological continuum from the diencephalic origin means that functionally advantageous photoreceptors could have differentiated at each transitional stage in evolution but how are we to apply the principle in the case of the lens of the lateral eye, which, given its separate origin from the ectoderm, is most unlikely to have evolved in morphological continuity with the lateral eye?

In spite of its simpler level of organization, the pineal eye (Fig. 3.9) should almost certainly not be regarded as a transitional form on the way to the evolution of the lateral eye. The two structures probably represent distinct morphological variants, founded on a common basic set of specialized cell types, which diverged from the start of their evolution as a consequence of their different points of emergence from the neural tube. In both, receptor cells face the lumen of the original ventricular cavity, but, whereas the pineal remains as a vesicle surrounding this cavity with receptors pointing forward, in the lateral eye infolding has resulted in orientation of receptors *away* from the lens. It has also resulted in a transformation of the interrelation of receptors with the pigment cells which are arranged between receptors in the case of the pineal (Studnicka, 1905; Eakin, 1973).

The structure of the pineal may, however, provide insights of possible relevance to the evolution of the lateral lens. The distal wall of the pineal optic vesicle is frequently specialized, reaching, in some species, degrees of elaboration which make it clearly recognizable as lens (Fig. 3.10). This identification is confirmed by the presence of crystallins (Eakin, 1973). This is of considerable interest because it demonstrates a possible sequence for lens evolution in which functional and morphological continuity are maintained. The structure of crystallins has so far offered few clues as to the evolutionary ancestry of lens cells (Bloemendal, 1981), but (echoing the diffuse arrangement of pigment cells between pineal receptors) they may well have originated from a cell type which was diffusely distributed in the neural tube and may initially have had quite different functions, perhaps supportive. Consistent with this supposition is the persistence of cilia in pineal lens cells (Eakin, 1973) and the widespread synthesis of lens crystallins in neural tissue in early embryos (Bloemendal, 1981). Parallelling pigment epithelial evolution from the

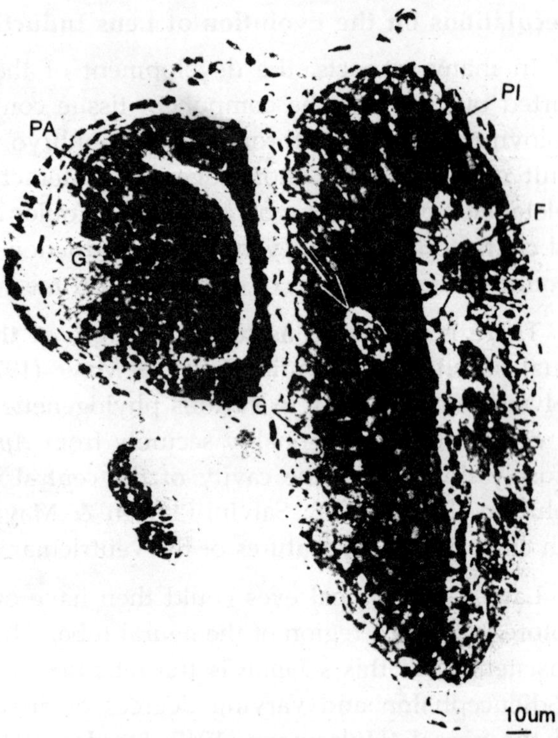

Fig. 3.9: Superficial pineal eye of *Entosphenus tridentatus* (PI) shows lens-type specialization distally (to right) where it is in contact with dermis, whereas there is no such specialization in the deeper parietal eye (PA). G = ganglion cells; IS and OS = inner and outer segments of receptor cells; F = fold in roof of pineal eye. (From Eakin, 1973; © 1973 by the Regents of the University of California, reprinted by permission of the University of California Press.)

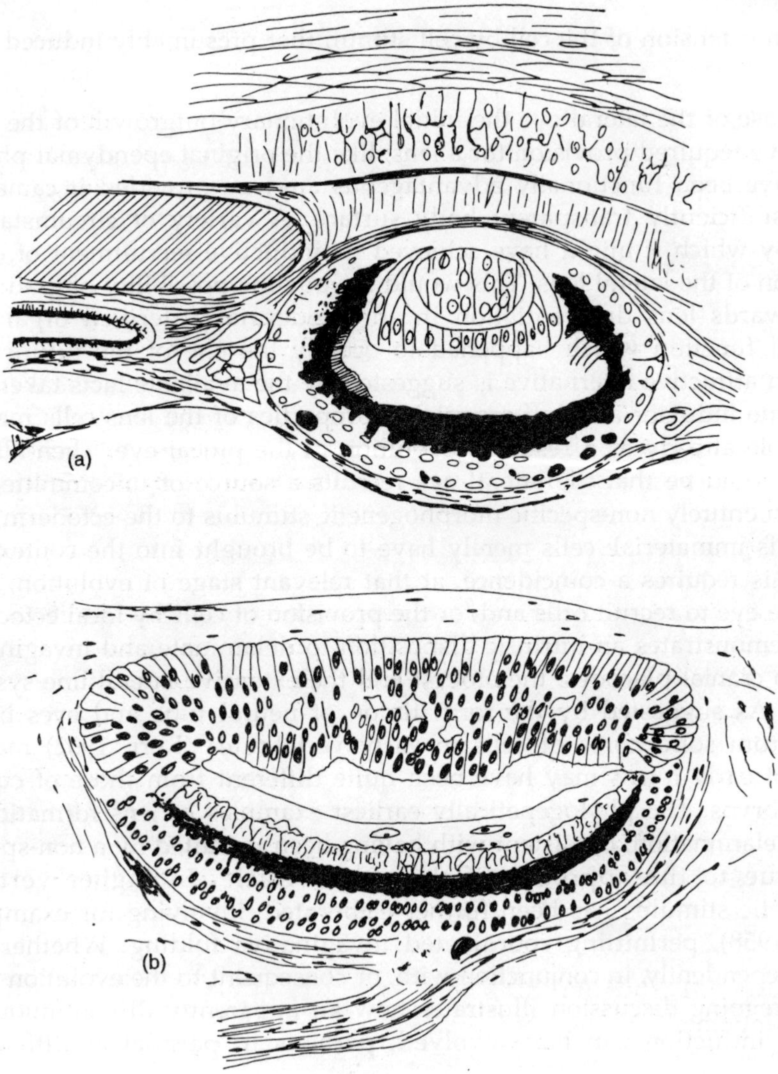

Fig. 3.10: (a) Pineal of *Anguis fragilis.* (From Studnicka, 1905, after Leydig, 1890). (*b*) Pineal of *Iguana tubericulata.* (From Studnicka, 1905, after Klinckowstroem, 1894).

stage seen in the pineal to that seen in the lateral eye, one can see, by comparing the pineal organs of various species (Studnicka, 1905), a continuous progressive concentration of lens-type cells at the distal surface (Fig. 3.9), culminating in a tendency towards total segregation by self-organization (Fig. 3.10). Of course pineal lens differentiation in extant vertebrates may owe some of its features to properties evolved in the context of the lateral eye. Eakin (1973) has shown that pineal lens induction is not dependent on contact with specific surrounding tissues: the inductive stimuli responsible for lens differentiation are likely to be the result of interactions between cells in the vesicle itself (the vesicle surfaces are initially in cellular contact (Studnicka,

1905) and an extension of the cell-by-cell stimuli that presumably induced the ancestor cells.

In the case of the lateral eye the initial evolutionary outgrowth of the optic vesicle need not have required provision for a lens; like the original ependymal photoreceptors it would have been functionally advantageous anyway provided it came sufficiently close to a sufficiently transparent body surface. A variety of circumstances can be envisaged by which it might have achieved its inductive recruitment of ectoderm for the formation of the lateral lens. This would include the possibility of gradual evolution *de movo* towards lens differentiation in an ectodermally derived organ of hitherto unconnected function which happened to occupy the site of the future lateral lens. However, an attractive alternative is suggested by the available facts taken as a whole. The eye-borne inductor and self-organizing properties of the lens cells may both have been available already as a result of evolution in the pineal eye. Then all that would be needed would be that the lateral eye recruits a source of uncommitted cells by, if necessary, an entirely non-specific morphogenetic stimulus to the ectoderm. The lineage of the cells is immaterial: cells merely have to be brought into the context of the lens inductor. This requires a coincidence, at that relevant stage of evolution, between the ability of the eye to recruit cells and/or the provision of cells by local ectoderm. Lateral ectoderm demonstrates an intrinsic disposition to delaminate and invaginate in that it gives rise to cranial placodes, the otocyst and the extensive lateral line system in early vertebrates. As suggested by the formatimon of neural tube and eyes by secondary cavitation from solid cores of cells in early vertebrates (Kerr, 1902) morphogenetic properties of early forms may have been quite different from those of current highly evolved embryos. The phylogenetically earliest examples of lens formation (Fig. 3.11) appear by delamination, consistent with having been recruited by a non-specific mitotic stimulus due to the approach of the optic vesicle. In higher vertebrates the morphogenetic stimulus has been further elaborated, involving for example local gut (Jacobson, 1958), permitting coordinated invaginative folding. Whether lateral lens evolved independently, in conjunction with, or consequent to the evolution of the pineal lens, the foregoing discussion illustrates how an apparently discontinuous event like lateral lens induction can have evolved, perhaps in parallel at different sites, in continuity.

In conclusion an attempt has been made to set out the sort of procedures available, and some of the difficulties encountered, in deducing evolutionary sequences using embryological evidence. Naturally there is no reason to anticipate any final definitive answers—that would require an ordered series of fossil embryos!—but it is likely that convergence of evidence into a best fit pattern may single out one scheme with overwhelming high probability. The sort of evidence that can be brought to bear includes attention to the plausibility of the required cellular and embryonic potentialities (these are likely to evolve relatively slowly and therefore to be few and common between species), satisfying the continuity principle at all points, and use of any general rules for the elaboration of new cell types and organ systems which may be found to apply throughout evolution.

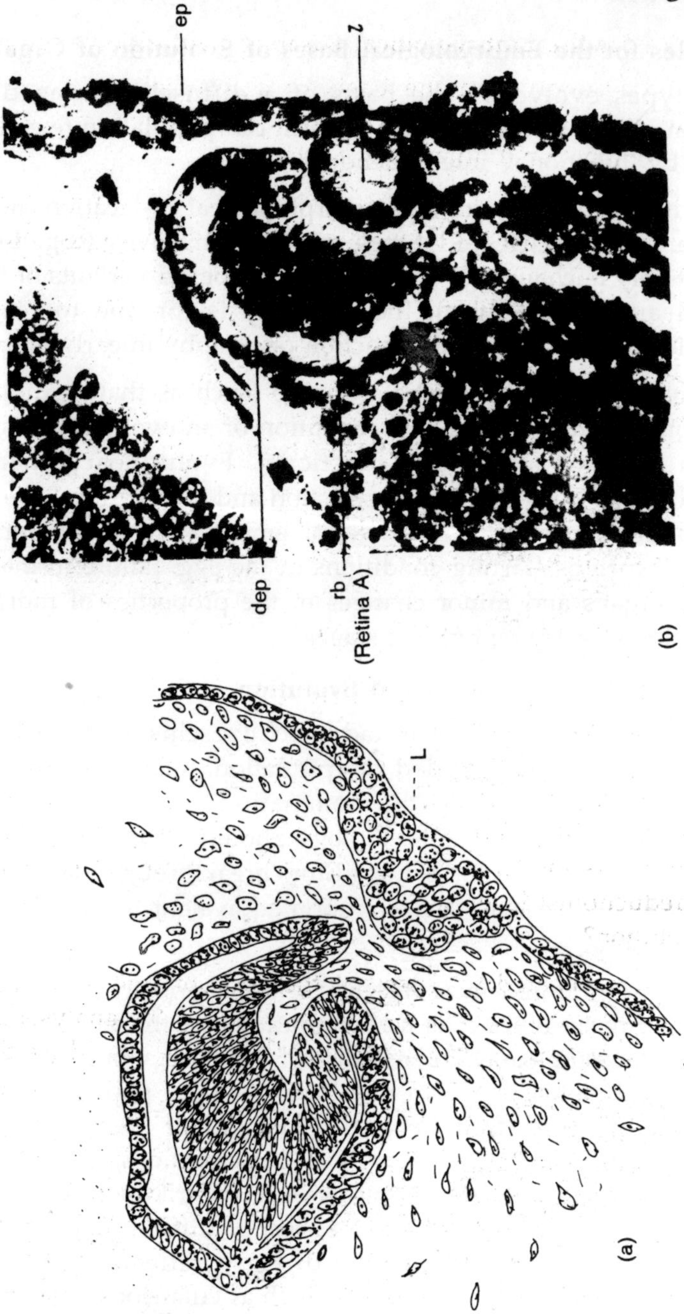

Fig. 3.11: Origin of latral lens in cyclostome embryo. (*a*) Delamination of ectoderm (*L*) in *Bdellostoma stouti*. (From Stockard, 1907). (*b*) Later lens vesicle (1) formation in *Petromyzon*: dorsal up (From Keibel, 1928). dep = pigmented epithelium; ep = epidermis;

Some General Rules for the Embryological Bases of Evolution of Organ-Systems

(*a*) New cell types evolve on the basis of a diffusely arranged cell by cell differentiation in an initially extensive and uniform cell population (e.g. an epithelium) 'induced' perhaps by 'functional' interdependencies.

(*b*) Later evolution primarily involves morphological re-location and segregation into discrete organs of specific cell types which themselves (together with their inductors) are relatively unchanging. Multiple morphological variants of the organ may occur (e.g. pineal and lateral eyes): the tendency is for one to reach definitive specialization while others change their function or atrophy in early embryogenesis.

(*c*) An embryological morphogenetic sequence (such as that leading from neural plate to eye) can only be understood as a repetition of an evolutionary morphological progression, each stage of which was functional. Evolutinoary change does not anticipate the final result: it is the result of selection and elaboration from the inevitable morphogenetic variation occurring in ancestral embryos at predifferentiation stages because of variables such as starting conditions in the egg, failure to mature of earlier evolved (vestigial) organs and minor changes in the properties of morphogenetically active cells prominent in early embryonic stages.

The Inescapable Gap between Genes and Evolution

Embryology's identity crisis, foreshadowed by schisms resulting from over-dogmatic theorizing such as surrounded recapitulation, intensified as the subject was overtaken by events in the form of the domination of biochemistry and genetics culminating in the triumphs of molecular biology. Should embryology use the Watson—Crick double helix as its model for a hoped-for 'breakthrough' or is it legitimate to operate at a less reductionist level of description equivalent to the framework created by Darwin for evolution?

Genetics has long appeared to represent the appropriately fundamental level for ultimate analysis of embryogenesis, as have mutations for the analysis of evolutionary change. If we knew the complete DNA sequence of a species, would we then necessarily completely understand embryology? In one important but poorly understood respect we would not: cell structure itself (including descent of organelles and 'maternal' factors) is inherited, independently of Mendelian considerations, by way of the template of the egg cell. Genetics has been disappointing in its contribution to the understanding of embryology: at the level of the final expression of differentiated states in cells, its relevance is obvious, and where pattern directly reflects properties of single differentiated cell types (as it does in situations such as cutaneous pigmentation patterns (Willier & Rawles, 1944; Fig 3.8)) genetic variation can directly account for complex pattern changes. But, in general, known mutations do not fit well with the requirements one would expect of controllers of elements of spatial pattern; as surveys such as that by Kalter (1980) in the mouse show, single mutations very often have multiple (Pleiotropic) effects at multiple locations and particular organs are disproportionately commonly involved. Disproportionately large groups of mutations share common teratological effects, suggesting that their effects are dictated by constraints resulting from embryological mechanisms. Homoeotic mutations in insects are exceptional. Their effect are clearly potentially valuable sources of segmental elaboration for insect evolution but, as discussed earlier, it is far from clear whether they involve a truly special class of pattern-controlling genes: they may be mutations of cellular expression whose effects are segmentally based as a result of the anatomical structure and homologies of insect segmentation.

Given the importance in embryogenesis of morphogenetic and functional interactions involving populations of cells, it is not surprising that the effects of many genes affecting pattern are buffered and transformed in a complex manner by embryonic processes (Grüneberg, 1963) which prevents resolution of their primary action as expressed in single cells. Integrated patterns of morphogenesis cannot be directly or explicitly programmed in the genome; they are the result of epigenetic interactions between differentiated embryonic cells progressively elaborated during embryogenesis and ultimately conditional on the organization of the egg cytoplasm.

Species differ most conspicuously in regard to differences in size and proportions, while cell types are, by contrast, few and stable in evolution. As we saw with the limb, it is the strategy for deployment of cells that generates anatomical complexity and variety: small changes to the initially simple starting conditions would be sufficient to cause major consequent anatomical modifications. Thus the gradual evolutionary modification of any particular embryonic feature, together with necessary harmonious changes elsewhere in the anatomy, would require small multidimensional genetic changes. Much may be achieved merely by rearrangement of genetic material (Wilson *et al.*, 1974). Embryology does not exclude the possibility of single mutations having global and relatively major ('macroevolutionary') effects which, in certain rare occasions, could be advantageous. While no aspect of embryogenesis can be entirely independent of the products of the genome, it may nonetheless be the case that control by genes is so indirect that analysis in such terms loses useful explanatory power. Pattern formation based on function (Fig. 3.8), in particular, enormously reduces the requirements for explicit genetic instructions, while also automatically ensuring adaptedness.

'A Hen is Only the Egg's Way of Making Another Egg' (Butler, 1878)

The key to the eventual identification of the basic parameters for evolutionary change in body form is to see evolution from the point of view of the parameters of variation available in development, rather than drawing inferences from the adult end result. Concepts such as induction, morphogenetic strategy and functional demand provide unifying explanatory terms which refer to processes at a level likely to be actually selected for in evolution. Precise definition of cellular behaviour and potentialities operating in the embryo, which may be unexpectedly different from the familiar adult picture, may lead directly to identification of switching events underlying changes of genetic expression. Detailed study of individual organ systems such as the eye or the limb, from an embryological view-point, suggests plausible evolutionary sequences and mechanisms, avoiding the multidimensional complexity of the whole organism and biologists' concern with the species concept, and in spite of the current theoretical heart searching by taxonomists and palaeontologists. At the very least it is clear from the foregoing discussion that the particular embryological theory that is adopted will have a major bearing on how evolutionary events and the basis of their inheritance are conceptualized.

REFERENCES

Ashburner, M. & Wright, T. R. F. (1978). *The genetics and biology of* Drosophila, vol. 2c. Academic Press: London.

Balinsky, B. I. (1927). Über experimentelle Induktion der Extremitatenanlage bei *Triton* mit besonderer Berucksichtigung der Inervation und Symmetrieverhaltnisse derselben. *Wilhelm Roux Arch. EntwMech. Org.* **110**, 71–88.

Beisson, H. & Sonneborn, T. M. (1965). Cytoplasmic inheritance of the organization of the cell cortex in *Paramecium aurelia. Proc. natn. Acad. Sci. USA* **53**, 175–82.

Berrill, N. J. (1961). *Growth, development, and pattern.* Freeman: San Francisco.

Bloemendal, H. (1981). *Molecular and cellular biology of the eye lens.* Wiley: New York.

Bohn, J. (1976). Tissue interactions in the regenerating cockroach leg. In *Insect development,* ed. P. A. Lawrence, Symposium of the Royal Entomological Society of London No. 8, pp. 170–85. Blackwell Scientific Publ.: Oxford.

Brower, D. L., Smith, R. J. & Wilcox, M. (1982). Cell shapes on the surface of the *Drosophila* wing imaginal disc. *J. Embryol. exp. Morph.* **67,** 137–51.

Bryant, P. J. (1970). Cell lineage relationships in the imaginal wing disc of *Drosophila melanogaster. Devl. Biol.* **22,** 389–411.

Bryant, P. J. (1975). Regeneration and duplication in imagina discs. In *Cell patterning,* Ciba Foundation Symposium No. 29, pp. 71–93. Associated Scientific Publishers: Amsterdam.

Bryant, S. V. French, V. & Bryant, P. J. (1981). Distal regeneration and symmetry. *Science* **212,** 993–1002.

Bryant, P. J. & Schneiderman, H. A. (1969). Cell lineage, growth and determination in the imaginal leg discs of *Drosophila melanogaster. Devl. Biol.* **20,** 263–90.

Butler, S. (1878). *Life and habit.* Trübner: London.

Chalmers, J. & Ray, R. D. (1962). The growth of transplanted foetal bones in different immunological environments. *J. Bone Joint Surg.* **44B,** 149–64.

Child, C. M. (1941). *Patterns and problems of development.* University of Chicago Press: Chicago.

Conklin, E. G. (1905). Mosaic development in ascidian eggs. *J. exp. Zool.* **2,** 145–223.

Conklin, E. G. (1933). The development of isolated and partially separated blastomeres of *Amphioxus. J. exp. Zool.* **64,** 303–76.

Coulombre, A. J. (1965). The eye. In *Organogenesis.* eds. R. L. DeHann and H. Ursprung, pp. 219–51. Holt, Rinehart & Winston: New York.

Counce, S. J. & Waddington, C. H. (1973). *Developmental systems: insects*, vol. 2. Academic Press: London.

Darwin, F. (1887). *The life and letters of Charles Darwin*, vol. 2, p. 273. Murray: London.

Davidson, E. H. (1976). *Gene activity in early development*, 2nd edn. Academic Press: New York.

de Beer, G. (1958). *Embryos and ancestors*, 3rd edn. Oxford University Press: Oxford.

Drastich, L. (1925). Uber das Leben der Salamandra-larven bei hohem und niedrigem Sauerstoffpartialdruck. *Z. vergl. Physiol.* **2**, 632–57.

Eakin, R. M. (1973). *The third eye*. University of California Press: Berkeley, Cal.

Ede, D. A. (1971). Control of form and pattern in the vertebrate limb. In *Control mechanisms of growth and differentiation*, eds. D. D. Davies and M. Balls, Symposium of the Society for Experimental Biology, No. 25, pp. 235–54. Cambridge University Press: Cambridge.

El Shatoury, H. H. (1955). Lethal no-differentiation and the development of the imaginal discs during the larval stage in *Drosophila*. *Wilhelm Roux Arch. EntwMech. Org.* **147**, 523–38.

Flint, J. M. (1910). Compensatory hypertrophy of the small intestine following resection of large portions of the jejunum and ileum. *Trans. Conn. State ed. Soc.* 283–335.

Freemann, G. (1981). The cleavage initiation site establishes the posterior pole of the hydrozoan embryo. *Wilhelm Roux Arch. EntwMech. Org.* **190**, 123–5.

French, V., Bryant, P. J. & Bryant, S. V. (1976). Pattern regulation in epimorphic fields. *Science* **193**, 969–81.

García-Bellido, A. (1967). Histotypic reaggregation of dissociated imaginal disc cells of *Drosophila melanogaster* cultured *in vivo*. *Wilhelm Roux Arch. EntwMech. Org.* **158**, 212–17.

García-Bellido, A. (1977). Inductive mechanisms in the process of wing vein formation in *Drosophila*. *Wilhelm Roux Arch. EntwMech. Org.* **182**, 93–106.

García-Bellido, A., Ripoll, P. & Morata, G. (1973). Developmental compartmentalisation of the wing disk of *Drosophila*. *Nature New Biol.* **245**, 251–3.

Gerhart, J., Ubbels, G., Black, S., Hara, K. & Kirschner, M. (1981). A reinvestigation of the role of the grey crescent in axis formation in *Xenopus laevis*. *Nature* **292**, 511–16.

Goldschmidt, R. B. (1940). *The material basis of evolution*. Yale University Press: New Haven, Conn.

Goss, R. J. (1961). Regeneration of vertebrate appendages. *Adv. Morph.* **1**, 103–52.

Goss, R. J. (1964). *Adaptive growth*. Logos Press: London.

Gruneberg, H. (1963). *The pathology of development*. Wiley: New York.

Harris, P. (1979). A spiral cortical fiber system in fertilized sea urchin eggs. *Devl. Biol.* **68**, 525–32.

Harrison, R. G. (1921). On relations of symmetry in transplanted limbs. *J. Exp. Zool.* **32**, 1–136.

Harrison, R. G. (1969). *Organization and development of the embryo*. Yale University Press: New Haven, Conn.

Herlitzka, A. (1897). Sullo sviluppo di embrioni completi da blastomeri isolati di uova di tritone (*Molge cristata*). *Wilhelm Roux Arch. EntwMech. Org.* **4**, 624–58.

Hinchliffe, J. R. & Johnson, D. R. (1980). *The development of the vertebrate limb*. Clarendon Press: Oxford.

Holtfreter, J. (1934). Über die Verbreitung induzierender Substanzen und ihre Leistungen im *Tritonkeim. Wilhelm Roux Arch. EntwMech. Org.* **132,** 307–83.

Holtfreter, J. (1947). Neural induction in explants which have passed through a sublethal cytolysis. *J. exp. Zool.* **106,** 197–222.

Horder, T. J. (1976). Pattern formation in animal development. In *The developmental biology of plants and animals,* eds. C. F. Graham and P. F. Wareing, pp. 169–97. Blackwell Scientific Publ.: Oxford.

Horder, T. J. (1981). On not throwing the baby out with the bath water. In *Evolution today,* eds. G. G. E. Scudder and J. L. Reveal, pp. 163–80. Hunt Institute for Botanical Documentation: Pittsburgh.

Hulsemann, M. (1967). Vergleichende histologische Untersuchungen uber das Vorkommen von Gliafasern in der Epiphysis cerebri von Saugetieren. *Acta anat.* **66,** 249–78.

Huxley, J. S. (1972). *Problems of relative growth,* 2nd edn. Dover Publ.: New York.

Jacobson, A. G. (1958). The roles of neural and non-neural tissues in lens induction. *J. exp. Zool.* **139,** 525–57.

Kalter, H. (1980). A compendium of the genetically induced congenital malformations of the house mouse. *Teratology* **21,** 397–429.

Keibel, F. (1928). Beitrage zur Anatomie, zur Entwicklungsgeschichte und zur Stammesgeschichte der Sehorgane der Cyklostomen. *Z. mikrosc.-anat. Forsch.* **12,** 391–456.

Kerr, J. G. (1902). The development of *Lepidosiren paradoxa. Q. J. microsc. Sci.* **46,** 417–59.

Kieny, M. (1964). Régulation des excédents et des déficiences du bourgeon d'aile de l'embryon de poulet. *Archs. Anat. micr. Morphol. exp.* **53,** 29–44.

Klinckowstroem, A. de (1894). Beitrage zur Kenntnis des Parietauges. *Zool. Jb.* (*Anat. Ont. Tiere*) **7,** 249–80.

Kramer, G. (1951). Body proportions of mainland and island lizards. *Evolution* **5,** 193–206.

Lawrence, P. A. & Green, S. M. (1975). The anatomy of a compartment border: the intersegmental boundry in *Oncopeltus. J. Cell Biol.* **65,** 373–82.

Lawrence, P. A. & Morata, G. (1976). Compartments in the wing of *Drosophila*: a study of the *engrailed* gene. *Devl Biol.* **50,** 321–37.

Lewis, W. H. (1904). Experimental studies on the development of the eye in Amphibia. I. On the origin of the lens. *Am. J. Anat.* **3,** 505–36.

MacCabe, J. A., Saunders, J. W. & Pickett, M. (1973). The control of the anteroposterior and dorsoventral axes in embryonic chick limbs constructed of dissociated and reaggregated limb-bud mesoderm. *Devl. Biol.* **31,** 323–35.

Maden, M. (1980). Structure of supernumerary limbs. *Nature* **286,** 803–5.

Nieuwkoop, P. D. & Sutasurya, L. A. (1979). *Primordial germ cells in the chordates.* Cambridge University Press: Cambridge.

Poodry, C. A., Bryant, P. J., Schneiderman, H. A. (1971). The mechanism of pattern reconstruction by dissociated imaginal discs of *Drosophila melanogaster. Devl Biol.* **26,** 464–77.

Postlethwait, J. H. & Schneiderman, H. A. (1971). Pattern formation and determination in the antenna of the homoeotic mutant *antennapedia* of *Drosophila melanogaster. Devl Biol.* **25,** 606–40.

Rappaport, R. (1970). An experimental analysis of the role of cytoplasmic fountain streaming in furrow establishment. *Devl. Growth Diff.* **12,** 31–40.

Reverberi, G. (1971). *Experimental embryology of marine and fresh-water invertebrates.* North-Holland: Amsterdam.

Roux, W. (1888). Beitrage zur Entwicklungsmechanik des Embryo. V. Ueber die kunstliche Hervorbrigung 'halber' Embryonen durch Zerstorung einer der beiden ersten Furchungszellen, sowie uber die Nachentwickelung (Postgeneration) der fehlenden Korperhalfte. *Arch. path. Anat* **114,** 113–53.

Rychter, Z. (1962). Experimental morphology of the aortic arches and the heart loop in chick embryos. *Adv. Morph.* **2,** 333–71.

Salvini-Plawen, L. V. & Mayr, E. (1977). On the evolution of photoreceptors and eyes. *Evol. Biol.* **10,** 207–63.

Saunders, J. W. (1977). The experimental analysis of chick limb bud development. In *Vertebrate limb and somite morphogenesis,* eds. D. A. Ede, J. R. Hinchliffe and M. Balls, pp. 1–24, Cambridge University Press: Cambridge.

Saunders, J. W., Gasseling, M. T. & Errick, J. (1976). Inductive activity and enduring cellular constitution of a supernumerary apical ectodermal ridge grafted to the limb bud of the chick embryo. *Devl Bill.* **50,** 16–25.

Saxen, L. & Toivonen, S. (1962). *Primary embryonic induction.* Academic Press: London.

Schmalhausen, I. I. (1948). *Factors in evolution.* Blakiston: Philadelphia.

Schneiderman, H. A. (1979). Pattern formation and determination in insects. In *Mechanisms of cell change,* eds. J. D. Ebert and T. S. Okada, pp. 243-72. Wiley: New York.

Simpson, P. (1976). Analysis of the compartments of the wing of *Drosophila melanogaster* mosaic for a temperature-sensitive mutation that reduces mitotic rate. *Devl. Biol.* **54,** 100–15.

Slack, J. M. W. (1977). Control of anteroposterior pattern in the axolotl forelimb by a smoothly graded signal. *J. Embryol. exp. Morph.* **39,** 169–82.

Slijper, E. J. (1942). Biologic-anatomical investigations on the bipedal giat and upright posture in mammals, with special reference to a little goat, born without forelegs. II. *Proc. Ned. Akada. Wetensch. Amsterdam* **45,** 407–15.

Smith, L. D. & Ecker, R. E. (1970). Uterine suppression of biochemical and morphogenetic events in *Rana pipiens. Devl. Biol.* **22,** 622–37.

Spemann, H. (1901). Ueber Correlationen in der Entwickelung des Auges. *Verh. anat. Ges Jena* 61-79.

Spemann, H. (1938). *Embryonic development and induction.* Yale University Press: New Haven, Conn.

Stanley, S. M. (1973). An explanation for Cope's rule. *Evolution* **27,** 1–26.

Stern, C. (1968). *Genetic mosaics and other essays.* Harvard University Press: Cambridge, Mass.

Stockard, C. R. (1907). The embryonic history of the lens in *Bdellostoma stonti* in relation to recent experiments. *Am. J. Anat.* **6,** 511–15.

Studnicka, F. K. (1905). *Die Parietalorgane, Lehrbuch der vergleichenden mikroskopischen Anatomie der Worbettiere,* ed. A. Oppel, Part V. Fischer: Jena.

Tank, P. W., Carlson, B. M. & Connelly, T. G. (1977). A scanning electron microscopic comparison of the development of embryonic and regenerating limbs in the axolotl. *J. exp. Zool.* **201,** 417–30.

Tank, P. W. & Holder, N. (1981). Pattern regulation in the regenerating limbs of urodele amphibians. *Q. Rev. Biol.* **56,** 113–42.

Thompson, D'A. W. (1942). *On growth and form.* Cambridge University Press: Cambridge.

Tickle, C. (1981). The number of polarizing region cells required to specify additional digits in the developing chick wing. *Nature* **289,** 295–8.

Tickle, C., Alberts, B., Wolpert, L. & Lee, J. (1982). Local application of retinoic acid to the limb bud mimics the action in the polarizing region. *Nature* **296,** 564–6.

Tobler, H. (1966). Zellspezifische Determination und Beziehung zwischen Proliferation und Transdetermination in Bein-und Flugelprimordien von *Drosophila melanogaster. J. Embryol. exp. Morph.* **16,** 609–33.

Tung, T. C., Wu, S. C. & Tung, Y. Y. F. (1962). Experimental studies on the neural induction in *Amphioxus. Scienta sin.* **11,** 805–20.

Vijverberg, A. J. (1974). A cytological study of the proliferation patterns in imaginal disks of *Calliphora erythrocephala* Meigen during larval and pupal development. *Neth. J. Zool.* **24,** 171–217.

Wallace, H. (1981). *Vertebrate limb regeneratio.* Wiley: Chichester.

Weiss, P. (1939). *Principles of development.* Holt: New York.

Wermel, J. (1935). Untersuchungen uber die Kinetogenese und ihre Bedeutung in der onto- und phylogenetischen Entwicklung (Experimente und Vergleichungen an Wirbeltierextremitaten). VI. Mitteilung: Veranderungen der Muskulatur. *Morph. Jb.* **75,** 452–8.

Wildermuth, H. (1968). Autoradiographische Untersuchungen zum Vermehrung-smuster der Zellen in proliferierenden Russelprimordien von *Drosopila melanogaster. Devl. Biol.* **18,** 1–13.

Willer, B. H. & Rawles, M. E. (1944). Melanophore control of the sexual dimorphism of feather pigmentation pattern in the barred Plymouth Rock fowl. *Yale J. Biol. Med.* **17,** 319–40.

Wilson, A. C., Sarich, V. M. & Maxson, L. R. (1974). The importance of gene rearrangement in evolution: evidence from studies on rates of chromosomal, protein, and anatomical evolution. *Proc. natn. Acad. Sci. USA* **71,** 3028–30.

Wilson, E. B. (1893). *Amphioxus,* and the mosaic theory of development. *J. Morph.* **8,** 579–638.

Wolpert, L. (1969). Positional iformation and the spatial pattern of cellular differentiation. *J. theor. Biol* **25,** 1–47.

Chapter—4

Mendelism and Evolution

1. Mutation and Selection

The essence of Mendelian heredity is that it is particulate. The genetic constitution is composed of discrete units. Each kind of unit can exist in a number of a discrete forms. The hereditary transmission of any one kind of unit is more or less independent of that of other units, the restriction of independence being a partial one, concerned with the phenomenon of linkage. The units are the Mendelian factors or genes, while their different forms are called allelomorphs or alleles.

The particulate nature of inheritance enables calculations to be made as to the proportions of offspring of different types in different generations after a cross. Like the atomic theory in chemistry, it is the basis of quantitative treatment.

The hereditary particles or genes are located within the visible chromosomes, whose manoeuvres distribute their contained genes equally to all cells of the body, and determine the quantitative details of Mendel's laws. Within a particular chromosome, each gene has its appointed place, which it keeps permanently (apart from rare rearrangements;). Of recent years, the study of the giant chromosomes in the salivary glands has converted *Drosophila* from an unfavourable into an exceedingly favourable cytological object. It is now possible in this genus to check genetic prophecies cytologically and cytological prophecies genetically, in a remarkably complete and detailed way. (For the cytological basis of heredity see Darlington, 1937; briefer treatment in M. J. D. White, 1937).

It used to be imagined that the precise arrangement of the genes within the chromosomes was biologically irrelevant. To-day, however, we know that this is not so. Genes (all or many of them) have somewhat different actions according to what neighbours they possess. This is the so-called position effect which, only recently discovered, will probably turn out to be widespread, although in some organisms (such as maize) it does not seem to occur. Where it occurs, it is likely to be of fundamental importance as well, since the rearrrangement of blocks of genes (sections of chromosomes) within the chromosome outfit, though considerably rarer than gene-mutation, is not infrequent in the long perspective of evolution; and this, through the position effects which it frequently causes, provides a large and previously unsuspected source of variability and potential evolutionary change. Its contribution, however, must be much less varied and much less abundant than that of genemutation (Muller, 1940).[1]

This does not affect the basic conception of the gene as particulate. Genes are in many ways as unitary as atoms, although we cannot isolate single genes. They do not grade into each other: but they vary in their action in accordance with their mutual relations. In this they are again like atoms: the chemical behaviour of a compound will be altered when we transfer an atom from one position to another in the molecule, even though the substantive constitution of the molecule remains unchanged. Thus the whole is not merely the sum of its parts: it is also their relation.

The discreteness of the genes may prove to be nothing more than the presence of predetermined zones of breakage at small and more or less regular distances along the chromosomes. For the independent hereditary behaviour of genes, from which their discreteness is deduced, is due to two facts. When the genes to be tested are in different kinds of chromosomes, their independence is due merely to the independent behaviour of the two chromosomes. But when they are both in the same kind of chromosome, their independence depends on what is known as crossing-over. Prior to the formation of reproductive cells, the two homolgous chromosomes of each kind pair together. Where they touch, they may break and exchange segments; in the daughter-chromosomes the kinds of genes and their order remain as before, but one or more blocks of genes from one chromosome will have exchanged places with precisely corresponding blocks of genes from the other. The breaks do not always occur in the same place. If there is more than one break in a chromosome-pair, the second break is at a considerable distance from the first; thus breaks can normally not occur on both sides of a single gene. What happens is thus that genes are separated from their erstwhile neighbours in a chromosome by these breaks; and it is the fact that breaks may occur at different places in the chromosome which makes it possible for any gene in a chromosome to be separated from its neighbours and thus to be inherited independently. Knowing this, we may put the matter the other way round, and say that the process of exchange of sections after breaking, and the fact that breakage only occurs at certain spots, determines what we call the gene, a gene-unit is thus a section of the chromosome between two adjacent sites of potential breakage at crossing-over.

The chromosomes may thus be looked on as "super-molecules" built up out of a series of regions, each region marked off by zones of potential breakage. The portions of these regions which we can recognize by their effects in inheritance are what we call genes. Rearrangement of the regions, as well as change within a single region, or the loss or duplication of a region or set of regions, can and normally will cause alterations in the action of the chromosome and its parts on the developing organism. Similarly the doubling of the whole chromosome-outfit, by producing a different relation between gene-outfit and cytoplasm, will also alter the characters of the organism.

The number of genes is much larger than was originally imagined. *Drosophila* is the only organism where adequate quantitative knowledge is available. Several recent estimates, based on different methods of approach (see summary in Gulick, 1938), vary from a minimum of over 2,000 to maximum of well under 13,000, with a probable number of about 5,000, for this minute insect. The size of a *Drosophila* gene must be between 10^{-8} and 10^{-5} μ^3 and probably between 10^{-7} and 10^{-6} μ^3, equivalent to some 10 medium-sized protein molecules (see also Lea, 1940). In some other organisms (*Lilium*) the genes may be larger, and in others more numerous (e.g. in man, perhaps 4 to 6 time more so than in *Drosophila*). It will be seen what astronomical possibilities of recombination and mutual interaction are afforded by an assemblage of this magnitude.

A gene-mutation will then be any intrinsic change in substance or structure, affecting the mode of action of one of these unit-regions.

One of the notable biological discoveries of the last few years was that of Muller, on the effect of X-rays in producing genemutations. The ordinary rate of mutation can by this means be multiplied a hundred-and-fifty fold, and a certain number of wholly new mutations, in addition to many already known, are produced. It is, however, interesting, in vie of our discussion, to note that X-rays also induce the rearrangement of chromosome-sections by translocation, inversion, etc. In view of the assertions of certain biologists that mutation is of its essence pathological, it should be mentioned that X-ray treatment can produce reverse mutation—i.e. cause a previously mutated gene to revert to normal (Patterson and Muller, 1930; Timoféeff-Ressovsky, 1934*a*, 1937), Comment is needless.

Timakov (1941) in wild *Drosophila* has detected a gene increasing mutation-rate at least 40 times, and possibly to a level higher than that induced by X-rays.

The fact that the genes and their arrangement normally remain constant, until altered by some kind of mutation (after which they again remain constant in their new form until a further mutation supervenes), accounts for the resemblance between parents and offspring. The fact of Mendelian recombination, by which new combinations of old genes are produced according to Mendel's second law (and to the rules of crossing-over), accounts for the great majority of the differences between parents and offspring, and between members of a family or population. But gene-mutation, though a rare event, appears to account for most that is truly new in evolution. Under the head of gene-mutation, position-effect due to very small sectional rearrangements can legitimately be included, since it involves a structural change and a novel effect; further, it cannot for practical reasons be excluded, since there is at present in many cases no possibility of distinguishing between it and true gene-mutation. Recombination also may in certain cases produce evolutionary novelty, for instance after a cross between two previously isolated types. Finally, as we shall see later, hybridization, with no subsequent recombination, may sometimes be responsible for evolutionary change (Chap. 6 § 9).

However, gene-mutations (including position-effects) appear to be the most important source of novelty in evolution, and we must now say a little more about them.

In the first place, no trace has been found in *Drosophila,* where analysis has been pushed to an extraordinary pitch of refinement, of any characters not dependent on chromosomal (and therefore mendelizing) differences (Muller, 1940). Secondly, although complete proof cannot be offered, the presumption, in the absence of evidence to the contrary, is that all mendelizing gene-differences owe their origin to mutation: up to the present we know of no other way by which they could have come into existence.

Finally, a number of mutations are known which are roughly neutral, or actually or potentially useful. In barley, Gustafsson (1941) induced by X-rays three mutations which increased yield, one of them markedly. Among those which are potentially useful a few cases may be mentioned here. Banta and Wood (1928) discovered a "thermal race" of the crustacean *Daphnia longispina,* which had its optimum 6°–8° C. higher and its thermal death-point 5°C. higher than the long established parthenogenetic strain from which it originated. It was also immobilized more quickly by low temperatures. Thus it was potentially adapted to a warmer environment than its parent strain. Its origin from sexually inbred individuals showed that it depended upon a recessive mutation arising during the long preceding period of parthenogenetic reproduction. Other more immediately useful mutations arose in the same way, e.g. some causing greater fertility and other greater longevity (Davenport, 1928).

Very important results were obtained by Johannsen (1913) in following up his classical researches on pure lines for seed-weight in beans. As is well-known (see Johannsen, 1926), he showed that the prime effect of selection in a mixed population was to isolate pure lines, and that further selection then had no further effect, in the absence of mutation. But he also showed that mutations might occur in pure lines, and might then be selected. During his experiments, two mutations, one for higher and one for lower seed-weight, were detected and "captured" by his selection for high and low seed-weight respectively.

Another interesting case is that of the variety of tobacco, originated apparently by mutation, described by Garner and Allard (1920). This did not flower at all in its place of origin (Washington); but when the daily period of light was reduced to twelve hours, it flowered and set seed better than the stock from which it had arisen. In other words, it was potentially better adapted than the parent stock to more tropical latitudes.

Still another case is the mutation described by McEwen (1937) in *Drosophila,* which abolishes the fly's phototropism. This would be adaptive in dark surroundings. It is noteworthy that this last effect is produced by the same recessive mutation that changes the body-colour to tan: thus here a useless visible effect is correlated with a potentially useful physiological effect (a "correlated character" in Darwin's usage).

Of mendelizing differences, alike in domestic and wild species, which are actually or potentially favourable, there is an abundance. We need only think of the genes producing small and large size in poultry (Punnett and Bailey, 1914); those producing the specific differences between the two species of snapdragon crossed by Baur (1923); the mimetic polymorphic forms of various butterflies; the different forms of heterostyled flowers such as primrose (*Primula*) and loosestrife (*Lythrum*); the single-brooded and double-brooded condition in silkworms; and so on. An interesting case of a *Drosophila* mutant establishing itself in considerable numbers in the wild is the vermilion-eyed type of *D. hydei* (Spencer, 1932). This mutant must be very delicately balanced in its selective relations. The recent establishment of other marked mutant types, like the black hamster, the black Tasmanian opossum, the *simplex*-toothed field-vole, etc., are discussed later. In our own species, the work of Blakeslee and his collaborators (see Blakeslee and Fox, 1932) has established the existence of remarkable differences, apparently mendelian, in sensitivity of taste and smell in regard to various chemical compounds and natural odours. These seem under present conditions to be, in themselves, somewhat selectively neutral. Later work (Fisher, Ford and Huxley, 1939) has shown that in chimpanzees not only are the same differences found, but tasters and non-tasters occur in about the same proportions as in man—close to 3 : 1. This appears to indicate a stable balance between the two conditions, and one depending upon some advantage, of unknown nature, enjoyed by the heterozygotes. The different blood-group genes would seem to fall into a somewhat similar category, though here the proportions vary markedly in different populations: some of these genes occur also in various lower mammals.

Many differences between "good" species have also been shown to depend on mendelizing gene-differences (see Goldschmidt, 1928, Chap. 15; Haldane, 1923*a*, Chap. 3, Lamprecht, 1941).

Further, an increasing number of characters once held to be non-mendelian are being shown to depend on mendelian genes (e.g. the multiple factors influencing the hooded pattern in rats; the distinctive characters of wild subspecies of the deer-mouse *Peromyscus*); and indeed wherever F_2 shows greater variability than F_1, inheritance must be particulate. Thus it may be legitimately argued that the majority of all inherited characters must rest on a mendelian basis. Even in the present incomplete state of our knowledge, there are strong presumptive grounds for this assertion, so that the onus of proof now lies on those who would maintain the contrary in any particular case.

In addition, initially deleterious factors can be rendered useful by genetic-evolutionary methods which we shall discuss in subsequent sections of this chapter.

Finally, mutations, while they seem to occur more readily in certain directions than in others can be legitimately said to be random with regard to evolution. That is to say, the directions of the changes produced by them appear to be unrelated either to the direction of the evolutionary change to be observed in the type, or to the adaptive or functional needs of the organism. Evolutionary direction has to be imposed on random mutation through the sifting and therefore guiding action of selection. It is, of course, possible that as the laborious technique of testing for mutation-rate is extended to more species, certain mutations may be discovered which show very much higher rates than others. However, the general agreement already found between organisms so different as a monocotyle-donous angiosperm, an insect and a mammal would indicate that in most species we may expect to find some mutations occurring at a rate of 1 in 10^5 individuals or even higher, and many genes with a mutation-frequency of abut 1 in 10^6. Occasional genes with much higher mutation-rates occur (see summary in Dobzhansky, 1937), and some genes promote increased mutation-frequency in other genes. In cotton, hybridization may increase the mutation-frequency of certain genes (Harland, 1936). Mutation-frequency must in some way be balanced against length of life; otherwise the chromosomes of long-lived species would become crowded with lethals before reproduction (Dobzhansky, 1937). Again, the mutation-rate for haemophilia in man is on the same scale as that for most *Drosophila* genes if computed per life-cycle, but much lower on the basis of time (Haldane, 1935*b*). Certainly species vary in mutation-rate: thus the fern *Nephrolepis exaltata* has produced many more mutants than any other species of the genus (Benedict, 1931). Zuitin (1941) in *Drosophila* finds that mutation-rate is increased by sudden environmental changes.

With mutation-rates of this order of magnitude, evolution must always be a somewhat slow process, judged in terms of years, but its speed in relation to geological time will be quite adequate. R. A. Fisher (1930*a*, 1932) has discussed the matter in a general way. He clears the ground by pointing out that "blending inheritance," which was currently postulated in Darwin's day, would be constantly annulling variability: to be accurate, the variance (in the absence of selective mating) would be halved in each generation. As a result, new genetic variations (i.e. what we should to-day call mutations) would have to be exceedingly abundant—far in excess of anything observed in actual fact—to produce the variance actually observed; and any variability available for selection to act upon would have to be of very recent origin. It was largely for these reasons that Darwin ascribed so high an influence to "the direct effects of environment".

The discovery that inheritance is almost entirely particulate and non-blending removes these difficulties, so that in point of fact the rise of Mendelism, far from being antagonistic to Darwinian views (as was claimed, notably by the early Mendelians themselves, in the years immediately following its rediscovery), makes a selectionist interpretation of evolution far simpler. In mathematical language, it indefinitely conserves much genetic variance instead of rapidly dissipating it, and thus amasses material on which selection can work.[2] Further if particulate inheritance and discontinuous mutation as they are known at the present time are the general basis of genetics and variation, selectionist views also gain support over those strictly to be called orthogenetic, in which the direction of mutation itself is supposed to determine the course of evolution, and over those to be called Lamarckian, in which the effects of use and function are supposed to be inherited. For no rate of hereditary change hitherto observed in nature would have any evolutionary effect in the teeth of even the slightest degree of adverse selection. Either mutation-rates many times higher than any as yet detected must be sometimes operative, or else the observed results can be far better accounted for by selection. A mutation with partial dominance occurring once in 10^5 individuals will, if selectively neutral, take a period of somewhat over 10^5 generations to establish itself in half the individuals of the species. If there were the faintest adverse selection against it, it could never increase at all. But if it conferred an advantage of only 1 per cent—i.e. if an individual bearing one such mutant gene has an expectation of reproducing itself which is only 1 per cent higher than those without the mutant gene, then it would establish itself in half the individuals of the species in a period of only about 10^2 generations (R. A. Fisher, 1930*a*; Haldane, 1932*a*). Fisher (1937*b*) has also made interesting studies on the form of the wave by which advantageous genes advance.

For ordinary natural selection involving a simple dominant with a selective advantage of 1 in 1,000 (i.e. where the ratio of dominant to recessive changes from 1 to 1.001 in each generation) it will take nearly 5,000 generations to increase the proportion of the dominant from 1 to 50 per cent, and nearly 12,000 more to raise it to 99 per cent. For an advantage of 1 in 100 the number of generations must be divided by 10. In the early stages of selection of a single mutation with constant effects, when the gene is still very rare dominants can spread much more rapidly than pure recessives, unless a certain degree of inbreeding occurs.

These results may be actualized in certain cases: e.g. in dominant melanism when conditions alter so as to favour the melanic mutant, the rate of change in the constitution of the population is of the order deduced. However, Haldane's detailed conclusions are not likely to be so directly applicable to evolutionary problems as was thought at the time, since we now realize that dominance or recessivity are themselves in large measure a result of evolution, produced in response to the deleterious nature of most mutations. The mutations that are of value for evolution will in most cases be of very small extent, of slight effect, and often at least of incomplete dominance or recessiveness. Further, we are now realizing that evolution will in general proceed, not by the selection of single mutations, but by the selection of mutations in relation to a favourable combination of existing small gene-differences, or in many cases by the selection of such new recombinations alone, to be followed later if occasion offers by appropriate new mutations. According to R. A. Fisher, this process will be considerably quicker than that of the selection of single recessives, which are the commonest obvious mutations found, and accordingly were, when Haldane wrote, usually considered to be the main source of evolutionary variance.

It will be observed that the amount of variance provided by mutation will, with a constant mutation-rate, vary directly with the size of the population. In a given time, therefore a rare species cannot lay hands on the same store of mutations as would be available to an abundant species. The problem of the relation of size of population to evolution is, however, much more complex than this (R. A. Fisher, 1930*a* Chap. 4, and

1937*a*; Sewall Wright, 1931, 1932, 1940). We must consider how much variance a population can hold, as well as how much variance it is provided with by mutation.

Many rare mutation must be extinguished by mere random loss: the individuals or gametes containing them fail to reproduce. There must always be a tendency for "minority" genes, which are present in low frequencies, to be lost from the germ-plasm by such accidental extinction if of no selective advantage. Even with a definite selective advantage such as 1 per cent, which is of the order of magnitude for rapid evolutionary change, the chances are strongly against a lone mutation surviving in the species (see, e.g., Haldane, 1939*b*). Mutations with a deleterious effect will of course be lost through selection, the rate of loss depending on the intensity of the effect.

Thus repeated mutation (i.e. a definite mutation-rate) together with a considerable-sized population, are necessary for new mutations to have an evolutionary chance.[3] Such abundant species as have been analysed prove to be carrying a surprisingly large number of recessive mutations in their germ-plasm Dobzhansky, 1937, Chap. 3).

In addition, the probability of mere accident playing a part in the actual survival of particular genes or gene-combinations is enhanced in small populations. This has been especially emphasized by Wright, who points out that we should expect to see, in the case of small species or isolated subspecies, certain types of useless or even deleterious change, which would not occur in an abundant form, becoming incorporated in the constitution through chance recombination.

Already in 1912 Lloyd had discovered instances of this process of accidental multiplication and decline of mutant genes, but without realizing its full theoretical implications; and by 1918 Muller had drawn general attention to its importance. Sewall Wright later christened the process "drift", and worked out its consequences in full detail (see Wright, 1940).

What may be regarded as the converse of the Sewall Wright phenomenon of drift in small populations is the impossibility of securing good results in artificial selection when only small numbers are employed. This, the general experience of poultry-breeders, has been confirmed by Hays (1940) in definite experiments designed to test the point. Using a flock never exceeding 50 birds, and often much smaller, he was unable in the course of eight generations to raise egg-production, though marked progress can be obtained by using large flocks. Apparently the numerous genes needed for the requisite multiple gene-combinations are not available in such small populations. In some characters involved in fecundity, indeed, the effect was contrary to the direction of selection—a result comparable with the deleterious changes sometimes seen in small populations in nature.

The smaller the size of natural population and the more perfectly it is isolated the more likely is drift to proceed to its limit, resulting either in the complete loss of a mutation from the group, or its fixation in all the individuals of the group—accompanied, of course, by the complete loss of its normal allele.

In larger and less isolated populations, however, drift will normally proceed only within limits, causing the frequency of a gene to fluctuate round a position of equilibrium. This equilibrium-frequency will be determined by the balance between the two opposing processes of mutation-frequency on the one hand and adverse selection on the other, while, as we have seen, population-size will also have an effect. Thus in large populations, slightly deleterious mutations may be present with reasonable frequency, especially when recessive, and will then constitute a reservoir of potential evolutionary change, since their unfavourable effects can generally be neutralized by appropriate combinations of modifiers.

In some cases, as with haemophilia and other sex-linked recessives in man, we know that the effect of a mutant gene is so deleterious that a comparatively high mutation-pressure must be postulated to account for its frequency. In other cases, changes in external environment will alter the amount or even the sign of the selection-pressure on the gene. Thus an increase in temperature would favour the spread of the "thermal race" of *Daphnia* mentioned on, while the lowering of temperature during the glacial period has doubtless led to autopolyploids replacing diploids at high latitudes in many plant species.

The same will be true of changes in the internal environment. Genes may have their expression altered by modifiers so as entirely to change their selective value.

Meanwhile it is important to realize that the frequency of mutant genes represents an equilibrium between mutation-frequency and selection, that variability represents a further equilibrium between recombination and selection, and that the size and structure of the population will have effects on both these equilibria.

We shall revert later to this last point. Here we will merely mention the important conclusion established by Sewall Wright (see Wright, 1940*a*), that the greatest amount of evolutionary potentiality is available to large species divided into partially discontinuous groups (subspecies etc.). The partial isolation between the groups favours diversity by local adaptation and also by drift and the establishment of non-adaptive recombination; while the fact that it is only partial implies that the variance provided by all the diversity taken together is potentially available to the spices as a whole.

Recent work has emphasized the importance of studies of population-structure for understanding the precise way in which evolution will operate in any particular species. Thus to take but two examples, Dobzhansky (1941) points out that certain theoretical calculations as to the relation between the mutation-rate and the number of lethals actually found in a population will only hold in unlimited populations. As the size of the normally inbreeding population is decreased, the number of lethals goes up. In *Drosophila pseudoobscura,* using this method, he was able to show that the size of inbreeding population-groups was quite different in California and in Central America. A region with smaller size of population-groups will show greater divergence between its constituent populations and these will each show greater variability in time; further, in such a region, the type as a whole can only change through the migration and selection of superior genotypes from small colonies. In general, the size of the constituent population-groups seems to be astonishingly small for an organism with such capacities for distribution.

In *Drosophila hydei* the situation is rather different (Spencer, 1941). This is a tropical species which has become widely established in U.S.A. as a hanger-on of urban man. Each city and town is the focus of a single population-group. Each city and town is the focus of a single population-group. Each such population-group passes through tremendous fluctuations in size, becoming quite small in winter. It is improbable that genetic equilibrium is ever reached within such markedly fluctuating groups, and the rapid increases in numbers give abundant opportunity for the spread of new genes even against selection-pressure. Analysis showed, as was expected, that different populations differed markedly in the type and number of mutant genes that they contained.

As Dobzhansky points out, we may say on the basis of such analysis that one of the most important recent evolutionary events has been the merging in the human species of small population-groups in a more or less freely interbreeding whole.

In general it seems clear that from the standpoint of mathematical theory, existing mutation-rates will in moderately abundant species suffice, with the aid of selection, for the distinctly slow processes of evolutionary change to be observed in fossils.[4]

This statement is a deductive one made on theoretical grounds from the standpoint of mathematical analysis. In the remainder of this chapter we shall deal with more concrete aspects of the relation between Mendelism and evolution.

2. Genes and Characters

A great deal of water has flowed under the bridges of biology since the early days of mendelian work, when mendelian factors were rigidly equated with mendelizing characters when only two states of a given gene were recognized, the dominant being supposed to represent its presence and the recessive its absence, and when all mutations and all mendelian genes were supposed to have considerable and obviously discontinuous effects.[5]

To-day the notion of mendelian characters has been entirely dropped (see, for instance, Sinnott and Dunn, 1932). The term may occasionally serve as a useful piece of shorthand notation, but is in point of fact a false conception. In the first place, a single gene may affect a number of characters, a phenomenon known as pleiotropism. Grüneberg (1938), in an illuminating analysis, points out that pleiotropic effects may be realized in three different ways. In the first place, a gene may exert a direct effect on two or more distinct processes. The example of the effect of the series of white eye-colour allelomorphs in *Drosophila*, which also exert an effect on the shape of the spermatheca, is probably an example of this category. Secondly, a gene may exert a direct effect on a single process, but in many different sites and conditions. This holds for the primary action of the gene studied by Grüneberg (1938) in the rat, which causes hyperplasia and abnormal growth of cartilage in the ribs, trachea, and elsewhere. Another and even more striking case is the array of anomalies in such different organs as eyes and feet, found in a particular strain of mice, which Bonnevie (1934) has shown are due to alteration in a single developmental mechanism, namely the causing of embryonic blebs of fluid at a particular stage of embryonic development.

The most interesting examples for our purpose, however, belong to the third category, of indirect effects. A gene exerts a primary direct effect, and this then causes numerous secondary effects. Grüneberg's gene in the rat has a hyperplastic anomaly of cartilage as its primary effect. But among the secondary effects are such varied "characters" as emphysema, hypertrophy of the right ventricle, blocked nostrils, and incompletely occluded incisor teeth.[6]

An equally good example is that of the frizzled breed of fowl. Here the primary effect is entirely on the feathers; as the secondary effect of the resultant abnormal heat loss, we find (in temperate climates) marked thyroid and adrenal enlargement, subnormal body-temperature, and much increased food-intake.

Such secondary effects are excellent examples of what Darwin called "correlated characters:, which may be of great evolutionary importance.

Furthermore, any given character represents the end-result of a great number of genes interacting with the environment during development, and is not inherited as such. What is investigated in any genetic experiment is the inherited basis for a constant character-difference. Thus a *character-difference* may be said to be inherited in mendelian fashion, while the *character* cannot but even so the differential effect of a particular gene on the character need not by any means always be the same. It may alter according to differences in the environment, and also according to differences in the remainder of the gene-complex. As an example of the first, we may take the well worked out case of "abnormal abdomen" in *Drosophila* (T. H. Morgan, 1915). This effect depends on a single partially dominant sex-linked gene: but it is only manifested in moist conditions. In dry conditions flies pure for the gene appear perfectly normal, while intermediates are produced by varying degrees of moisture (see Gordon and Sang, 1941, on the similar case of *antennaless*).

An equally striking botanical case is that of a type of albinism in barley (Collins, 1927), dependent on a single gene. When grown below 6.5°C. the plants entirely lack chlorophyll, while above 18°C. they are quite normal. Between these limits the mutants produce a graded amount of chlorophyll. Baur's classical case (1923) in *Primula siensis* has now been shown to be due to faulty experimentation. For other plant cases see Lawrence and Price (1940). In Himalayan rabbits and Siamese cats (both simple recessives) black pigment is produced only below a certain threshold temperature. Normally only the extremities fall below this; but Iljin (1927, 1930) has experimentally produced pale extremities, and black on the body. Another *Drosophila* example is *short-wing*, a sex-linked recessive which at 27.5°C. markedly reduces wing-length and affects eye-development. However, the effect falls away with temperature and is absent at

14°C. (Eker, 1935). Thus environmental changes may either mask or bring out the results of genetic difference. We must therefore distinguish carefully between the *nature* of the gene and its *expression*. The gene itself can only alter by mutation; but its expression can be affected in a number of ways.

The most revolutionary change has come in regard to the way in which the expression of a gene can be altered by other genes. The discovery of this fact has given us the two fundamental concepts of genic balance and the gene-complex. Thus the internal or genetic environment, of a gene may produce effects upon its expression which are as striking as those induced by the external environment, and of course very much more important from the point of view of evolution.

By genic balance we imply that individual genes act, not absolutely, in virtue solely of their inherent qualities, but relatively, in virtue of their interaction with other genes. The concept was first reached by studies on sex. It was at one time supposed that in *Drosophila* and other forms with male heterogamety, one X-chromosome automatically determined maleness, and two femaleness. It has since been shown, however, that it is the balance of the X-chromosomes to the autosomes (A) which is operative. A ratio of 1 X to 2 A produces maleness, of 1 X to 1 A produces femaleness; while one of 1 X to about 1.5 A produces intersexuality: Sterile "super-males" and "super-females" are produced by ratios of 1 X to over 2A and under 1A respectively. Here we can deduce that sex-determination is effected by the quantitative ratio between sets of male-determining and female-determining genes, though we do not know how many separate genes are involved in each set.

The principle, however, appears to be of universal application: the effect produced by any gene depends on other genes with which it happens to be co-operating. The effects of modifying genes are the most striking examples.

A classical case of the kind is the alteration in the hooded pattern of rats by modifiers (Castle and Pincus, 1928). The basic gene remains the same, but its effects may be reduced to a few specks of black on the head, or progressively extended over the whole back and most of the belly, by the agency of accessory genes. In cotton (*Gossypium*) differences in leaf-shape have been evolved in a precisely similar way (Silow, 1941).

Extending this concept, we reach that of the gene-complex. The environment of a gene must include many, perhaps all other genes, in all the chromosomes. This gene-complex may be altered in numerous ways by mutation or recombination so as to modify the effects and mode of action of particular genes, whether well-established ones or new mutations. We can thus distinguish between the genetic and the somatic environment of genes.

Further, it is now known that a gene can exist in a great variety of allelomorphic forms (alleles), up to a dozen or more being known for single loci. The effects of these usually differ in a quantitative way (though occasionally in a qualitative way as well), and the steps between the various alleles may be very slight. Multiple alleles are, in general, taken to represent different states of a homologous material unit. They thus constitute one type of gene-differences with quite small effects. Many modifiers and cumulative factors such as those involved in quantitative characters also have small effects. In many cases the actual origin of such small differentials by mutation has been observed. Further, where a gene's effect is small, the variations of expression, due to environment and to other genes, may readily cause an overlap with the phenotypic expression of an allele or of another gene with similar type of expression. Thus though genetic variability must be discontinuous, its expression in measurable characters may become continuous.

Even mutations which in one gene-complex are pathological, in another may be perfectly harmless, and in yet another advantageous. A striking example of this is provided by the work of M. Gordon (1931) on generic hybrids between the viviparous freshwater fishes *Platypoecilus* and *Xiphophorus*. Some strains of the former possess a gene for the production of large pigment-cells with black pigment, responsible for a certain type of spotted pattern. When, however, this gene interacts with certain genes in the sword-tail (*Xiphophorus*), the pigment-cells produce cancer-like melanotic tumours (Kosswig, 1929).

Equally striking and curious results may occur as the result of the interaction of two gene-complexes in species-hybridization. As an example of this, we may cite an intergeneric pheasant cross recently described (Huxley, 1941*b*). The cross was between a Lady Amherst pheasant (*Chrysolophus ampherstiae*) and an Impeyan pheasant (*Lophophorus impeyanus*), and the hybrid was a male. The coloration of the males of both parent species is not only brilliant but varied. Thus the Lady Amherst cock has a black-and-white extensible "cape" on the head, a striking regional pattern on the body, and elaborately barred central tail feathers; while the Impeyan cock is distinguished by brilliant patches of burnished bronze, green, and blue-back on its upper parts, with white rump and buff tail. The hybrid, however, has most of its upper parts uniformly black, with mere traces of green and bronze iridescence, but neither regional patterning nor brilliant colouring. The lower parts and central tail feathers are mottled with brown, grey and white in various ways. This simple colour-scheme, by the way, cannot be considered to have any reversionary or "atavistic" significance whatever: it is simply that the delicately balanced gene-systems responsible for the two elaborate patterns have cancelled out, so to speak, to produce a wholly different and almost uniform coloration. It may perhaps be mentioned that in other characters the hybrid is intermediate (e.g. the shape and size of the cape), and in still others shows obvious dominance (e.g. the blue face-skin of the Impeyan).

What we may call partial gene-complexes may also arise in relation to the separate chromosomes into which the total gene-complex is divided. This has been demonstrated by Mather (1941) as regards what he terms polygenic characters—i.e. quantitative characters dependent on the co-operation and interaction of numerous genes. In *Drosophila*, the number of ventral abdominal hairs can be changed by selection so as to trangress the limits of normal variability in both plus and minus directions (cf. Castle's hooded rats). But the effect of selection is exerted in two main stages. During the first two generations, a marked change is effected, which Mather interprets as being due to recombination of whole chromosomes. Then, after a period of relative stability for two or three further generations, a further and more marked change is produced, continuing for a number of generations; this appears to be due to recombination of originally linked genes forming polygenic combinations for hair-number. Different polygenic combinations for this character have arisen in homologous chromosomes in different straits, each combination being balanced in that it contains both plus and minus modifiers of the character. Furthermore, a number of such combinations will tend to co-exist in a species with considerable out-crossing, since the delicacy of the balance (see below) is improved when the genes for a polygenic character are heterozygous. When selection is practised, crossing-over provides new and extreme combinations. (Cf. the more fully isolated partial systems of Darlington).

No such balanced polygenic combinations can be detected in the modifiers of abnormal mutant characters, such as bar eye. Mather suggests that they will arise by natural selection in respect of wild-type characters, in order to prevent too great deviation from the normal, while at the same time affording the possibility of change under selection, through crossing-over providing a limited number of extreme recombinations. Such polygenic combinations, like other features of genetic systems are thus a compromise between immediate individual fitness and long-term evolutionary plasticity.

To sum up the evolutionary bearings of recent discoveries about gene-complexes, we may say that evolution not only need not occur by a series of sharp single steps, but is not likely to do so; each such step is immediately *buffered*, as it were, by ancillary changes in genes and gene-combinations which can act as modifiers for the major mutating gene and adjust it more or less completely to the needs of the organism, though final adjustment may have to wait upon further mutation. In any case, what evolves is the gene-complex; and it can do so in a series of small, if irregular steps, so finely graded as to constitute a continuous ramp.

When we reflect further that it is theoretically possible for a gene to alter its character radically by mutating step by small step from one member of a multiple allelomorph series to another, we shall see that the discontinuity inherent in Mendelian genetics is no obstacle to the visible continuity revealed in paleontological evolution. Discontinuous germinal changes are perfectly capable of producing continuous changes in somatic characters. Nor, as we shall set forth more fully later, is the pathological character of many mutations at their first appearance necessarily a bar to their final evolutionary utilization by the species.

The divergence of two stocks will always involve the accumulation of different genes in the two lines, each buffered by special modifiers and adjusted in its own way to other genes; and this will inevitably lead to a certain amount of disharmony on crossing, the F1 or later generations being less fertile or less viable, or both (see discussion in Muller, 1939, 1940). The fact of internal adaptation within the gene-complex thus automatically helps to bring about the inter-sterility of species.

3. The Alteration of Genic Expression

Let us take some examples of mutations, at first deleterious, being rendered innocuous. One of the most striking cases occurs in the meal-moth *Ephestia kuhniella.* Here a red-eyed mutant is known which shows considerably lowered viability. But when the recessive gene for red eyes is combined with another recessive gene for transparency of eyes, the double recessive is as viable as the normal wild type (Kühn, 1934).

A somewhat similar example comes from *Drosophila.* The mutation *purple* (eye-colour) causes the duration of life to be considerably reduced. Another mutation, *arc,* affecting wing-shape, produces nearly as great a reduction. But the two in combination cause much less reduction than either separately (Gonzalez, 1923). The figures (for both sexes together) are as follows:

Genes	*Length of life (days)*
Purple	24.54 ± 0.18
Arc	26.81 ± 0.29
Purple/arc	33.71 ± 0.34
(Wild-type	39.47 ± 0.28)

Considering what severe effects are exerted by the to genes separately, the favourable result of their combination is very striking.[7] Brierley (1938) has worked out a means of determining the "selective index" of any gene-combination, and has obtained some suggestive preliminary results, also in *Drosophila,* on the general viability interactions of numerous genes.

In a number of cases, the restoration of viability occurs gradually in the mere course of maintaining the mutant stock. The classical analysis of this phenomenon is that of the eyeless mutant of *Drosophila.*

Eyeless is due to a 4th-chromosome recessive gene. Its characteristics on its first discovery were that it considerably reduced the size of the eyes, in some cases to complete absence, decreased fertility markedly, and had a depressing effect on viability. After, however, a stock for eyeless had been inbred without any artificial selection for a number of generations, it was found that practically all the flics had normal eyes and showed little reduction in either fertility or viability. On outcrossing to the normal wild type and re-extracting the recessives in F2, it was found that these once more manifested the original characters of eyeless, though in even more variable degree (T.H. Morgan, 1926, 1929).

The explanation of these facts is that the manifestations of eyeless are readily influenced by other genes, and that in general those modifiers which make for normal viability and fertility also make for normality in eye-size. Thus natural selection acting upon the recombinations of modifiers present in the stock speedily saw to it that the combination making for the manifestation of reduced eyes was eliminated. In competition with its wild-type allelomorph, eyeless would be eliminated; but in stocks pure for eyeless, the genes to be eliminated will be the plus modifiers of the mutation. In broadest terms, there has been a selection of the most favourable gene-complex.

A similar genetic modification of recessive mutations towards wild-type expression was found by W. W. Marshall and Muller (1917) in *Drosophila melanogaster* for the wing-characters *balloon* and *curved*. Still another example from *Drosophila* is the sex-linked mutation *vesiculated,* affecting the wings: this can be brought back to normal expression by means of autosomal modifiers (Evang, 1925). A very similar botanical case is recorded by Harland (1932) for chlorophyll deficiency in cotton; in a way this is even more striking, since the original manifestation of the genes (in this case three pairs are involved) was markedly lethal. Here again inbreeding and selection led to the production of a reasonably viable form, while outcrossing of this to normals caused the reappearance of lethal segregants. The genetic mechanism is similar to that operative in the classical case of Castle's hooded rats and the alteration of their pattern in either plus or minus direction by an accumulation of modifiers (Castle and Pincus, 1928), though the selective implications are of course different.

Selection of this type, it now appears, is a constant and indeed normal process. It has become almost a commonplace in animal used for genetic analysis to find that mutant types, which at first are extremely difficult to keep going, after a few generations become quite viable. This has repeatedly occurred in *Gammarus,* for instance (Sexton, Clark and Spooner, 1930), and Mr. E. B. Ford tells me that it has often occurred in his cultures of other mutant strains of the same species. A recently-described example from *Drosophila* is that of white eye in *D. obscura* (Crew and Lamy, 1932). This recessive mutant was at first very delicate but its viability improved progressively on inbreeding. A precisely similar course of events was observed in the hairless mutant of mice (Crew and Mirskaia, 1931), showing that the phenomenon occurs in mammals. In plants, we have referred to cotton: an analogous case has been found in the nasturtium (Weiss, quoted by R. A. Fisher, 1931).

In all cases, the explanation is basically similar to that for eyeless. The experimenter, however, will also attempt to keep the mutant character sharp: he will therefore be selecting for combinations which keep the viability up without altering the visible expression of the gene, so that the process may take a little longer.[8] R. L. Berg (1941) finds that, owing presumably to this form of selection, dominance become more intense in laboratory stocks than in the wild.

A converse effect is found when a gene, which in one species or variety is harmless, becomes deleterious on outcrossing. The explanation is that the expression of the gene in its normal situation has become so conditioned by favourable modifiers that it exerts no ill-effects; on outcrossing, however, it finds itself in a genic environment lacking some or all of these modifiers, and consequently expresses itself in ways unfavourable to viability. We have mentioned a case of this sort in Gordon's fish crosses. A striking example, particularly relevant to our present discussion, comes from Stockard's work (1931, 1941) on dogs. The St. Benard breed shows various symptoms of hyperpituitarism that simulate the pathological condition known as acromegaly. Matings between St. Bernards give normal litters; but when the St. Bernard is crossed with the Great Dane, a breed that may be regarded as a simple giant type with no hyperpituitary characters, a considerable proportion of the F1 (and later) offspring show serious disturbances during their growth, notably hydrocephalus and paralysis of the hind limbs, these effects being clearly of genetic origin.

Man, it seems has pushed the St. Bernard breed as far as it can go in the direction of large size, heavy jowl, and other effects of extreme or one-sided pituitary action; and in the process has amassed those combinations of modifiers which will protect the organism against the harmful effects of its exaggerated glandular development. When the breed is outcrossed, the protective genes are diluted to a greater or lesser extent, with corresponding ill-effects. The modern show type of bulldog has similarly been produced by selection for genes causing abnormal thyroid structure and function; here the breed has been pushed still further towards the glandular limit, since a considerable proportion of males are partially or wholly sterile.

An example in which genic expression is altered without noticeable effects on viability is that of the hybrids between *Drosophila melanogaster* and *D. simulans*. About 50 per cent of these lack bristles present in both parent species—i.e. certain combinations of modifiers from the two parent species suppress the expression of certain genes controlling bristle development (Biddle, 1932).

Excellent examples involving artificial election have resulted from the work of R. A. Fisher (1935, 1938), who by repeated back-crossing introduced dominant or semi-dominant genes from domestic breeds of poultry, into the unselected gene-complex of the wild jungle fowl. Polydactyly varies in its single-dose expression[9] in domestic breeds. Punnett and Pease (1929) found it to behave usually as a complete dominant, but sometimes as a recessive and sometimes as an irregular partial dominant. To explain these facts, they postulated one dominant inhibitor preventing the expression of the gene for polydactyly, and a second capable of suppressing the action of the first. Hutchinson (1931) pointed out that a simpler explanation is provided by Fisher's theory, according to which there is a single gene which remains constant, but whose dominance-relations differ in different gene-complexes. In the wild gene-complex, moreover, as Fisher showed, the homozygote can be definitely distinguished from the heterozygote by possessing larger extra toes, with more bones. What has happened is that an originally intermediate single-dose expression has, in most domestic breeds, become more complete. The gene for barred plumage behaves in a somewhat similar way. Most interesting are the results with the gene for crest on the head. In the wild gene-complex the crested gene in single dose produces crest alone; in double dose, however, it produces not only an unusually large crest but a cerebral hernia of deleterious character. In the Japanese silky fowl, no hernia is ever produced, and the effect of the crested gene is the same in single and in double dose. Thus firstly, the gene has become fully dominant in domestication in place of intermediate; secondly, its effect on hernia has been suppressed by modifying factors (cf.); and thirdly, in the wild gene complex, its two effects are of different type, the harmful hernia being fully recessive, the neutral crest being partially dominant.

Another aspect of the adaptation of genic expression to the needs of the organism concerns the stability of expression of genes. Plunkett (1932), for instance, has analysed the fact, well known in general terms, that wild-type characters are usually much less modifiable by changes in environment than are those determined by mutant genes. His analysis was for the most part restricted to temperature-effects in *Drosophila,* but the principle can be widely generalized. The explanation is based on the fact that genes are in most cases concerned with the rates of processes (see Goldschmidt, 1938*a*). The curves expression the rate of the processes are in general obliquely S-shaped, tending to a final horizontal equilibrium-position (Ford and Huxley, 1929). In wild-type genes, the flattening is normally completed before the imaginal state (or corresponding definitive stage) is reached, whereas in the majority of mutant genes, the curve is still in the ascending phase at this stage. Thus quite small disturbances of the curve will have marked effects in mutant genes, but very slight ones on wild-type characters.

There can be no doubt that selection has been at work to adjust the rates of gene-controlled processes so as to produce this result in wild-type genes, thus conferring a high degree of stability on the characters concerned. As Plunkett further points out, the evolution of complete dominance, with which our next section deals, can be regarded as a special case of this principle, viz. that in general natural selection favours the genotype which produces the most stable and therefore uniform phenotype.

Where special circumstances demand the contrary effect, that different conditions shall be met by quite distinct phenotypes, selection has often operated to produce plasticity of genic expression. This plasticity, however, is usually of a special type, operating by some sort of switch mechanism, so that two or a few contrasted phenotypes, each of them relatively stable, are produced. The classical case is that of the environmental control of caste in social hymenoptera, whereby the same genotype can be made to produce either neuter or fertile females, and intermediates are rare aberrations. The same sort of mechanism seems to be at work in regard to the winged and wingless condition of aphids, and in environmental sex-determination.[10]

4. The Evolution of Dominance

R. A. Fisher (1928, 1931, 1934) has extended this concept of the alterability of gene-expression by modifiers to account for dominance in general, or at least for many features of dominance as found in nature. His argument runs as follows. Mutation is always throwing up new genes; the majority of these will inevitably be deleterious, since in a delicately-adjusted system like the gene-complex most changes are likely to be for the worse unless compensated. Further, we know as an empirical fact that the majority of mutations are repeatedly produced. Obviously the great majority of mutant genes will be carried in single dose, so that it will be an advantage to minimize any activity shown by them while in the heterozygous state. Thus, even when a harmful mutation at its first appearance shows considerable single-dose expression, i.e. manifests some or all of its effects when in the heterozygous state, then, if it be repeatedly produced (which is the case with most mutations), the way is open for it to be forced into recessivity by selection action on the rest of the gene-complex. If it is relatively abundant (and recent studies of wild populations—e.g. C. Gordon (1936), Dubinin and others (1936), and Sexton and Clark (1936*a*)—have shown the surprisingly high incidence of various recessives which they carry in the fly *Drosophila* and the shrimp *Gammarus* respectively), selection may get to work on the homozygous condition, and render it also inactive. In such a way, as Fisher points out, mutations may be reduced to the rank of specific modifiers, normally inoperative, but exerting effects in abnormal gene-situations.[11]

In support of this view, we will cite some of the array of facts that show how readily the degree of dominance of a gene may be altered by the presence of other genes. The classical case is that of horns in domestic sheep (Wood, 1975). The difference between horned and hornless breeds depends on a single gene-difference, but whereas a single dose of the gene for horns will produce horns in rams, a double dose is necessary in ewes. The same gene is thus dominant, in the internal environment of males, but recessive in that of females. In fowls, Landauer (1937) found that the gene for frizzled plumage, whose effects are normally incompletely dominant is converted into an almost complete recessive by the presence of a particular recessive modifier in double dose. Dunn and Landauer (1234, 1936), with the gene *rumpless*, which reduces the tail and posterior end of the body, were able to go further and to show that this could be converted either into a dominant or a recessive by crossing to different stocks, followed by selection for dominance or recessivity respectively. In mice, on the other hand, the gene for black, which is normally a complete recessive, can be converted into an incomplete dominant by modifiers (Barrows, 1934). In *Drosopila virilis,* the dominant gene for rounded win converts the gene for ruffled bristles from a recessive into an incomplete dominant (Lebedeff, 1933). Mather and North (1940) describe gene in mice whose only known effect is to modify the dominance-relations of the *agouti* gene.

A case of some historical interest is that described by Federley (1911) of the behaviour of white spotting in the larvae of the moth *Phgaera*. In *P. anachoreta* an unspotted variety is found, and this behaves as a simple recessive to the normal spotted condition. In *P. curtula,* however, only the unspotted condition exists. In the F1 of a cross between the two species, using the spotted form of *P. anachoreta,* spotting is expressed in an intermediate form, i.e. its dominance has been partially abolished.

Twenty-five years ago, this fact seemed so remarkable that an authority such as Sturtevant (1912) was disinclined to accept it: to-day, however, such alteration of gene-expression by modifiers has become a commonplace.

We need not multiply examples. It is clear that the dominance of a gene can be radically modified according to the genic environment in which it happens to find itself.

The next step is to show that this undoubted fact of the modifiability of the degree of dominance has been utilized in the course of evolution to make most commonly-recurring mutations recessive, so as to reduce the degree of their heterozygous expression, including that of the decreased viability which accompanies most mutations at their first appearance.

In cotton, Harland (1933) and Hutchinson and Ghose (1937) have studied the mutation *crinkled dwarf*. This occurs in the Sea Island variety of *Gossypium barbadense*, and is there a complete recessive. When crossed to unrelated strains of the same species, it is not completely recessive but shows a low degree of single-dose expression.

When, however, the mutation was introduced from *G. barbadense* into the related species *G. hirsutum* (upland cotton) there proved to be a complete absence of dominance of the normal type. The F1 is intermediate, so that at first sight we might imagine the single-dose expression to be about 50 per cent; but the fact that the F2 gives a large and unclassifiable range shows that the degree of dominance must be under the influence of a number of modifying genes. This is confirmed by the results of back-crossing the F1 species-hybrid bearing the crinkled dwarf gene to various strains of *G. hirsutum*. In certain lines, complete or almost complete recessivity of crinkled dwarf was re-established, while in others the single-dose expressivity was rendered accurately intermediate, the heterozygote class being clearly separable from either homozygote. As Hutchinson and Ghose have clearly shown, the results entirely support R. A. Fisher's views. Later work (see summary by Harland, 1941) has shown that it also occurs as a very rare mutant in *G. hirsutum*. The *barbadense* crinkled show intermediate single-dose expression in the F1 with *hirsutum*, but is recessive-when transferred to a pure *hirsutum* gene-complex. When, however, *hirsutum* crinkled is transferred to a pure *barbadense* gene-complex, it behaves as an intermediate.

Harland (1941) develops the thesis that the dominance of normal alleles in cotton may be built up in a considerable variety of ways. Thus the character *petal spot* in both *barbadense* and *hirsutum* is based primarily on a main gene which, however, exists in different allelic forms in the two species. But whereas in *barbadense* the action of this main gene must be reinforced by a number of plus modifiers to produce its full effect, in *hirsutum* the main gene is stronger, and requires no (or few) modifiers. As corollaries of these facts we find (1) that in *barbadense*, the modifiers exert some positive action (a small spot) even in the absence of the main gene; (2) the gene for *barbadense* spot transferred to the *hirsutum* gene-complex has a very weak effect, and the petals are barely spotted; (3) the gene for *hirsutum* spot transferred to the *barbadense* gene-complex is reinforced by the modifiers there present, and produces a spot which is larger and more intense than any previously known; (4) crosses of the spotted forms of the one species with the unspotted of the other give a graded F2 with all intensities of spotting; (5) the F2 from the unspotted forms of the same two species produces some spots as large as normal *barbadense*—i.e. due to recombination of modifiers only, presumably from both species. In *G. arboreum* yet a third method of producing the spotted character is found: there is no main gene, but spotting depends on the genes that in *barbadense* act as modifiers, but here must be called a multiple factor series or polygenic combination.

Haraland suggests that if a character is of advantage to the species, it can be more readily retained, in spite of recessive mutation, if its expression depends on several genes. However, this conclusion does no seem justified. The advantage of the method of using a single main gene together with modifiers would rather seem to lie in keeping the expression of the character relatively constant in the normal range of environmental conditions, but retaining a considerable reserve of potential variability to meet new or extreme conditions. The single-gene control will give greater stability, the purely multifactorial control (by "modifiers" only) will give greater plasticity. Single-gene control will permit the character to be more readily lost under the influence of selection or by drift. Presumably in correlation with this, in *hirsutum,* where spot depends on one gene only, the spot character has been lost in all but a few rare varieties.

To return to crinkled dwarf, the same principles apply. Normality (non-crinkled) in *barbadense* depends on a single strong allele, like spot in *hirsutum.* Normality in *hirsutum* on the other hand, depends on a weaker main non-crinkled allele, together with a number of modifiers. These not only encourage the dominance of normal over crinkles, but make the pure crinkled forms more normal, both in appearance and viability. By rigorous selection, new combinations of modifiers have been obtained which render the *hirsutum* crinkled practically indistinguishable from normal. When the two species are crossed, using the normal of one and the crinkled form of the other, F2 ranges from forms which are phenotypically normal through all grades of crinkling to "super-crinkled" types which are almost lethal. The petal spot experiments demonstrate two types of dominance. In *hirsutum* the Fisher effect is clearly operative, with modifiers aiding a weak "normal" gene, and also modifying the recessive towards normality. Recessive modification is much harder where a nearly dominant main gene exists, as in *barbadense.* This may then be due to an extension of the Haldane effect, by selection of "stronger" normal alleles. Harland and Atteck (1941) consider that this also operates for *crinkled dwarf* in some species, but the evidence is not decisive. They further point out that the Haldane effect is likely to be more important in self-fertilized forms, where the Fisher effect cannot so readily be produced (Haldane, 1939*a*). Where both types occur, as in *Gossypium,* doubtless "accidental" events such as the time of occurrence of suitable mutations, will determine which mechanism evolves (see also Silow, 1941).

An important empirical fact which was among those that led Fisher to promulgate his theory is that of the behaviour of multiple alleles. In almost every case so far investigated, the wild-type allele shows complete dominance over all the lower members of the series, whereas these when crossed with each other show intermediate expression. Thus the normal red colour of the eye in wild *Drosophila* is completely dominant over white, ivory, eosin, cherry, and all the other members of the white allelomorphic series: but white crossed with eosin, or ivory with cherry, gives an F1 with eyes intermediate in colour between the two parents.

The exception proves the rule: and the exception to this rule, to which attention was first drawn by Ford (1930), concerns the effect of this same series of genes upon an internal character of apparently no selective value. Dobzhansky (1927) had shown that the genes of the white-eye series affect the shape of the spermatheca to a small but constant degree. Whereas, as we have seen, the genes are all recessive to wild-type as regards their effects on the eye, they show intermediate expression as regards this secondary effect on the spermatheca. The body-colour genes of the ebony-sooty series show the same effect on the spermatheca, and the same differential expression as regards their effect on the external and the internal character (except that ebony body-colour is not wholly recessive). The most obvious explanation is that selection has been operative in modifying the expression of the disadvantageous external character, whereas on such effect was called for, or has been produced, with regard to the harmless secondary internal effect. We may also compare the different dominance-relations of the deleterious and harmless effects of the *crest* factor in fowls.

It is worth, nothing that this differential expression of the gene as regards two characters affected by it cannot be reconciled with any rigid form of the "inactivation" theory of recessiveness. This extension of the old Presence and Absence theory, which is obviously unternable in its original form, claims that the degree of recessivity corresponds to the degree of partial loss or inactivation suffered by the gene in mutating to one or other of its recessive allelomorphs. It is clear from what we know of actual deficiency-mutations, in which a portion of the chromosome is missing, that loss may produce effects of the same nature as gene-mutations (see, e.g., Mohr, 1920, who showed that that loss of the *white* locus produces an *ultra-white* effect more intense than *white* itself), so that the inactivation theory may sometimes apply. But the demonstration that some genes can become dominant or recessive according to the gene-complex shows that it can at best have a partial application. It is also noteworthy that the true absence of a gene in *ultra-white* produces some effects (spermatheca shape) which are not recessive!

As R. A. Fisher (1931) points out, the majority of the characters of most domestic breeds, or at least of the most obvious characters, especially of pattern and colour, depend on mutations recessive to the wild type. It would seem clear that man has here taken advantage of two facts, first that more or less recessive mutations are commoner in nature and secondly that they can be readily fixed by mating two similar individuals, in order to utilize striking and more or less recessive characters during his selection. Furthermore the natural tendency to concentrate for breeding purposes on individuals showing a character in more extreme form will, in the case of genes originally largely recessive, then automatically encourage the production of complete recessivity. If, however, the gene had more than intermediate single-dose expression, selection would tend to make it more dominant. Thus in general man's artificial selection will tend to encourage either complete dominance or complete recessivity, though for reasons quite other than those operative in nature. But, as mentioned, the distinguishing characters of domestic breeds in most species are usually recessive.

The chief exception occurs in poultry, where the majority of "domestic" characters are partially or wholly dominant. Fisher (1931) suggests that this difference is due to the fact that in the countries of their origin, the domestic forms, even to-day and more so in earlier times, would frequently be mated by wild cocks. In such a case, recessive, could not readily be fixed, whereas partial dominants would at least reappear in every generation; thus dominants would tend to be bred into the race by a natural selection of man's selective processes. A further effect would be that the degree of their dominance would be increased, through the new varieties being almost wholly heterozygous, and through man selecting the most striking individuals from which to breed. We have seen that his experimental tests have confirmed this hypothesis.

Marchlewski (1941) has recently confirmed Fisher's theory in dogs. Here, black was originally dominant over yellow; however, in the dingo and in various domestic breeds, yellow has become dominant through the selection of modifiers.

The recessivity of characters in other domestic species is not so nearly universal as Fisher was at first inclined to think (Castle, 1934, on various species; J. A. F. Roberts and White, 1930, and J. A. F. Roberts, 1932, on sheep). It seems clear, however, that man, by his breeding methods, has modified the single-dose expression of wild-type genes in his domestic animals, accentuating the recessivity of those with low, and the dominance of those with high single-dose expression; and that while this will result in most species in a preponderance of recessive breed-characters, in poultry it will tend to a preponderance of dominant ones.

It should be stated that Wright (1934*a*) does not agree with Fisher's views on the evolution of dominance, but wishes to ascribe recessivity to partial inactivation of the gene. An alternative hypothesis for the origin of the recessive character of most mutations has been given by Haldane (1930; and see 1939*a*). This is based on an interesting view as to the mode of action of genes, namely, that different multiple allelomorphs produce different amounts of some substance, but only up to a certain saturation value: beyond this they can produce no phenotypic effect. Thus any mutation in a minus direction below this level can be detected, but those in a plus direction cannot. In consequence a number of different multiple alleles of different strength, but all above that needed to give the saturation value, may readily accumulate without being phenotypically detected. If now minus mutations occur, Haldane suggests that, in order (I speak teleological shorthand) to prevent their visible and viability effects from being manifested in the heterozygote, those higher alleles will be selected which in combination with the mutation will not fall below saturation level.

In other words, higher members of the series will be selected, and visible dominance will be the result. Ford, however, has pointed out that the saturation level itself will be determined in relation to the residual gene-complex, so that even here the action postulated by R. A. Fisher may be operative, though in some cases in addition to the Haldane mechanism. In any case both suggestions involve selection acting upon other genes than the mutant. Since the above was written, Ford (1940*b*) has shown that the Fisher effect can be artificially produced. The currant-moth, *Abraxas grossulariata*, has a single-gene wild variety (*lutea*) with yellow instead of white ground-colour, which normally gives an intermediate F1 with wild type. By four generations of plus and minus selection, Ford has conferred both complete dominance and complete recessivity upon the gene.

Sewall Wright (1929, 1934*a*) attacks Fisher's general conception on the ground that the selection-pressure available will be inadequate to achieve the results envisaged. However, there seems little doubt that dominance of the "normal" wild-type allele has been evolved; and Plunkett and Muller independently (see Muller, 1935) have shown how the need for stability of gene-expression in development will secondarily result in the evolution of dominance.

Whatever the precise method employed, it seems clear that dominance and recessiveness must be regarded as modifiable characters, not as unalterable inherent properties of genes. Dominant genes, or many of them, are not born dominant: they have dominance thrust upon them. Mutations may *become* dominant or recessive, through the action of other genes in the gene-complex. The evolution of dominance is thus seen to be in large measure an adaptation to the deleterious nature of most mutations.

R. L. Berg (1941) points out that the intensity of dominance will be selectively balanced against the accumulation of deleterious recessives which it makes possible, and that it will tend to be decreased in species consisting of incompletely isolated groups. The extra plasticity thus conferred upon such species will be in addition to that deduced by Wright.

As another example of an adaptation of the genetic mechanism itself, Fisher (1930*a*) cites the plasticity conferred by sexuality. For one thing, it will permit evolutionary advance by the combination of new mutations. If several favourable mutations occur in a population in a given time, then in a sexual cross-fertilizing species they can be combined. But if the species is asexual, they will almost certainly remain isolated, each confined to one line; for them to be combined, one mutant must be selected until it has become the main type, and only then will a second favourable mutation have a chance of becoming combined with the first.

In the second place, it will permit recombination to throw up new gene-combinations and so to use the existing genetic variance of the species to alter the type quickly in relation to changed conditions. Thus it promotes both progressive specialization and plasticity in response to the changes and chances of the environment. In addition, as Fisher (1932) stresses, it has a function to perform in relation to the deleterious nature of most mutations. For, by allowing recombination, it permits mutations to appear in homozygous form, and thus facilitates the elimination of the more deleterious. Elimination will be greater when the frequency of homozygosis is increased by inbreeding or self-fertilization. Thus variations in the type of sexual reproduction will alter the emphasis of its evolutionary function (Darlington, 1939): evolutionary plasticity will be more encouraged by crossbreeding, evolutionary stability by inbreeding. Inbreeding will also promote both the rejection of unfavourable and the spread of favourable mutations.

Recombinational plasticity will be especially valuable when conditions vary and become less favourable. This is doubtless the reason why so many organisms adopt some method of asexual reproduction (which is more efficient *qua* reproduction) so long as environmental conditions are favourable, but resort to a sexual process as soon as they become unfavourable. This is so, for instance, with many protozoa, rotifers, lower crustacea, and aphids.

The biological meaning of this has been clearly brought out by careful genetic studies on *Paramecium* and other ciliates (Jennings, 1929). It has been found in general that conjugation causes an increase of genetic variability in the resultant population, and that while (as in higher organisms) the majority of the new type may be regarded as unfavourable, some are actually or potentially better-adapted than those prevailing before conjugation. Thus conjugation will in many cases provide an increased chance of throwing up a recombination better able to cope with unusual and unfavourable conditions. Precisely similar results have been obtained with lower crustacea by Banta and his colleagues (see Davenport, 1933).

Attention is elsewhere drawn to the action of the impulse to migrate in unfavourable conditions. This also confers plasticity on a species, but in this case by increasing the range of environmental opportunities available to a given hereditary constitution, instead of increasing the range of hereditary constitutions available to cope with given environmental conditions. This, however, can hardly be called an adaptation of the genetic mechanism.

On the other hand, the peculiar reactions of the crossing-over mechanism to temperatures may well, as Mr. E. B. Ford has suggested to me, fall into this category. In *Drosophila*, the best-investigated case (Plough, 1917), crossing-over is least at temperatures close to the optimum for the species, and increases rapidly both with increase and with decrease of temperature. Increased crossing-over will, of course, have the effect of increasing the recombination of the genes located in a single kind of chromosome, and this will have a considerable effect in a form like *Drosophila* where the chromosomes are few in number. Unfavourable temperatures will thus increase the genetic variance available to a population.

The discovery of the position-effect allows us to deduce certain ways, previously quite unsuspected, in which the evolutionary mechanism must itself have evolved. If, as now seems established, it is the case with some or all genes that interaction with near neighbours in the same chromosome affects their expression in an important way, then it is clear that all the genes within a given chromosome must be delicately adjusted to each other so as to produce a harmoniously functioning whole. Any given gene must be adjusted to its neighbours within a certain chromosome-distance either way; the genes at the limit of this range will be adjusted both to the gene at our hypothetical starting-point and to others further away, thus conferring a certain organization on the chromosome as a whole.

This involves a new conception of chromosomes. Up till quite recently, it was possible and usual to regard them as mere vehicles for the carriage and distribution of the hereditary constitution, without any functional organization of the genes they contained. These were assumed to be arranged at random, like coloured beads picked up haphazard by a blind man and threaded on a string; and their positions in the chromosomes were not supposed to have any relation with their effects on visible or other characters.

With the discovery of the position-effect, however, this assumption, as a hard-and-fast principle, has gone by the board. Although many genes affecting one character are scattered irregularly through the chromosomes, and genes affecting different characters are often contiguous, yet some degree of non-random arrangement does occur (Morgan, Schultz and Curry, 1940). Basically, and in origin, their arrangement doubtless is random, and what we know of the frequency of sectional rearrangements shows that genes must often change their neighbours in an essentially accidental way. But this randomness must then be given a functional polish: neighbouring genes must be adjusted to each other by new mutation and by recombination. To continue our metaphor, the blind man's necklace is looked at, and colour disharmonies are got rid of by choosing new beads of the same general colour but slightly different shade.

The same general type of adaptation to position-effects has been necessary as with dominance in relation to gene-mutation. Indeed, the "Weakening" of genes in abnormal positions (Dobzhansky, 1936) indicates that a disproportionate fraction of the single-dose expression of dominant genes is determined by their relations with their immediate neighbours. In addition, functionally-balanced groups of genes affecting polygenic characters will be evolved within the separate chromosomes (Mather, 1941). We must to-day consider chromosomes not as being purely mechanical gene-vehicles, but to a certain degree as organic gene-arrangements.

5. Types of Mutation

So far, under the head of mutation, we have been considering only gene-mutation or point-mutation, i.e. the substantive alteration of a definite unit-region of the chromosome outfit. But various other kinds of mutation are also known to occur and we must devote a brief section to these.

In the first place, there are *genome-mutations,* involving one or more whole sets of chromosomes and therefore of gene-outfits. The normal diploid complement of chromosomes of a species may become doubled (autotetraploidy). Or reduction may fail to occur, and a diploid instead of a haploid gamete may be formed. Or, as a result of the union of a normal gamate with a diploid one, however, formed, auto-triploid forms with three genomes may result.

Tetraploidy in nature may also result from a cross between two species, when the corresponding chromosomes from the two parents, do not pair before meiosis and the hybrid is therefore originally sterile. If, however, the chromosomes of a cell divide but not the cell-body itself, all descendants of this cell will be tetraploid, and the two members of each kind of chromosome can act as mates at reduction. The result will be that the gametes have complete genomes from either parent. This is known as allotetraploidy, and its actual origin has been observed in *Primula kewensis,* the fertile hybrid between *P. floribunda* and *P. verticillata*. In this case the original hybrid was sterile, and the fertile type, with larger leaves and flowers, arises sporadically in cuttings, from a cell in which chromosome-doubling has occurred. Allotetraploidy is almost confined to plants, because of the favourable conditions provided by their vegetative growth for the rare chromosome-doubling to occur and to give rise to reproductive tissue, and because of their lack of the sex-chromosome mechanism normal to animals, which would not function after chromosome-doubling (Muller, 1925). For polyploidy in animals see Vandel (1937).

The presence of six, eight, and more genomes in a strain or species is also known, again almost entirely in plants; presumably the condition is usually consequent upon species-crossing (Darlington, 1937).

In all genome-mutations, the genome-units may be approximate only, sometimes with loss and sometimes with gain of one or a few chromosomes (aneuploidy).

Triploid and other anisoploid strains with an odd number of genomes are relatively rare in nature, and cannot normally reproduce themselves sexually, since the chromosomes of one genome cannot find mates. But allopolyploids and other strains with an even genome-number can maintain themselves.[12] Such polyploidy has undoubtedly been of considerable evolutionary importance in plants. One method of species-formation is by allotetraploidy after crossing. But apart from this, polyploid of any kind, so long as not excessive, by multiplying the number of gene-pairs of each kind in the hereditary constitution, confers long-ranges potential variability and plasticity on the species. For different gene-pairs may mutate in slightly different ways, giving a gradation of new recombinations. High autopolypolidy, however, by virtually suppressing the chance of manifestation of recessives, reduces plasticity. It appears that the phenomenon has been of importance in the evolution of higher plants, where series of related forms with two, four, six, and higher numbers of genomes often occur. Polyploidy has also undoubtedly contributed to the evolution of many cultivated plants, notably the cereals and cotton.

The second type of chromosome-mutation is that of the addition or subtraction of single chromosomes. This again appears to be much commoner in plans than in animals. It has been thoroughly investigated in the Jimson weed, *Datura stramonium*, by Blakeslee and his collaborators (1928). The commonest case is where one kind of chromosome is represented three times instead of twice in the hereditary outfit. Such *trisomic* mutants, as they are called, show greater differences from normal than do ordinary gene-mutants. They also show reduced viability due to the quantitative upset of gene-balance which they cause.

In *Datura*, these types cannot become fixed, since no pollen with an extra chromosome is viable. Thus all viable pollen-grains with contain n chromosomes, while the ova will be either n or $n + 1$.

It would seem, however, that in the evolution of some plants, the condition has become stabilized; but this is always in polyploid forms, where imbalance is not so readily brought about.

The third type of chromosome-mutation is the sectional involving only parts of chromosomes. For its occurrence and evolutionary bearings in *Drosophila*, see Muller (1940). One form of this is known as *deficiency* and involves the loss of a portion of a chromosome. This is known to occur not uncommonly in *Drosophila*, but is here and probably elsewhere of little evolutionary significance, since homozygous deficiency is usually lethal.

The converse is known as *duplication*, when a portion of a chromosome comes to be repeated, occurring instead of once, either in the form of a translocation to another chromosome, or of a "repeat" within the same chromosome, often immediately adjacent to its original position. Small "repeats" of this type have been shown by the salivary gland technique to be not infrequent in *Drosophila*, and are of considerable evolutionary importance. They are of immediate importance, since the alteration in genic balance would usually produce definite but not deleterious effects. They are of much greater ultimate importance, since they constitute the chief method by which the number of genes is increased, thus providing duplicate factors, and the opportunity for slight divergent specialization of homologous genes, giving great delicacy of adjustment. In this respect they would appear to be a good deal more important than the earlier-detected and more spectacular process of duplicating whole genomes by autopolyploidy.

The next type of sectional chromosome-mutation includes all cases involving spatial rearrangement of sections of two kinds of chromosome. The most important form of this is *reciprocal translocation* or segmental interchange. When this occurs, two chromosomes break and exchange fragments. The precise mechanism need not concern us. It can be induced with greater frequency by X-rays, and appears to occur where chromosomes, actually or almost touch each other. It is known to occur or can be deduced to have occurred in a number of plants, and in *Drosophila* and certain other animals.

In *Datura,* over forty so-called "prime types", which differ from each other merely by rearrangement of segments of the chromosomes, and which appear to owe their origin to reciprocal translocations, are known to occur in nature; they do not show visible differences. Different prime types differ in their geographical distribution.

Owing to certain peculiarities of chromosome behaviour, these prime types in *Datura* tend to persist as such, even after a cross. This is in effect a form of isolation and should eventually give opportunities for mutation and selection to produce visible differences between the various chromosomal types.

After crossing between two prime types, the hybrid type will, owing to certain chromosomal peculiarities, be reproduced as such in later generations, without rearrangement due to crossing-over, as well as the two pure types. If now a recessive lethal mutation occurs in one of the chromosomes which have suffered segmental interchange, the prime type containing that chromosome cannot be reconstituted, as a double dose of the lethal is *ex hypothesi* fatal. Since lethals are relatively common types of mutation, one may readily occur also in corresponding portions of the chromosomes derived from the other prime type. If so, and if we are dealing with an inbreeding group, we shall have a condition of "balanced lethals", and only the hybrid chromosome-combination will be capable of survival.

Wherever much segmental interchange has occurred, followed by long-continued intercrossing between the resultant prime types, we may expect to find balanced-lethal and therefore permanently hybrid combinations. And once these come into existence, they can differentiate still further by the accumulation of gene-mutations.

This, in actual fact, is what appears to have occurred in the evening primroses, *Oenothera* (see Renner, 1925, and Cleland, 1928; and summary e.g. in Dobzhansky, 1937). One known species (*O. hookeri*) is of normal chromosomal behaviour, but all the others present balanced-lethal chromosome combinations of greater or lesser extent. The chief evolutionary significance of these phenomena would appear to lie in its providing a special method of species-formation. It is, however, of historical interest since occasion crossing-over will give apparent mutations (really recombination of large blocks of genes); and on the basis of these de Vries advanced his mutation theory.

Small translocations of various types seem to occur quite frequently in *Drosophila.* They have probably been of some, though secondary, importance in initiating the differentiation of species (Dobzhansky and Tan, 1936).

As final form of sectional chromosome-mutation we have *inversion,* in which one segment of a chromosome becomes reversed within the chromosome as a whole. Quite large or very small portions of the chromosome may become inverted. Here again the frequency of the process may be accelerated by X-rays.

Crossing-over, of a type which will give viable offspring, cannot occur in the inverted section of a chromosome paired with a normal mate. This being so, inversion may produce a distinct type, homozygous for the inverted chromosome, in addition to the normal; in some cases, in fact, hybrids between the two types will not be able to reproduce so freely, because of the death of cross-overs. The resultant isolation of the two types of chromosome will permit their differentiation. More than that, selection will tend to erect barriers to intercrossing, so that the resultant waste due to the reduced fertility of the hybrids may be avoided. In consequence, the two types may develop into distinct species.

We may here again mention the curious and unexpected phenomenon of the position effect, according to which the mere fact of rearrangement of genes produced by inversion, etc., may cause a difference in their visible effects, thus simulating mutation (Dobzhansky, 1936). Indeed, studies like those of Muller (Muller, Prokofyeva, and Raffel, 1935) make it probable that a large number of the genetic changes in *Drosophila* previously ascribed to gene-mutation are in reality due to such "position-effects", produced by inversions of very small sections of a chromosome.

It has been suggested by some authors that what are normally called gene-mutations are in reality only the effects of small rearrangements. However, Mackenzie and Muller (1940) have recently demonstrated that there is a real distinction between the two types of mutation, since ultra-violet radiation can produce true gene-mutation, but not the chromosome-breakage needed to effect sectional rearrangements, however small. This is *prima facie* evidence that the substantive changes due to true gene-mutation do (as is to be expected) play a part in nature, in addition to the organizational changes due to rearrangement of pre-existing units.

From the point of view of evolution, however, the significance of such changes will be very similar to that of true or substantive mutation; the changes produced will be inherited according to Mendelian laws, and will be of small extent.

Muller (1930) has also pointed out that if two homologous chromosomes with different but overlapping inverted regions are brought together by crossing, crossing-over will result in a new type of chromosome containing one region in duplicate. Such small duplications will have visible effects, and may also be employed as sources of evolutionary change. Recently the discovery of the giant chromosomes in the salivary glands has converted *Drosophila* from a very bad to by far the best material for detailed chromosomal study, permitting the cytologist to produce a detailed map of the visible structure of its chromosomes and to detect even minute inversions and other rearrangements. It is as if an astronomer armed only with Galileo's telescope had been suddenly equipped with a 50-inch reflector.

The use of this method has shown that rearrangements of segments of the chromosomes are far commoner, and have played a much larger part in evolutionary processes, such as the differentiation of species, than was previously supposed. We have just mentioned the important role of small inversions within a single species. When we compare related species (e.g. Dobzhansky and Tan, 1936; Dobzhansky, 1937), we find they are distinguished by numerous characteristic differences in segmental arrangement, *Drosophila pseudoobscura* and *D. miranda*, for instance, are so closely related that they will mate ad produce healthy (but sterile) offspring. The chromosomes of the one are approximately homologous with those of the other, but some segments have been translocated to other chromosomes, and numerous segments have been inverted, so that some sections of certain chromosomes show "profoundly different patterns". Other sections, however, remain approximately similar. On the other hand, it is probable that such changes only pave the way for full separation, the later stages of speciation being effected by a series of single gene-mutations (Muller, 1940).

In barley, however, Gustafsson (1941) finds that induced sectional rearrangements occurring simultaneously with induced gene-mutations are most likely to give favourable results, as providing a new internal environment for the new gene.

But in spite of the frequency of these larger types of mutation, it would seem that gene-mutation, together with the "pseudomutation" due to position effects, is the most important source of evolutionary change.

6. Special Cases: Melanism; Polymorphism' Fluctuating Populations

Before proceeding further in our main argument, however, we must turn aside to discuss certain special cases which illustrate various points concerning neo-mendelism and selection.

(a) Melanism in moths

The first of these concerns the phenomenon of melanism in moths, which has played a prominent art in recent evolutionary discussions. The facts may be summarized as follows:

In the first place, there is no doubt that melanism among many species has become much more frequent during the last century, and that this change has been associated with industrialization: the predominance or abundance of melanic occurs in large cities and in industrial areas.[13] In some cases the entire population of an area has become melanic. Descriptions of the historical course of events have been given for Britain by Harrison (1920*b*, 1932), and for the continent by Walther (1927), and by Hasebroek (1934). A summary of their genetic basis is given by Ford (1937).

As regards their origin, Harrison (1928, 1935) claimed that he had been able to cause melanic forms to appear, in two species belonging to different genera, by means of incorporating lead and manganese salts in the food, and that the induced melanism behaved as mendelian recessive, as does the naturally-occurring but very rare melanism of these species; (in a third species in which natural melanism is dominant, and has shown industrial spread, he abandoned the work after only preliminary results).

However, repetition of the work on a large scale by Hughes (1932) and Thomsen and Lemcke (1933) has failed to confirm these results. R. A. Fisher (1933*a*) has also criticized Harrison's views as involving a mutation-rate much higher than any obtainable by X-ray treatment. Further, all industrial melanism is due to dominant genes (see below). It would seem best to assume that some error has been at work, and that no true induction occurred. If so, then melanic mutations must, like other recurrent mutations, have been thrown up sporadically for a long period, but have spread owing to the altered selective conditions of an industrial environment. Numerous cases of sporadic melanism which have not become more frequent recently are known in moths as in other groups. An interesting case is *Boarmia extersaria*. This shows industrial melanism in Germany; but in Britain it does not enter industrial areas, and the melanic type has remained sporadic. A survey of industrial melanism reveals that the intensity of darkening varies from species to species. As regards its genetic basis, no case of recessive melanism is known to have shown industrial spread. Industrial melanism always depends either on a single dominant gene, or on multiple factors each exhibiting complete or partial dominance (Ford, 1937).

As regards its physiological effects, numerous authors have shown that dominant or partially dominant melanism confers extra hardiness and viability. Ford (1940*b*) has shown in *Boarmia repandata* that in highly unfavourable conditions (feeding on alternate days) the ratio of melanics emerging (on an expectation of 50 per cent) goes up from about 54 to about 70 per cent. He has also (1937) pointed out that, in spite of this physiological advantage, dominant melanic forms in non-industrial areas have not shown any spread or increase in frequency.

He accordingly concludes that recessive melanism is due to genes which have been forced to become recessive by selection of modifiers, on account of their deleterious effects on viability. Dominant melanism, on the other hand, has favourable effects on viability, but in normal conditions is kept in check by counter-selection operating through natural enemies, the type forms being definitely cryptic in coloration, while the melanics stand out sharply against the normal background. A balance is thus reached, with a low percentage of melanics.

In industrial areas, however, the counter-selection in favour of the type is not so strong, since the background is darker. It has been suggested that in some cases ecological selection would here be reversed, and the dark forms become better protected by background resemblance. Detailed counts by Harrison, however, have shown that in some species at least this is not the case. In industrial areas it is further to be expected that many natural enemies of the adult would be reduced in numbers or absent, so that selection for concealment would be less stringent. As a still further possibility, it appears probable that in the chemically unfavourable environment of industrialism, the greater hardiness of melanics would have an increased advantage.

In any case, according to Ford's general hypothesis, the balance between opposed selective forces is in industrial areas tilted in favour of the melanic variety, with the result that this has speedily increased and in some cases has completely ousted the original type. The strength of the selection acting against melanics on account of their coloration is shown by the fact that in one species, although the melanic is more cold-resistant, and has spread in industrial areas, it is not found so far north as the normal.[14]

It would appear that on no other hypothesis can the lack of spread of dominant melanics in rural areas and their spread in industrial areas be reconciled. If so, we have one of the most striking demonstrations of the efficacy of selection.

(b) Genetic polymorphism

Genetic polymorphism, or the co-existence of two or more genetically-determined and well-defined forms ("phases") of a species in the same area, presents certain peculiar problems. We speak of polymorphism when the difference between the various forms is sharp, or at least expressed as a bi- or multi-modality in a frequency curve of variability; when the equilibrium between them is relatively stable; and when the frequency of the least abundant is high enough to make it certain that it is not due merely to mutation-pressure (see Ford, 1940*a*). Polymorphism must be clearly distinguished from normal variability, however extensive, which will be grouped in a single unimodal frequency curve. The existence of separate forms or distinct modes is an essential characteristic of polymorphism.

Since we are here concerned only with genetic polymorphism, we can neglect all such cases as those of the social hymenoptera, the seasonal forms of certain butterflies, etc., which are determined environmentally. We can also neglect the particular type of genetic polymorphism involved in the genetic determination of two sexes, since this is primarily maintained not by a selective balance but by the inherent nature of the genetic-reproductive mechanism: the same applies to heterostyly in plants. Paulian (1936) applies the term *genetic* to certain cases of polymorphism in male insects where the forms are discontinuous and not to be explained by simple allometry (heterogony). Until, however, the developmental basis of this phenomenon has been ascertained, it is better not to assume that it must be genetically determined.

The interest to the evolutionist of genetic polymorphism within a freely interbreeding population is that, as R. A. Fisher (1936*a*) was the first to point out in general terms, it must always involve a balance of selective advantages between the different types. For, *ex hypothesi,* mutation-pressure alone will not account for the facts, and it can readily be shown that in the absence of selective balance, one type would rapidly oust the other from any considerable representation in the population.

There are two distinct methods by which this balance is actually effected—genetical and ecological.

(i) In the case of genetical balance, the heterozygote is more viable or enjoys some other selective advantage over either of the homozygotes. (For simplicity's sake we will consider only cases involving dimorphism: trimorphism will occur, as in certain species of foxes, when the heterozygote differs in appearance from either homozygote, or in other cases, as in *Papilio polytes* and *P. memnon,* when two interacting gene-pairs are involved (Ford, 1937), and polymorphism when two or more non-interacting gene-pairs are involved.) This may occur in two ways. Either the gene itself is less viable or even lethal in homozygous condition; or it is closely linked with a recessive lethal, which exert no effect in single dose but is lethal when homozygous.

Owing to the difficulty of proving a negative, no certain case of the former conditions is known, though the mutant *curly* in *Drosophila melanogaster* is a possible example. In the conditions of ordinary laboratory cultures, this maintains itself indefinitely, giving a dimorphism with wild-type. It does this because it is almost fully lethal when homozygous, but usually rather more viable than wild-type when heterozygous. As, however, it is located within an inversion, the possibility of linkage, either with a lethal or with the genes for the increased viability, is not excluded. Of linkage with a lethal, however, two examples have been thoroughly worked out. In the case of the butterfly *Argynnis paphia*, over most of its range an occasional second form of female occurs, known as *valesina*, in which the ground colour is dull green instead of rich brown. This form is usually rare, but in some localities constitutes 5 to 15 per cent of the female population. Goldschmidt and Fischer (1922) showed that *valesina* depends on a dominant gene closely linked with a recessive lethal; they were able to break the linkage, thus obtaining broods with all females of the *valesina* type. In China and neighbouring areas almost all the wild females are *valesina*. This well shows the relativity of the term *normal* as applied to organisms in nature.

A similar situation exists in the American Clouded Yellow Butterfly, *Colias philodice* (Gerould, 1923). Here a white female variant exists, and normally constitutes 4 to 20 per cent of wild females. As with *Argynnis*, the rarer form is due to a dominant gene linked with a recessive lethal, and Gerould was able to separate the two genes. In one area the white type is the more abundant. The situation as regards selective advantage is complex. The lethal must have some advantage (*a*) when heterozygous, as it is (in *Colias*) more widely spread in the population than the dominant colour-gene; (*b*) in association with the dominant colour-gene, since the linkage, though not very close, survives in nature in most areas. The dominant colour-gene must also have some advantage in association with the lethal to balance the wastage arising from homozygosity. Where it is homozygous in nature, the advantage must come from association with some other gene.

Numerous cases, as yet unanalysed genetically, are known in nature where polymorphis, as with *Argynnis paphia*, exists in one area of the range of the species but not in another. A familiar example is the common red squirrel, *Sciurus vulgaris*, which is always red in some areas, e.g. Britain, but both red and black in mountainous parts of Europe (Lühring, 1928). Other cases are mentioned. In some it would appear that a new area has been colonized exclusively by individuals of one type, which presumably then is not linked with a full lethal, but either is linked with a gene somewhat reducing viability, or is the type which is less advantageous than the lethal-linked form. In other cases the lethal linkage may have been broken, as with *Argynnis* and *Colias*.

Highly polymorphic species exist in nature among land-snails, such as the common *Cepaea nemoralis* and *C. hortensis* (A. Lang, 1908), with their vast range of ground-colour and degree of banding; grasshoppers (grouse-locusts, etc., Nabours, 1925; grasshoppers, Rubtzov, 1935[15]); and certain fresh-water fish such as *Lebistes* (Winge, 1927). In these cases the polymorphism appears to depend on the phenomenon of close linkage within each chromosome or of the obligatory association of many whole chromosomes to produce a similar effect to close linkage (Fisher, 1930*b*; Diver, 1932). In these circumstances, a recessive lethal will prevent the free recombination of any favourable mutations in the same chromosome. Thus, since recessive lethals are common types of mutation, whole chromosomes will have to compete with each other, instead of selection being able to act so as to produce an approximation to a single "best" combination of genes. Further, since homozygotes cannot live, there must be at least two different forms of each of the chromosomes which contain lethals; the different combinations of these will of necessity give a variety of forms; and this variety will itself be subjected to selection so as to give the best balance, and the least waste through excess mortality of one or some of the forms. In the land-snails, the interesting fact has been discovered that the frequency of the different types of bandling has remained about the same since the neolithic period (Diver, 1929), showing that the balance is an enduring one. The problem, of course, remains as to why the close linkage has become a characteristic of the species in the first instance, since no obvious advantage inheres in such a condition. Haldane (1930; and see discussion in Ford, 1934) has suggested that it is due to translocation of segments from one chromosome to another. The unusual phenomenon of the dominance of the mutant types over a "universal recessive" would then also be accounted for, since the mutants would possess the translocated segment in duplicate, both in its original and its new position. Further, if, as is to be expected and as appears actually to be the case, the translocation is less viable when homozygous, we could dispense with lethal genes as an explanation of the selective balance reached.

The common sea-anemone *Metridium senile* exists in a number of strinkingly different colour-varieties. D. L. Fox and Pantin (1941) enumerate eight, including white, red, brown, grey, and various combinations of these; in addition, there is much variation in the intensity of the colours. The different forms occur in different proportions in different localities. The various colour-types depend on the interrelation of (1) lipochrome, giving colours from red to yellow; (2) brown melanin, restricted to the ectoderm; (3) black melanin, restricted to the endodern. There seems no doubt that the main types are genetically determined, and differ in their metabolic properties, and also that the colour is here adaptively non-significant, but correlated with some basic physiological difference. Fox and Pantin conclude that selection is weak as between the different colour-varieties, and that this will account for the existence of the numerous phases. We have seen, however, that a selective balance is required for this, and experiments on the physiological and ecological differences between the varieties should yield interesting results. There are the additional complications that asexual reproductions occurs, and that single individuals can persist for great lengths of time, perhaps even indefinitely.

In the Mexican fresh-water fish *Platypoecilus maculatus,* M. Gordon (1939) has found well over 120 patterns in a state of nature, mostly dependent on the recombination of 15 gene-pairs. This is a remarkable degree of polymorphism for a wild species, especially as collections dating from 1867 indicate that it is a balanced one. Furthermore, as in the other examples we have cited, related species and genera (*Xiphophorus*) present much parallel variation. Some of the excessive variation, however, is apparently due to the fixation of "accidental" characters by drift. A quite different type of balance occurs in variable plant species which consist of numerous ecotypes.

(ii) Ecological balance, on the other hand, depends either on a diminution in the degree of selective advantage due to increase of frequency of one or all of the polymorphic forms above a certain level, or on an alternation in the amount or type of selection due to alteration in the environmental conditions. The best examples of the former concern mimetic butterflies. Either all the forms are mimetic, or one is non-mimetic and the other or others mimetic. If a mimetic form happens to become too abundant relative to its model, the protection afforded by the resemblance will diminish. Where certain mimetic forms are wholly absent, the corresponding models are also fond to be missing. In any area, a balance will thus be struck, depending on relative abundance of models, intensity of predation, and viability factors.

Among the best-analysed examples are those of *Papilio polytes* and *Hypolimnas dubius*. In the former case, only the females are polymorphic, existing in three forms, two mimetic and one non-mimetic. The species is a successful one, able to live outside the range of its models: it is then, of course, monomorphic, all females being non-mimetic.

It seems clear that where models are available, mimicry confers a definite advantage, but one which diminishes rapidly with increased frequency of the mimetic forms.

In *Hypolimnas dubius,* both sexes are alike, and are dimorphic, with two mimetic forms. For the details of the genetic basis of the condition, readers are referred to Ford (1937). In general the equilibrium due to ecological selective balance may be broadly compared with the effects of mass action in chemistry.

As a matter of fact, *Papilio polytes* appears to illustrate a combination of genetic with ecological control, since the two mimetic forms, both of which are dominant, are more viable in the heterozygous than in the homozygous condition.

Various authors have regarded as a theoretical difficulty the fact that such enormous differences in pattern, and even shape and habit, obviously involving many independent characters, can be controlled by a single gene. It should, however, be clear that once a mutant type is established, conferring even a small mimetic advantage, the residual gene-complex can undergo evolution of the same nature as what we have discussed in previous sections. But here the extent of such modification may be pushed much further, with a result that is best described, not as a modification of the visible effects of the original gene, but as an addition of various new effects, all of which, however, are dependent for their expression on the presence of the original gene. Presumably this change in the gene-complex will depend more on new mutation and relatively less on recombination of previously existing genes than in e.g. alteration of dominance.

We may put it in another way by saying that the original gene-difference comes to act as a switch controlling the action of numerous mutant modifying and modifiable genes, precisely as occurs in the case of the primary sex-difference in regard to genetically sex-limited characters. There is no reason for, and every reason against, postulating the sudden origin of the whole pattern by one mutation. Further, the frequent superficiality of the characters by which mimetic resemblance is achieved shows that the resemblance cannot have arisen though parallel mutations occurring and being preserved owing to similarity of conditions (see Punnett, 1915; Cott, 1940). The apparently cryptic colour-polymorphism of certain stick-insects and mantids may have a similar ecological basis. The egg-colour, polymorphism of the cuckoo, *Cuculus canorus,* is largely related to the risk of ejection by the host (Jourdain, 1925; Huskins, 1934). The different egg-colour strains are presumably balanced in relation to host-abundance.

In the ruff (*Machetes pugnax*) polymorphism is confined to males in the breeding season, and the number of distinctive types is extraordinary. No other birds rivals it Ford (1940*a*) suggests that the cumulative effect on females of display stimulation by numerous males at a common courting-ground, (Darling, 1939; Huxley, 1938*a*) promotes maximum variability. Mayr and Rand (1937) cite a peculiar and striking dimorphism in tail-coloration in the bird *Rhipidura brachrhyncha*.

Polymorphism is much commoner than usually realized, e.g. in birds and mammals only, it exists among squirrels, foxes, bears, cats, owls, herons, hawks, skuas, etc.

Next we come to cases where changes in external environment determine the ecological balance, not those in internal conditions (relative abundance of the separate types) We have mentioned the common squirrel (For plan ecotypes).

Elton (quoted by Ford, 1934) has found that in the red fox (*Vulpes fulva*), which shows trimorphism the "cross" type being apparently a heterozygote between red and silver, the rare silver type changes in its relative frequency in a regular way during each of the 10-year cycles of abundance to which the species is exposed. In this species, Cross (1941) finds a rough polymorph-ratio cline, red being commoner to the S., silver and cross to the N.

The arctic fox (*Alopex lagopus*) is dimorphic in winter (blue or white). Only the white type occurs in Kamchatka, and only the blue in certain Alaskan islands; while on the Alaskan mainland a N-S gradient is found, the white type decreasing in frequency with latitude.

Here any selective advantage afforded by the white coat in winter must presumably be offset by some disadvantage, probably connected with viability, for the blue type to be able to exist in numbers at all. The primary basis of the dimorphism (as in the red fox) is thus a genetic balance. But environment may somewhat alter this equilibrium, so that the balance is in part also ecological.

A remarkable case where the relation with environmental conditions is more direct is that of the hamster (*Cricetus cricetus*), as described by Kirikov (1934) and Timoféeff-Ressovsky (1940). About 150 years ago the naturalist-geographer Lepekhin noted that in a certain region of Russia black hamsters were unusually prevalent. Since then, the statistics of the Russian fur-markets have enabled biologists to trace the steady spread of the black type, until to-day, throughout a broad zone along the northern border of the range of the species, black forms are in the majority and in some areas are present to the total or virtual exclusion of the typical greys. This area coincides with the sub-steppe (wood-steppe) climatic zone, and is cooler and moister than the steppe region proper, which forms the main home of the species. In the steppes the black type occurs only as an occasional aberration. It seems clear that the black form enjoys some selective advantage in the sub-steppe area, while it is at a disadvantage in the drier steppes.

In the bird *Coereba*, the recent replacement of the typical by the melanic form on certain West Indian islands appears to be similar. In other cases, although the actual process of spread has not been followed, we can be certain that it is taking place. For instance, in the brush opossum (*Trichosurus vulpecula*), melanic variants are very rare on the Australian mainland. In the Tasmanian subspecies, however, they are common. Pearson (1938), from an examination of many thousand skins, was able to plot a contour ("phenocontour") map of the relative frequency of the black type, with contour lines ("isophenes") marking regions of a given frequency. In the first place, he was able to show that neither an isolated small island off the north-west, nor another on the north-east on the curse of the submerged land-bridge from Australia, contained black animals except at aberrations; this demonstrates that the abundance of the black type must have arisen after the isolation of the subspecies in Tasmania. The north-west corner of Tasmania contains only black opossums, while on parts of the east coast the proportion is under 25 per cent. There is no correlation of the frequency of blacks with climatic gradients. The suggestion that the black type appeared (whether by new mutation or by the crossing of two carriers of the black gene in single dose) in the north-west, and is gradually extending eastwards, is confirmed by conditions on the narrow-necked Tasman peninsula. This shows a markedly lower frequency of blacks than the adjacent zone of the main island: it seems clear that the extreme narrowness of its neck has hindered the spread of the black gene.

It is interesting to note that Tasmania is cooler and moister than Australia, so that the similarity to the case of the black hamster is very close. Humidity also favours melanism in *Coereba*.

Another case where a phenocontour map of a mutant type has been plotted, and examination of the map shows that spread of the mutant is occurring and is being impeded by geographical barriers, is that of the *simplex* condition of the teeth in the field-vole *Microtus arvalis* in north-central Germany (Zimmermann, 1935). The aberration here consists in the absence of the last ridge of enamel on the molar. Here the *simplex* condition occurs in over 90 per cent of individuals in Schleswig-Holstei, with zones of decreasing frequency to west, south, and east. The mountains of central Germany have proved a complete obstacle, while certain large rivers have obviously hindered the spread of the character. This case is genetically slightly more complex, since various gradations in the expression of the character occur, and not only is the character more frequent in the presumed centre of origin, but also more extreme in type. It is further worth noting that the gradient in *simplex*-frequency bears no relation to another character-gradient or cline within the species, namely, the gradual east-west darkening and reddening of the coat-colour across the north German plain.

In the last two cases, the polymorphism may be regarded as transient, since the species or subspecies appears to be moving from a condition in which a given character is rare, maintained only by mutation-pressure, to one where it is universal, again apart from the rare and sporadic occurrence of its allele. This is theoretically to be distinguished from true polymorphism, in which a state of balance between the contrasting types is indefinitely maintained, but it will not always be possible to distinguish the one condition from the other (Ford, 1940*a*). Thus Southern (1939) has shown that the bridled variety of the guillemot (*Uria aalge*) increases northwards from 0.5 per cent or lower to well over 50 per cent of the total population, in a fairly regular gradient. There is, however, no ways as yet of telling whether the two types are indefinitely balanced against a climatic gradient, as with the arctic fox, or whether the bridled type is spreading at the expense of the "normal". The owl *Megascops* shows a centrifugal cline between red and normal (grey) forms in U.S.A. (Hasbrouck, 1893); while that between dark and light fulmar petrels is rather more complex.

Mather (1941) has recently propounded a possible genetic explanation of clines in such apparently non-adaptive characters. He suggests that the gene responsible is linked with one of two balanced polygenic combinations within homologous chromosomes, affecting some other character, and that this is related to some environmental condition. The mean development of the character will be adjusted by the relative frequency of the two combinations, and thus the frequency of the linked gene will also vary in relation to the environmental gradient concerned. One may add that the same result would be obtained if one of the genes in the polygenic combination were pleiotropic and also produced the visible effect (e.g. bridling).

A frequent condition found in nature is that of regional or geographical polymorphis, when two or more contrasted forms are confined to different regions. Thus in the moth *Spilosoma mendica* (Ford, 1937), the normal condition is for the male to be dark, the female white. In Ireland, however, the males as well as the females are white, and this condition (variety *rustica*) is known sporadically on the European continent. Such a state of affairs may represent the final stage of a transitory polymorphism, in an area which favours the spread of an alternative type (cf. the brush opossum), or be due to colonization by one only of the two forms. Elsewhere, the condition may represent the end of a dimorph-ratio cline. Thus, in *Accipiter novaehollandiae* such a cline culminates at either end in an area exclusively inhabited by one of the two forms. Again, the white form of the palearctic moth *Leucodonta bicoloria,* abundant in the east, decreases westwards and is absent in the extreme west (Suomalainen, 1941).

We may conclude by referring to the floral dimorphism of higher plants, even though that is solely or mainly environmental or developmental in its determination, not genetic. Floral dimorphism is concerned with the relation between cross- and self-pollination. (See Kerner and Oliver, 1902, for a good general but somewhat out-of-date account; Uphof, 1938, for cleistogamy; for gynodioecism; Mather, 1941, for heterostyly).

Strictly speaking, the term dimorphism should perhaps only be employed when whole plants of distinct type are found. This occur in most cases of gynodioecism, e.g. in the common plantain *Plantago lanceolata,* the viper's bugloss *Echium vulgare,* etc., where some plants produce normal hermaphrodite flowers, while others have only female (pistillate) flowers, which are sometimes smaller than the hermaphrodite ones; in a few cases of cleistogamy, such as the balsam *Impatiens noli-me-tangere,* where in addition to plants exclusively producing the normal showy flowers, others may occur bearing only the inconspicuous and permanently closed cleistogamous flowers adapted solely to self-pollination (as well as still others with both types); and in heterostyly. Mather (1941) has an interesting discussion on the evolution of heterostyly (see also Mather and de Winton, 1941). He points out that an illegitimacy reaction appears to be compulsorily associated with heterostyly, and is the chief bar to inbreeding. Homozygous thrum plants, which normally do not occur in nature, are less viable than the heterozygotes or the recessives (pms). This reversal of the usual relations of dominance to viability must be due to the accumulation of deleterious mutations, in the region adjacent to the thrum gene, which will only exist in nature in a heterozygous condition; this is similar to the accumulation of loss mutations in the Y-chromosome, which has led to its almost total inertness.

If conditions demand greater stability and therefore increased inbreeding, in some cases selection may reverse the intensity of the illegitimacy reaction. Since commercial seed-raisers prefer selfing, this has occurred with cultivated species, e.g. *Primula sinensis.* Here the fertility of illegitimate relative to that of legitimate pollinations has almost doubled since the experiments of Hildebrand and Darwin in 1864 and 1877. Another possibility is the selection of mutants giving homostyle plants which are then capable of self-pollination. This has occurred in nature.

In a few cases of heterostyly, such as certain species of *Primula,* the two types of flowers also have different-sized corollas. In some cases the "pin", in others the "thrum" type is thus distinguished. The meaning of these last conditions is obscure. Kerner and Oliver (1902) suggest that the smaller size of corollais associated with the type where autogamy is more prevalent, and cross-pollination accordingly less essential; but this Dr. Mather informs me is not in accord with experimental facts.

It is also quite logical to use the term dimorphism for types in which two distinct types of flower are produced on the same individual plant, as occurs in most cleistogamous forms (species of *Viola, Glechoma, Lamium, Oxalis, Helianthemum, Juncus,* many grasses, etc.). Further, it is perhaps even legitimate to extend the concept to cover the frequent combination, in one and the same flower at different times, of definite devices to secure cross-pollination and self-pollination. This would then constitute what we may perhaps call a dimorphism in time, since in the great majority of cases the plant produces flowers which are adapted to ensure cross-pollination, but if this does not occur, it transforms these same organs into what is virtually a few type of flower adapted for self-pollination.

Furthermore, in all these cases, a selective balance is at work. However, the balance is a complex one. In plants whose flowers change from exogamy to autogamy, as well as those which produce both showy and cleistogamous flowers, it used to be supposed that all that was involved was what the experimental embryologists style "double assurance" (*doppelte Sicherung*) to secure pollination. Cross-pollination was assumed to be in some way better, but if, through lack of suitable insects or other reason, it chanced not to be effected, then the plant fell back on its second line of defence, in the shape of self-fertilization. This in itself would constitute a selective balance of an ecological nature. The advantages accruing from cross-fertilization are offset against the disadvantage of its not being always possible; the disadvantage of having to produce a second type of flower (or to develop new adaptations within the original flower) is offset against the advantage of assured seed-production.

Ecological factors may further complicate the picture. Thus certain species produce solely or almost solely cleistogamous flowers when growing in deep shade, where few insects are likely to visit them, but go to the reverse extreme when grown in sunny open localities. In *Viola,* cleistogamous flowers are much more abundant in high latitudes; this is due to a photo-periodic response (Borgström, 1939). Along these lines, the course of evolution of the cleistogamous condition is readily envisaged. Cleistogamy in the strict sense implies a special type of flower which not only does not open, but shows other adaptations; usually the petals and stamen are reduced in size, and the pollen-grains are not liberated from the anthers but germinate *in situ.* This condition has doubtless followed on one of "pseudocleistogamy", where, in certain ecological conditions, the normal flowers simply fail to open and self-pollination occurs.

This, however, is not the whole story. Autogamy, it is now realized, is not merely a *pis aller.* It has certain advantages, in perpetuating unchanged a vigorous and well-balanced genetic constitution once this has been evolved. This stability, however, will only be advantageous so long as conditions also remain unchanged: in addition, an entirely stable type loses the possibility of invading new environments. Thus the provision of devices for both cross- and self-fertilization constitutes a balance between the advantages of plasticity (with its disadvantage of wasteful production of less well-adapted recombinations) and of stability (with its long-term disadvantage of absence of adjustment to new conditions). Some forms are exclusively of one or the other type; but in a large number, probably the majority, of flowering plants, the two have been brought into balance, with consequent floral dimorphism of one sort or another.

A third type of selective advantage concerns the degree of waste of gametes associated with a particular mode of reproduction. This wastage is especially marked in monoecious plants, and its implications are discussed in reference to the selective balance involved in gynodioecism.

It is difficult to evaluate the precise shares of these various advantages and disadvantages in floral dimorphism. We must be content to observe that a selective balance is involved, and that the dimorphism is thereby maintained.

In other cases the dimorphism is between sexual and vegetative reproduction. Dr. W. B. Turrill informs me in a letter that in the lesser celandine, *Ranunculus ficaria*, the following types of reproduction occur: normal amphimixis, apomixis, abundant vegetative multiplication (of two types; one variety is wholly or almost wholly vegetative in its reproduction), and plants with only male, only female, and only hermaphrodite flowers. Different populations show different proportions of these various types. Here, in the field of reproduction, we may perhaps have something akin to the balance of ecotypes in many plant species.

We may in fact conclude that polymorphism always involves a selective balance, whether it is determined genetically, or environmentally, or internally by the processes of normal ontogeny, as when two or more kinds of persons or organs, adapted to different functions, are formed by the same colony or individual (for social hymenoptera).

Finally in view of its peculiar evolutionary interest as inevitably involving a selective balance and as in some cases leading by way of ecological regional differences, to sharp geographical differentiation, genetic polymorphism deserves the most intensive study, especially in cases where the ratios of the types are geographically graded, since here we may hope not only to measure the intensity of the selective forces at work, but also to discover something as to their nature.

(c) Selection in fluctuating populations

Elton (1930) has pointed out that the customary assumption of a population approximately constant from year to year is very far from the truth for many, if not most, species. A stable "balance of nature" does not exist. Fluctuation in numbers, rather than constancy, is the rule. This fluctuation may be broadly progressive towards increase or decrease, it may be irregular, or it may be cyclic and regular.

Animal species subject to such regular or periodic fluctuations include lemmings (*Lemmus*), snowshoe rabbits (*Lepus americanus*), mice, voles, jerbils and other Muridae, foxes (*Vulpes, Alopex*), lynxes (*Lynx*), and other fur-bearing carnivores, certain birds, such as ptarmigan (*Lagopus*) in Labrador and nutcrackers (*Nucifraga*) in Siberia, some invertebrates, land and marine, and probably certain antelopes and other larger mammals. The period of the fluctuation from crest to crest of abundance varies from 3 to 4 years in the smaller rodents, to 10 to 11 years in the snowshoe rabbit, and probably a good deal longer in certain ungulates. The difference in abundance between crest and trough may be very great. In the snowshoe rabbit the ratio of high to low population numbers in extreme cases must reach at least 100: 1. An interesting point made by Rowan (1931) for Canadian birds is that migrant species are not affected by these cycles. In years when the grouse population has been reduced to a minimum, the migrants are present in normal numbers. This fact must, in combination with others peculiar to migrants, have important evolutionary consequences restricted to migrant forms. However, it seems not to be of universal occurrence.

Elton has pointed our certain important evolutionary consequences of these facts. In the first place, both the intensity and the type of selection will vary continuously during the cycle. During the period of rapid increase, when numbers are low and conditions favourable, the intensity of selection will be very low. During the peak period, intraspecific selection due to pressure of competition will be high. Since the catastrophic fall in numbers is normally due to infectious disease, selection during this period will mainly concern disease-resistance. And in the subsequent period of unfavourable environmental conditions (for all these cycles seem to have an external determination) selection will be concerned with resistance to cold and hunger and similar aspects of the struggle for existence. To use Elton's metaphor, the species is put through a series of examinations, with easy times between-whiles, and the different examinations test different capacities. Apart from special resistances, such varying selection will promote a general elasticity of response.

Elton also made the suggestion that periodic fluctuations would allow greater scope for chance in evolution, since if a rare mutation or gene-combination happens to be present in the much-reduced minimum population, it will be automatically reproduced in the same proportion during the period of rapid increase when the struggle for existence is light and the intensity of selection low.

His views have been criticized on mathematical grounds by Haldane. However, the proof of the pudding is in the eating, and the studies of Ford and Ford (1930) make it clear, first that selection-intensity may actually be relaxed during a period of rapid increase, and secondly that when it is once more tightened up, the resultant type may differ from that obtaining in the proceeding period of abundance. They observed a sharply isolated colony of the small Greasy Fritillary butterfly, *Melitaea aurinia*, for thirteen years, and obtained records and specimens for a total period of forty-nine years. The population was increasingly abundant from 1881 to 1897; it then decreased, and became scarce by 1906 and extremely rare from 1912 to 1920. A rapid increase to abundance then took place to 1924, from when until 1930 it showed a progressive slight further increase.

Variability was slight during the first relatively stable period of abundance. During the period of rapid increase after scarcity, however, (I quote from Ford, 1934), "an extraordinary outbrust of variation occurred. Hardly two specimens were alike and marked departures from the normal form of the species, both in size, shape and colour, were very common. A high proportion of these were deformed in various ways, the amount of deformity being closely correlated with the degree of variation."

With the colony entered on its second period of abundance, the abnormal types and extreme variants practically disappeared, and the population settled down again to a uniform type. This, however, was not the same as before, but recognizably distinct.

It seems clear that the relaxation of selection during the recovery period allowed an excess of variability; and that when it again became rigorous, the new stable type was slightly different, owing to the accidental incorporation of different genes. R. L. Berg (1941) has demonstrated a similar effect in micropopulations of *Drosophila melanogaster,* aberrations increasing with decreased intensity of selection.

Sewall Wright (1932), in discussing such problems in more general terms, concludes that there must be available to most species a number of gene-combinations all of about the same survival-value; he compares them to peaks, separated by "valleys" of intermediate combinations which are less favourable. Normally it is difficult or impossible for selection to shift the type from an established peak to another, although this might be equally satisfactory if reached: but when the intensity of selection is reduced (or when low size of population promotes the accidental survival of genes and gene-combinations), many "valley" combinations are realized, the species can cross dryshod to other peaks, and it will be a matter of chance on which Ararat the type eventually remains perched when the rising tide of selection again floods out the valleys.

Elton draws a further interesting conclusion from the facts of periodic fluctuation. He concludes that it will have promoted the migratory impulse which is so strong in so many types of animals when in unfavourable conditions. At first sight, the existence of this impulse seems hard to explain on any selective hypothesis, since, in the more spectacular mass emigrations, such as those of locusts, lemmings, or certain butterflies, all, or all but a negligible fraction of the migrants perish, while the population is renewed from among those which do not manifest the impulse and stay in their original habitat. One would thus suppose that migratory tendencies would be strongly antagonized by selection.

However, although such migrations are exceedingly striking and have thus received a disproportionate share of attention from biologists, they are, in fact, but extreme and in a sense abnormal manifestations of a much more widespread phenomenon, namely, an impulse to react to unfavourable conditions by changed behaviour, notably by movement away from an environment which has become unfavourable. This does not normally result in mass migration on a vast scale, but in an irregular movement that disperses the population over a wide area. When lemmings are scarce in the Arctic, snowy owls (*Nyctea nyctea*) descend into north temperate latitudes in search of food (Elton, 1927). When the cedar-cone crop fails in Siberian forests, the Siberian nutcrackers

(*Nucifraga caryocatactes macrorhynchus*) leave their usual haunts and may reach western Europe (Formosov, 1933). And many quite small and inconspicuous movements of animal population are going on all the time.

When migration is of this less extreme type, many individuals which would otherwise die will survive temporarily in regions beyond the normal range of the species and be able to return later to their original habitat, while others may survive by reaching and remaining in other parts of the normal range. In addition, some individuals may be able to survive and to remain in areas outside the normal habitat of the species, either by adopting slightly different habits and so colonizing different habitats within the original geographical range, or by colonizing areas outside this range. This extension of habitat may in the first instance be dependent on a non-inherited modification of behaviour, mutation and selection later stepping in to fix the change genetically (the "organic selection" of Baldwin and Lloyd Morgan); or genetic variants may find themselves in surroundings to which their constitution is better adapted than was the normal environment of the species (pre-adaptation). In either case, migration will have been advantageous to the species as well as to the individual.

Elton (1930) draws an illuminating comparison between the sexual process and the migratory urge. Both are extremely widespread, and both confer additional evolutionary plasticity on a species. The sexual process enables the species to exploit to the fullest extent the mutations, old and new, which are carried by the species or which crop up during its evolutionary career, by making possible every kind of recombination of them. The migratory impulse, in relation to unfavourable conditions, has a precisely analogous effect, in increasing the range of environmental conditions with which any genetic variation that exist can be brought into contact. The two are complementary and often mutually reinforcing processes, and both have their most important function in times of stress.

Fluctuations in numbers, both irregular and periodic, thus may have important evolutionary consequences.

7. Mutation and Evolution

There remains the difficulty that most mutations so far investigated are deleterious. If mutations are the raw material of evolution, it is clearly not enough that they should be as it were sterilized and rendered innocuous; some of them must sometimes be, or become advantageous. However, this also is not so serious as at first sight appears. Since the gene-complex is an elaborately co-ordinated system, any changes in it are much more likely to act as defects rather than as improvements. Further, the larger the change the less likely is it to be an improvement; and inevitably the geneticist will detect large changes more readily than small. The detailed analysis of the last ten or fifteen years, however, has revealed large numbers of gene-differences with extremely small effects, down almost to the limit of detectability. It is not only possible but highly probable that among these are to be sought the chief building-blocks of evolutionary change, and that it is by means of small mutations, notably in the form of series of multiple allelic steps, each adjusted for viability and efficiency by recombinations and further small mutations, that progressive and adaptive evolution has occurred. Indeed, in cases where fertile species-crosses are possible, this contention has been definitely proved, as for instance by the prevalence of multiple-factor (polygenic) differences, each factor with only a small effect, as the basis for specific differences in *Antirrhinum* (Baur, 1932), *Phaseolus* (Lamprecht, 1941), cotton (*Gossypium;* Silow, 1941), etc., and in wolf-dog crosses (Iljin, 1941). Many sub-specific characters have a similar genetic basis, e.g. in the plant *Camelina sativa* (Tedin, 1925), in deermice (*Peromyscus*), gipsymoths (*Lymantria*), etc. (see Dobzhansky, 1937). Specific differences in *Drosophila* depend on many single genes, often grouped in polygenic systems (Mather, 1941).

The very large number of genes with small effects involved in the inheritance of quantitative characters has been stressed by "Student". With reference to Winter's experiments (1929) on oil-content in maize, R. A. Fisher (1933*b*) writes: "all commercial varieties must be segregating in hundreds, and quite possibly in thousands, of factors." With this amount of available variance, Winter was able to select high and low lines differing *sixfold* in oil-content. Silow (1941) estimates that in cotton (*Gossypium*) the closest species differ in over half their genes.

It must be admitted that the direct and complete proof of the utilization of mutations in evolution under natural conditions has not yet been given. Even the case of industrial melanism, apart from its concerning the results of man's interference, will not be complete until the induction of melanic mutations has been finally disproved. On the other hand, a complete and direct demonstration is inevitably very difficult to provide. The mutations concerned will normally have small effects. Thus the species concerned must be easily bred, and should have been subjected to detailed genetic analysis: otherwise small mutations will not be detected. The species must then be followed through a period of evolutionary change, and during this period selection must be proved to have been operative on certain mutations.

Thus it is inevitable that for the present we must rely mainly on the convergence of a number of separate lines of evidence each partial and indirect or incomplete, but severally cumulative and demonstrative.

The different partial lines of evidence may be summarized as follows:—

(1) *The existence of small mutations.* This has been proved in every organism subjected to detailed genetic analysis. While most of these are deleterious, it should be remembered that reverse mutations to wild type have frequently been demonstrated, both "spontaneously" and under the influence of the same agency (e.g. X-rays) used to induce the original mutation. Thus it cannot be maintained that the process of mutation is of its nature deleterious, since the "abnormal" can mutate to the "normal" Position-effects due to small sectional rearrangements must for our purpose be included in this category.

(2) *The existence of mendelizing variations of small extent constitution the differential characters of subspecies and species.* This has been shown in many cases, though it can,

of course, only be demonstrated for species where fertility and segregation occur after an inter-specific cross. However, the cases of this are fairly numerous (see Haldane, 1932*a*, Goldschmidt, 1928, Dobzhansky, 1937).

This point is important, since the presumption is very strong that all mendelizing variations owe their origin to mutation.

(3) *The existence of selection-pressure against small unfavourable mutations.* All cases of reduced viability in culture and of elimination of deleterious mutants in nature fall under this head. One of the best proofs is the low incidence of mutant genes in the sex-chromosome of wild-caught individuals as compared with their incidence in the other chromosomes (autosomes), as discovered by C. Gordon (1936) and Dubinin and his co-workers (1934, 1936) in *Drosophila*.

Recessive mutations in the autosomes will not exert their effects unless in double dose, and they cannot occur in double dose unless two individuals heterozygous for the gene happen to mate, which will be a very rare event. Sex-linked mutations, on the other hand, will immediately exert their effects on a number of males, since these possess only a single sex-chromosome, so that any recessive genes located in this chromosome can exert their effects in single dose, not masked by their normal allelomorphs. If the effects comprise reduced viability, selection will at once be brought into play and will tend to eliminate the gene from the constitution of the species.

Thus both the recessivity of most mutations and the scarcity of sex-linked recessives are consequences of selection.

A special case is the proof by Gerould (1921) that the normal grass-green larvae of the butterfly *Colias philodice* enjoy a selective advantage over the blue-green recessive mutant type in relation to the attacks of bird enemies, no doubt on account of their close resemblance in colour to the food-plant.

(4) *The existence of mutations which from the outset are non-deleterious, and especially of those which are potentially favourable.* In view of the recurrent nature of mutation, it is extremely unlikely that in an experiment mutations should arise which are markedly favourable at the outset, in normal conditions; for in most cases such mutations would long previously have been incorporated in the constitution of the species.

Of mutations which appear to be potentially favourable—i.e. capable of being immediately utilized by selection in certain conditions—there are numerous examples. We may mention the mutations in seed-weight of beans found by Johannsen; those modifying hooded pattern in rats the mutation altering temperature-resistance in water-fleas by Banta; and that in tobacco adjusting flowering to a different rhythm of light and darkness. Some of the mutant genes found by Dubinin *et al.* (1936) in wild *Drosophila* might readily increase (or even become the "normal type") under slightly altered conditions. The higher variability of abundant species demonstrates this process in action.

Zimmermann (1941) has found numerous recessive genes in the heterozygous state among populations of wild rodents, not only house-mice (*Mus m. musculus*), but also field mice (*Apodemus*) and voles (*Cleithrionomys*). Though some of these were for gross abnormalities, and others for partial albinism or spotting, still others determined characters which might readily be utilized in normal evolution, e.g. a darker type of agouti. In one case (*Cleithrionomys g. glareolus*) a dominant gene was found, changing the normal red of the back to the brown character typical of the alpine subspecies *C. g. nageri*.

An interesting example from domestic animals is that of frizzled fowls. These have a peculiar plumage, with upcurled feathers which do not retain heat well, and are at a great disadvantage in temperate climates. The condition depends on mendelian genes (F. G. Benedict, Landauer, and Fox, 1932). In tropical climates, however, as in West Africa, the breed is extremely common: here the frizzled plumage is an advantage, since it enables the birds to keep cooler than normal birds (Haldane, 1935; Landauer and Dunn, 1930; Landauer, 1937; Landauer and Upham, 1936).

(5) *The existence of genetic polymorphism within a species*. This, as we have seen, can only occur where there is a selective balance. Since both the visible and the lethal characters involved are known to mendelize in all cases properly investigated, the presumption is that they always do so, and the further presumption exists that they owe their origin to mutation.

Beautiful examples of the action of selection in causing the spread of favourable mendelian characters are seen in those cases of genetically-controlled polymorphism where one type, when freed from its linked lethal, has ousted the other, as well as in those where a mutant or rare allele spreads in certain environmental conditions, as in the industrial melanism of moths, and in other cases.

Polymorphism must be distinguished from normal variability, however large, grouped in a single normal curve of error, or at least one without sharply defined modes. When, however, wide normal variability exists, it appears, in so far as it is genotypic, to depend on mendelizing factors and their recombinations, since when a cross is made between extreme variants, the F2 is much more variable than the F1. The adaptive reasons for the existence of high variability of this unimodal sort are unknown, though it would appear that in some cases they must exist. Possibly it supplies the same kind of plasticity in relation to a wide range of environmental conditions as is found in plant species with numerous integrading ecotypes.

(6) *The effect of variation of conditions in altering the incidence of selection on* (*a*) *mutations, or* (*b*) *naturally-existing genetic differences*. In some cases mutations, which in what may be described as normal conditions are deleterious, may become advantageous in other conditions.[16] A good example is that of the vestigial-winged mutant of *Drosophila*, studied by Spencer (1932). In conditions near the optimum, vestigial is much shorter-lived than wild-type. But if vestigials and normals are kept together without food and water, the vestigials survive longer. Thus in environments which occasionally become very unfavourable the vestigial type might even oust the normal. It is worth noting that the advantage of disadvantage concerns the viability characters: the size

of the wings would thus be a "correlated characters" of no immediate selective value. On the other hand, reduced size of wings may have a direct selective value in certain conditions. Thus L'Heritier, Neefs, and Teissier (1937) also working with vestigial, found that flies with this character survived better than wild-type *Drosophila* when subjected to constant wind. This, as they rightly conclude, has a bearing on the prevalence of insects with reduced or functionless wings on oceanic islands.

Variation in the environment often leads to selection of certain types from among the range occurring naturally. This may refer either to continuous or discontinuous variations. An example of the former is the case described by Harrison (1920*a*) of the selection for depth of pigmentation in the moth *Oporinia autumnata*. The relative abundance of lighter and darker forms in a dark pinewood and an adjacent light birchwood was quite different, and so, but inversely, was the intensity of selection, as revealed by the number of wings left by birds. The result was that in the dark environment the darker types had become sixteen times the commoner, while in the light environment the lighter types were six times more abundant than the darker.

As an example of selection between sharply-delimited types we may mention the experiments of extreme interest carried out by Sukatschew (1928) on pure lines in dandelions (*Taraxacum*). He found that altering the density of total numbers of plants per plot might completely alter both the survival of the seedlings and the fertility of the survivors, so that a pure line which was inferior in one set of conditions would oust the rest if the conditions were changed.

We may also consider selection as between related species. Here, similar results to those with varieties of dandelions have been obtained by Timoféeff-Ressovsky (1933) with the competition between the larvae of two species of *Drosophila* at different temperatures; by Tansley (1917) on the varying results of competition between two species of bedstraw (*Galium*) according to the type of soil on which they are growing; and by Beauchamp and Ullyott (1932) on the decisive effect of temperature in bringing about the selection of one or other of two species of competing planarian worms.

Sukatschew's detailed analysis is entirely in accord with the elaborate ecological work of the Stapledon school, showing the effect of varying intensity of grazing on the survival of different species and strains of pasture plants.

In bacteria, the alteration of type with culture-medium appears not to depend upon any lamarckian or modificatory effect, but upon the selection of variants (to use a non-committal term), though the method of origin of these is still obscure. Again, selection has different effects on different pure lines of yeast according to conditions (Gause, 1934).

The diminution in size and other changes which occurred over a period of 150 years in a stock of horses placed on Sable Island, Nova Scotia, appear almost certainly to be due to selection in relation to the somewhat unfavourable conditions (Gates

1930). This doubtless has a bearing on the evolution of dwarf forms of large mammals on islands or near the limit of their range, for instance the very small Spitsbergen race of reindeer (*Rangifer tarandus*), the pigmy elephants, now extinct, of Malta and other Mediterranean islands, the Corsican subspecies of the red deer (*Cervus elaphus*) etc.

(7) *The interaction of two or more unfavorable mutations to produce a neutral or beneficial effect*. We have spoken of the cases of the mutations for red and transparent eyes in *Ephestia* and for purple and arc in *Drosophila*. Another case, of a rather different kind, is that of the recessive *facet-notched*, which produces a notch at the free end of the wing in *Drosophila*. Its allelomorph *facet*, also a recessive, produces irregular ommatidia in the eye, together with a slight irregularity of the wing-types. When, however, the heterozygous combination of the two is synthesized, it is found to produce no visible effects. The combination of the two recessive alleles restores the appearance and viability of the wild type (Glass, 1933, who cites other cases of the same phenomenon). Such a state of affairs might lead to the establishment of genetically-conditioned trimorphism.

(8) *The effect of selection of the gene-complex in altering the expression of mutations, and especially in abolishing their unfavourable action while retaining other effects*. We have dealt with numerous cases of this phenomenon earlier in the chapter, both as regards natural and artificial selection. The most striking cases are *eyeless* in *Drosophila*, the crest character in fowls and the restored viability of red-eyed meal-moths when the mutant gene responsible is combined with another recessive gene.

(9) *The existence of genetically-determined adaptations*. Once more the presumption is that these, if genetic, have arisen by mutation. If they are truly adaptive, the presumption is that they have arisen by selection.

(10) *The correlation between the incidence and type of genetically-determined variations in different parts of the range of a species with variation in the conditions and with the incidence and type of selection*. Some of the best examples concern polymorphic mimetic butterflies. We have first the adjustment of the frequencies of the different mimetic forms to the frequency of their respective model species; secondly, the adjustment of the pattern of the separate forms to the geographical variation of the models, or to the replacement of one species of model by another species; and thirdly, the relaxation of close mimetic resemblance in areas where the mimic outnumbers the model. None of these phenomena can be explained except on selectionist grounds. Another equally good example concerns the replacement of the "normal" type of the species in certain parts of the animal's range by a type which remains rare in other regions, e.g. black *Coereba*, black hamsters, black opossums, voles with *simplex* type of teeth, melanic moths, bridled guillemots, etc. In all such cases, thee are general grounds, and in some cases particular for believing that selection is at work.

These various lines of evidence all converge to support a neo-mendelian view, some showing that small mutations occur, others that selection is active, that some mutations are potentially beneficial, that through selection of the gene-complex, mutations can be adjusted to the needs of the organism, and that adaptation are genetically determined and vary in type and accuracy with direction and intensity of selection.

Three further general points of considerable importance must be mentioned. In the first place, R. A. Fisher (1930*a*) has provided mathematical proof of the interesting theorem that the combination of mutations to provide adaptive improvement in an organ or process in which the harmonious adjustment of many independently varying characters is required, is much more readily effected by mutations with small than by those with large effects.

Secondly, both Fisher (1932) and Haldane (1932*a*) have shown the enormous superiority, in the light of existing knowledge, of selection to other suggested agencies of evolutionary change, such as true orthogenesis.[17] Even if genes were to mutate repeatedly in the same direction, this could have no evolutionary effect unless they had no influence on viability and general fitness. A reduction of one-tenth of one per cent in viability would result in adverse selection which would override mutation at the highest rate ever yet observed in nature. Similarly, if a mutation caused an increase of viability of only 0.1 per cent or over, its spread would of necessity be mainly due to favourable selection. The same argument applies to the slow accumulation of lamarckian effects postulated by some believers in the inheritance of acquired characters: if this is so extremely slow as to escape detection in the course of an experiment covering a few generations, as they often assert, it would be overridden by selection-pressure whenever any but the most trifling differences in viability existed.

Thirdly, variability varies inversely with selection-pressure. The butterfly *Papilio dardanus* provides a striking example. This possesses several polymorphic female forms. Most of these are mimetic and highly invariable, except when, for special environmental reasons, they are able to live outside their models' range. The rare form *dionysos*, however, is non-mimetic, and highly variable (Ford, 1936).

In the field we are discussing, of the relations between genetics and evolution, perhaps the most important single concept of recent years is that of the adjustment of mutations through changes in the gene-complex. Before this had been developed by R. A. Fisher and his followers, notably E. B. Ford, the effect of a mutation was assumed to be constant. A given mutation, we may say, made an offer to the germplasm of the species, which had to be accepted or declined as it stood. And the data on laboratory mutants at the time indicated that the great majority would have to be declined.

To-day we are able to look at the matter in a wholly different way. To continue our metaphor, the offer made by a mutation to the species is not necessarily a final offer. It may be merely a preliminary proposal, subject to negotiation. Biologically, this negotiation is effected in the first instance by recombination and secondarily by mutation in the residual gene-complex. It can lead to a marked alteration in the effects of the mutation, which may make the proposal acceptable to the organism.

Ten years ago evolutionary change, on the neo-mendelian view, depended on the co-operation of two processes only—the presentation of ready-made building-blocks by mutation, and the utilization of certain of them under the influence of selection. To-day we have been brought to realize that a third process is at work—change, primarily recombinational, in the residual variability of the gene-complex; and this can shape the building-blocks so as to enable them to fit in better with their neighbours and with the general plan of the building.

Thus evolutionary change, in so far as Darwinian, is not due simply to the co-operation of mutation and selection; a third, intermediate agency is involved in the shape of the residual variability of the species. Adjustment intervenes between presentation and acceptance.

REFERENCES

1. A special case of position effect is the modification of variegation (mosaicism) of various sorts in *Drosophila,* exerted on various genes if translocated into proximty with heterochromatin ("inert" regions of the chromosomes). Schultz's summary (1941) makes it clear that this is caused by a change in the nucleic acid metabolism of the translocated regions, and that this exerts its effects in early stages of development, by causing a process akin to inactivation of the genes involved. The degree of this inactivation decreases with the distance from the point of breakage and re-union with the inert region.
2. A certain number of rare mutant genes will be lost to the species by accidental elimination. In addition, genetic variance will be reduced by the selective elimination of deleterious mutant genes.
3. R. A. Fisher (1937*a*) points out that the *number* of the rare approximately neutral genes carried by a species increases roughly as the logarithm of its population-size. Such genes, however, will only cause observable variability of any magnitude if they can increase their *frequencye,* as will occur if they are slightly favourable or if changed conditions cause them to become so. Thus, as Ford (1940*c*) points out, increased variability ascribable to large population-size depends on genes actually engaged in causing evolutionary change, and the observed fact of such increased variability demonstrates the spread in nature of genes with small advantageous effects.
4. In a stock like that of the horses, which shows a functional evolution that by geological standards must be called rapid, the time needed to effect a change of specific magnitude is of the order of 100,000 generations, and to effect one of generic magnitude of the order of 1,000,000 generations (Wells, Huxley and Wells, 1930).
5. A valuable summary of the modern outlook, which treats certain aspects of the problem more fully than is possible in a single chapter, is given by Ford in his little book *Mendelism and Evolution* (1934).
6. Waddington (1941*a*) points out that certain ontogenetic events act as "epigenetic crises", in that quite slight modifications of their course will have a considerable effect on a number of characters. Thus alterations in the pupal contraction of *Drosophila* are involved in mutant characters of legs, wings, bristles, etc.
7. As the stocks used in this experiment were not "isogenic" in regard to their residual gene-complex, an alternative explanation is possible by which the increased viability may have been due to modifiers and not to the specific combination of the two main genes.
8. Another method by which viability may be improved is by mutation in the primary gene concerned. As an example, we may take the work of Mohr (1932). Two stocks of vestigial-

winged fruit-flies (*Drosophila melanogaster*) were maintained for a long time as inbred cultures. In both of them, the wings eventually became almost normal. Analysis showed that this was due to the selection of a less extreme allele of vestigial which had presumably arisen by mutation from full vestigial, and had then been favoured by selection because of its less extreme effects. An interesting point is that these nearly normal alleles were not identical in the two cases, but represented different steps in the multiple series.

9. The term *dominant* has unfortunately been employed very loosely, some authors using it to mean that the heterozygote is indistinguishable from the homozygote, while others call a dominant any gene whose effects in single dose can be distinguished at all: e.g. Bowater (1914) called the melanic form of the moth *Aplecta nebulosa* a dominant, although the heterozygote, as he himself goes on to state, has an intermediate expression; and *all* the so-called dominant mutations in *Drosophila melanogaster* are either lethal in double dose, when, of course, the visible effect of the homozygote cannot be determined, or their heterozygous expression is less extreme than their homozygous (e.g. abnormal abdomen).

   For this reason, and because it is in many ways unsatisfactory to have positive and negative terms, like dominant and recessive, to denote gradations in what is really a single scale of positive efffects, I would suggest that some other term, such as *single-dose expression* (or *heterozygous expressivity*) would be more suitable. Full single-dose expression would then be equivalent to true dominance: zero single-dose expression to recessivity; and truly intermediate expression to absence of dominance in which the heterozygote is intermediate between the two homozygotes.Timoféeff-Ressovsky (1934b) deals with expressivity from a rather different angle.

10. The same result can of course be secured genetically, either by special chromosomal mechanisms, as in genetic sex-determination, or by a selective balance resulting in polymorphism, as especially well illustrated by butterflies with polymorphic mimetic forms.

11. As showing the intensity of the selection acting against certain recessive mutations, C. Gordon (1935) liberated 36,000 *Drosophila melanogaster* in England, where they are not endenic. The population originally contained 50 per cent of the recessive gene ebony (25 per cent pure wild-type flies, 25 per cent ebony and 50 per cent heterozygous for ebony). After 120 days (5 to 6 generations) the frequency of the ebony gene had falled to 11 per cent. From the data it appears that some heterozygotes were selectively eliminated in each generation, as well as the homozygotes; this is in accordance with the fact that ebony has a slight single-dose expression.

    A further important fact is that in nature recessisves are almost wholly absent from the X-chromosome of *Drosophila*, where, of course, they are expsed (in males) to selection in single dose.

    Dubinin and others (1936) found more than one detectable mutant (recessive) allele in each wild fruit-fly! They also showed that various mutant genes altered in their frequency during three years, some becoming more and others less frequent.

12. Autopolyploids will originally produce many inviable gametes, owing to the aggregation of chromosomes by fours instead of by pairs before meiosis. But there is evidence to show that this condition, too, may be adjusted by selection, leading to reasonably true-breeding forms (Darlington, 1937; Müntzing, 1936).

13 It has been claimed by Harrison that melanism is also commoner in coastal areas: Ford (1937), however, shows that this conclusion is certainly not universal.

   Hardy (1937) states that slight but definite darkening has occurred in the house-sparrow (*Passer domesticus*) in the Liverpool area. It will be of interest to see whether this change, too, will show progressive spread. Sporadic melanism has occurred in the passerine West Indian bird *Coereba flaveola,* producing four separate melanic island subspecies. In some cases the replacement of the normal by the black form has been followed during recent historic times. Furthermore, almost all island forms of the genus are somewhat darker than the mainland forms (Lowe, 1912). Two cases of recent spread of melanic mutants in mammals are considered later.

14. In certain cases, the melanic form has spread from its original industrial area into surrounding non-industrial country, with a decreasing frequency-gradient. If the selective balance in favour of non-melanics in non-industrial areas is sight, this is to be expected as the result of mere population-pressure: of Summer's views on population-pressure in subspecies of *Peromyscus.*

15. Parallel variations occur here as in *Cepaea,* but in much more striking fashion, since they affect a large number of related species and indeed genera.

16. When this is so they fall conveniently under the heading of pre-adaptation.

17. Haldane (op. cit., p. 142 qualifies this view by pointing out that the paths open to selective guidance are limited by the nature of variatin. This restrictive or subsidiary orthogenesis we shall discuss later.

Chapter—5
Hybrids in Evolution

In the last chapter, we learned that biologically distinct, reproductively isolated species are most likely to evolve in an environment which is unstable in terms of the evolutionary time scale, that is, over periods of hundreds or thousands of years. In such an environment, the conditions which favour populations having a particular set of adaptive characteristics will be constantly shifting back and forth from one region to another. That the populations of plants and animals themselves will also be shifting about in this way is a foregone conclusion.

Given these unstable conditions, we would expect to find many examples of previously separated populations coming together and having opportunities for their individuals to hybridize with each other. The evolutionary consequences of such hybridization will be explored in the present chapter.

Evolutionary Definition and Immediate Effects of Hybridization. Biologists have used the term hybridization in many different ways. Geneticists apply the term to crossing between pure lines which differ by a single gene, as in the experiments of Gregor Mendel. Plant breeders speak of hybrid corn which is the result of crossing inbred lines differing in respect to many genes. We also speak of hybridization between breeds or races of domestic animals, and between species, such as the horse and donkey. If the evolutionist is to use the term hybridization, he must define it in a way which will be reasonably precise and meaningful to him. Since evolution consists chiefly of changes in adaptive complexes of genes, the most meaningful definition of hybridization to an evolutionist is crossing between populations having different adaptive gene complexes. Such populations may be either different races or subspecies of the same species, or different species which are separated by variously developed isolating mechanisms.

Hybridization between populations having different adaptive gene combinations can increase greatly the size of the gene pool with respect to genes having different adaptive values, provided that the hybrids can give rise to segregating progeny in later generations. Most of the segregates produced from such hybrids will, of course, have a lower adaptive value than that of either of the parental populations, particularly in the environments to which these parental populations are adapted. A small proportion of them may, however, prove to be better adapted to certain environments. Natural hybridization between differently adapted wild races is in many ways comparable to the artificial hybridization carried out by animal and plant breeders with widely different varieties of domestic animals or cultivated plants. If, for instance, we cross two widely different varieties of *Iris* or *Gladiolus,* we get in the second generation an enormous array of types, having every conceivable combination of the characters of plant size, earliness, flower size, texture of petals, and flower color which were present in the parents, plus many characteristic which appear to be new. Most of these progeny are in many ways less vigorous or less attractive than their parents. The skilled breeder can, nevertheless, select from this mass of variants one out of a hundred or even one out of a thousand individual which in various ways is superior to either parent. Similarly, even though the great majority of the progeny of hybrids between two natural varieties or species are poorer in every respect than their parents, the forces of natural selection are strong enough to pick out that tiny fraction of them which may be better adapted to an available environment, and to increase greatly the frequency of such gene combinations in the environments to which they are adapted.

There are, however, two drawbacks to the success of such hybrids. In the first place, most hybrids between species are wholly or partly sterile, so that segregating offspring from them are difficult or impossible to obtain. Secondly, the genotypes found even in the second generation of a wide cross are highly heterozygous, and so will not breed true. Hybridization will, therefore, be an important factor in evolution only if the three following conditions are met. 1. There must be one or more environmental niches available to which certain hybrid derivatives are better adapted than are either of the parental populations. 2. In the progeny of partly sterile hybrids, fertility must be restored while valuable new gene combinations resulting from hybridization are still retained. 3. New combinations must become sufficiently constant genetically so that they can maintain themselves under natural conditions.

When these three factors are taken together, they make us realize that the successful contributions of hybrid progeny to evolution will always be rare, and in many groups will be highly exceptional events. Evolution, however, does not always progress through gradual accumulation of the kinds of changes which occur most commonly in populations. At many times during the course of evolution, combinations of events which occur with a very low frequency have effects far more significant than might be imagined on the basis of their infrequency. With this consideration in mind, ways will be considered by which populations can overcome these three restrictions or limitations to the success of hybrid progeny.

Hybridization in Old and New Environments

Anyone familiar with animal and plant communities in nature realizes the fact that if a particular set of environmental conditions have existed for a long time, every available environmental niche is filled with well-adapted species of animals and plants, so that populations with new adaptive characteristics have "no place to go." On the other hand, when such communities are disturbed in any way, the equilibrium between their species is upset, so that new kinds of plants and animals have a chance to enter them. This is most likely to be true if the disturbance is accompanied by a change in climate.

In our modern world, the commonest cause of such disturbances is man himself. We are hewing down forests; ploughing and bulldozing land; destroying natural predators; introducing domestic animals, which devour vegetation much faster and more indiscriminatingly than do the native animals; and spreading weeds, pests, and diseases at a rate which is many times faster than any which the world has seen in its long history. The destructive effects of these disturbances are familiar to everyone and are the subject of numerous books and other polemics. Since construction is always slower than destruction, the evolutionary changes which these disturbances have promoted are much more subtle and hard to recognize. A small-scale example of them is industrial melanism. If we study the wild animals and plants which have been able to profit from man's disturbances, we find many more extensive examples. At least in plants many of these involve hybridization.

When the early colonists came to North America from Europe, they cut down many of the forests and made open hillsides and cow pastures. Among the plants which took advantage of this change were the hawthorns. These shrubs, which are sun lovers, were originally confined to stream banks, open glades, rocky hillsides, and similar gaps in the otherwise continuous primeval forest. Thus, they not only were uncommon, but also were separated from each other in different kinds of habitats. A number of distinct, sympatric species were present, but we cannot now be sure how many. As pastures were made hawthorns invaded them, and since their seedlings were protected from grazing cattle by their strong spines, they were able to spread rapidly. Consequently, species which had previously been separated from each other by ecological isolation were able to come together. They created swarms of hybrids which are now so complex that in spite of repeated attempts, no botanist has been able to classify them in any reasonable fashion. Some of these hybrids are fertile when cross pollinated, while others have evolved ways of producing seeds by asexual means.

Many groups of weeds, particularly in the grass family, contain species which on the basis of cytogenetic evidence are clearly of recent origin and are found only in habitats disturbed by man. A genus related to wheat and known as goat grass (*Aegilops*), has been carefully analyzed by several geneticists. Its species originated in the Middle East, where man first cultivated wheat and barley about 10,000 years ago. Several of these species have almost certainly evolved since that time, and some of them are clearly of hybrid origin. A similar origin is equally probably for a wild relative of corn, teosinte, which grows in Mexico and Central America.

The evolutionary history of the animals which have invaded the new habitats created by man is harder to follow. Nevertheless, a good deal of evidence indicates that many new subspecies have evolved as a result of this activity, while some species may have evolved in response to it. The house sparrow, or "English sparrow," as it is called in most of North America, arrived here with the early colonists, and is now one of our most common birds. While most of the American house sparrows are indistinguishable from their English ancestors, subspecies have evolved when this bird has invaded areas such as the arid southwest and Hawaii, in which the climate and vegetation are very different from that to which the species was adjusted when it evolved in its original European home. For various reasons, the origin of species in connection with human disturbances in prehistoric times is harder to trace in animals than it is in plants. There are, however, several species which are now found only in habitats associated with man or created by him, and which may therefore have evolved in response to this activity. Among them are the house mouse and the species of *Drosophila* most familiar to geneticists, *D. melanogaster*.

Although the animal examples given above are not known to involve hybridization, and in the case of the house sparrow certainly do not, several examples are known of animal species which are sharply distinct from each other in undisturbed areas, but which have formed hybrid swarms in habitats disturbed by man.

One of the best of these is that of the spotted towhee (*Pipilo erythrophthalmus*) and collared towhee (*P. ocai*) in Mexico (Figure 5.1). The spotted towhee, a bird familiar to people living in our southern states and California, has a black and white head and back and rufous flanks. The related collared towhee, found in the mountains of Mexico, is mostly green and brownish olive. In the mountains of southeastern Mexico the two species are found together in their typical form, and little or no hybridization occurs between them. But in many parts of western Mexico, particularly on certain isolated mountain summits, are found hybrid swarms of intermediate birds. On these mountains, the original oak forests have been partly or completely cleared and replaced by scrubby second growth. This disturbance of the habitat has apparently favoured the hybrids.

Although human disturbance is by far the commonest cause of evolutionary instability in modern populations of animals and plants, other kinds of habitat disturbance have been taking place ever since life appeared, and in the case of the more recent ones clear evidence is available of their effects on plant and animal populations. In the northern hemisphere, the effects of the last ice age are the easiest to follow.

The fourfold advance and retreat of ice caps which covered hundreds of thousands of square miles of land created habitat disturbances of many kinds. The presence of the ice itself brought about local lowering of temperatures, although at least during some phases, particularly the retreat of the glaciers, relatively warm habitats must have existed at no great distance from them. This made possible the sympatric coexistence of populations with very different temperature requirements. In addition, the advancing ice scraped huge masses of soil and rock from the territory which it covered and deposited them along its southern margins, thereby creating a whole series of completely new habitats. Finally, the retreat of the glaciers laid bare tens of thousands of square miles of territory in which the earth's surface had been so profoundly altered that it consisted entirely of a vast mosaic of new habitats. These included not only terrestrial habitats with greatly altered soil conditions, but also innumerable new aquatic habitats in the form of lakes and swamps which were formed from the basins scooped, out by the ice or dammed up by masses of glacial debris.

Many groups of animals and plants are known to have evolved in response to these changes. In some groups of mammals this evolution is documented by the fossil record. For instance, several species of bears and of bisons appeared and became extinct during the glacial epoch, as did the well-known woolly mammoth. In plants, active evolution of races and species in the glaciated areas is found in many groups. Two of the most conspicuous examples are willows and birches. In eastern North America, the species of birch trees found south of the former glacial margin are all clearly distinct from each other. They occupy for the most part stable forest communities and are apparently ancient. In the glaciated areas populations of birch trees, particularly those related to the white or canoe birch, are extremely variable and hard to classify into species, indicating that they are unstabilized and still evolving. The same is true of populations of birches in northern Eurasia, from Siberai west to Scotland. Evidence from both phenotypic variability and chromosomal numbers and behaviour indicates that their evolution is associated with extensive hybridization.

A good example in fishes is the group of whitefishes and ciscoes found in lakes, streams, and occasionally shallow seas in northern Eurasia and North America. These fishes are mostly confined to drainage basins which were greatly altered by the ice age. Species have been particularly hard to recognize in the group, and different authorities have interpreted them rather differently. The most careful recent studies have been carried out by Swedish specialists, and have been supplemented by artificial hybridization and observations on progeny reared in hatcheries under controlled conditions. These studies have shown that many lakes and streams contain two or three species which are very hard to tell apart, but which can nevertheless be recognized as distinct when carefully studied, and pass the test of sympatry in most places where they occur together. Furthermore, each of these species is found in several lakes over a rather wide area. Some of the lakes, however, contain populations intermediate between these widespread and otherwise distinct species. These intermediate populations are probably hybrid swarms. Apparently, variability has been considerably increased by hybridization between species in some of the lakes which these fishes have occupied since the glaciers retreated and made these habitats available.

In addition to human disturbance and glaciation, several other kinds of environmental upheavals which have occurred in geologically recent times have almost certainly stimulated evolution in the regions where they occurred. In plant populations this evolution appears to have been accompanied by hybridization. In the southwest, from Texas to southern California, the advance of deserts accompanied by earthquakes

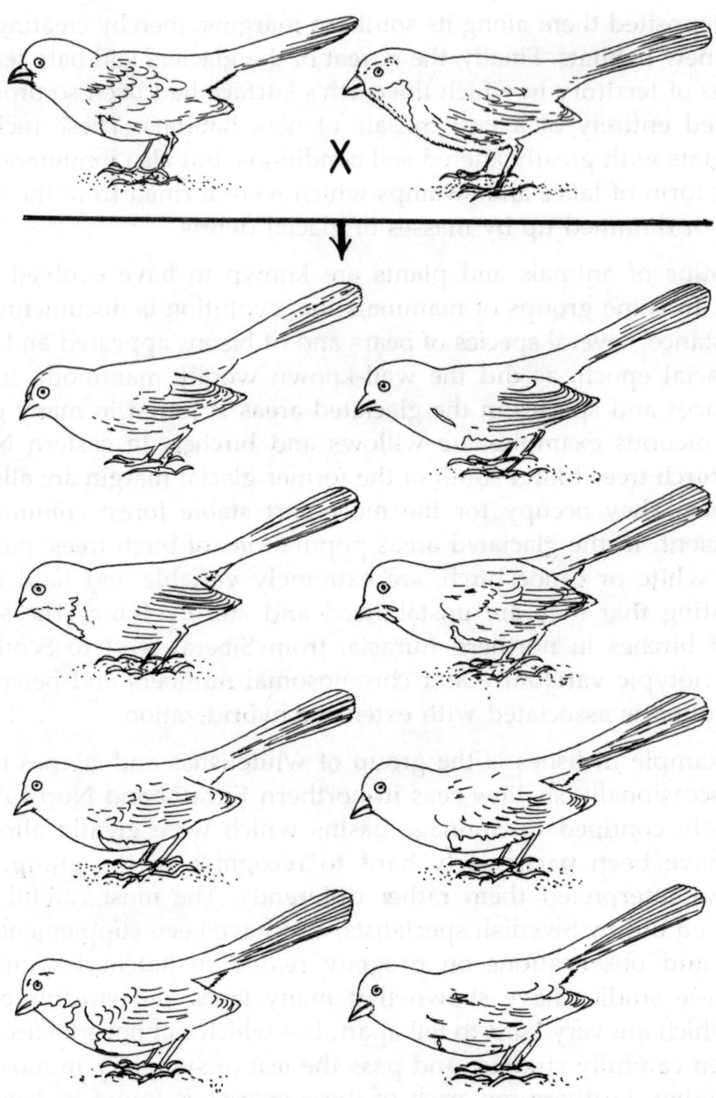

and other mountain building movements have had such effects. In Florida, the advance and retreat of shallow seas have been chiefly responsible. In Hawaii, entire islands have been built up by volcanic activity and have wholly or partly subsided into the sea. On individual islands, opportunities for evolution (including hybridization) have been created by recent volcanic eruptions. These eruptions have built up new craters consisting of raw lava, cinders, and unweathered soil, side by side with ancient, mature volcanic slopes. Biologists desiring to find examples of evolution in action have learned to look for them among the populations found in all of these different kinds of disturbed areas. They have rarely been disappointed.

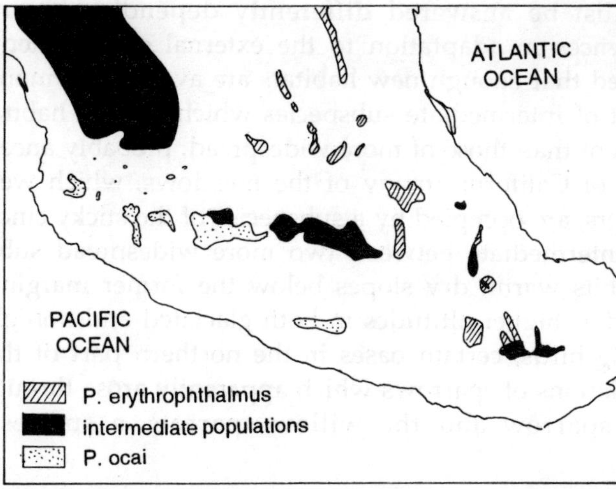

Fig. 5.1: Two species of towhee, *Pipilo erythrophthalmus* and *P. ocai*, together with their hybrids, and a map of their distribution in central Mexico. In the eastern part of this region they occur together with little or no hybridization. In the western and central areas, however, intermediate and strongly segregating populations (indicated in black) are found in many places. From Sibley, University of California Publications in Zoology, 50:109–194, 1950; and Evolution, 8:255, 1954.

Stabilization of Hybrid Progeny

The two other conditions which hybrid progeny must meet in order to achieve evolutionary success, i.e. restoration of fertility and stabilization of intermediate progeny, will be considered together. From the genetical point of view the both involve the following principle. If a particular difference between two genotypes is governed by several pairs of non-allelic genes, then an intermediate condition with respect to the character can be achieved in two different ways: 1. heterozygosity at all of the gene loci concerned, or 2. homozygosity at all of them, but in a different combination from that possessed by either parent. For example, if the parents differ with respect to four gene loci controlling the same character, and are homozygous at each of these loci, their respective genotypes may be written as follows: *AABBCCDD* and *aabbccdd*. (In this case the use of capitals versus small letters is merely for convenience, and does not indicate dominance or recessiveness.) The intermediate F_1 hybrid would then have the constitution *AaBbCcDd*. This hybrid would, of course, segregate toward both parental conditions at each locus. These segregants, however, would include six homozygotes which would have an equal contribution of genes from both parents, and would, consequently breed true for the intermediate condition: *AABBccdd*, *AAbbCCdd*, *AAbbccDD*, *aaBBCCdd*, *aaBBccDD*, and *aabbCCDD*. These new, intermediate homozygotes would, therefore, outnumber by three-fold the two parental homozygotes. Since they would certainly be formed, the critical question is: Would they be adapted to any of the new environments which might become available?

This question must be answered differently depending upon the character concerned. For differences in adaptation to the external environment the answer is often positive, provided that enough new habitats are available. Among higher plants, several examples exist of intermediate subspecies which occupy habitats intermediate between and more recent than those of more widespread, probably ancestral subspecies. In the Sierra Nevada of California many of the meadows, which were formed after the retreat of the glaciers, are occupied by a subspecies of the sticky cinquefoil (*Potentilla glandulosa*) which is intermediate between two more widespread subspecies. One of these subspecies inhabits warm, dry slopes below the former margin of the glaciers, and the other is found at higher altitudes in both glaciated and non-glaciated parts of the mountains. Among birds, certain oases in the northern part of the Sahara desert are occupied by populations of sparrows which apparently arose through hybridization between the house sparrow and the willow sparrow, a species native to the Mediterranean region.

In respect to those genetic differences which form the principal isolating mechanisms between species of animals: ethological isolation, hybrid weakness, and developmental hybrid sterility, successful intermediate combinations derived from hybridization are not known. One might expect them to be rare, since the genes which cooperate to produce these barriers form complex, integrated systems. Consequently, most of the recombinations between them would be badly maladjusted within themselves, regardless of the external environment. Successful recombinations of the genes controlling such barriers must not, however, be regarded as impossible. Experiments to find out whether they can occur need to be performed.

Among plants, recombination of the genes which control isolating mechanisms can produce new genotypes which are partly isolated from both of their original parents. This has been clearly demonstrated in respect to both floral isolation and segregational hybrid sterility. In respect to floral isolation, good examples can be found in the beard tongues (*Penstemon*) of the western United States. This genus contains a large number of species which are distinguished from each other principally in the shapes, sizes, and colours of their flowers. Four of the best known species in southern California are shown in Figure 5.2. One of them, found in pine forests in the mountains, has large, two-lipped, widely gaping corollas which are bright blue in colour. A second, known as scarlet bugler, which is widespread in the drier parts of southern California, has scarlet corollas, tubular in shape. A third species, the showy penstemon, lives on brush-covered hills, a relatively recent habitat which is intermediate between those of the mountain penstemon and the scarlet bugler. It has blue or slightly purplish flowers, intermediate in shape between the other two. In a few places, natural hybrids between the mountain penstemon and the scarlet bugler have been found, and these are much like the showy penstemon in a number of respects. The three species are normally visited by the following pollinators: the mountain penstemon by large carpenter bees (*Xylocopa*); the scarlet bugler by hummingbirds; and the showy penstemon by solitary wasps (*Pseudomasaris*).

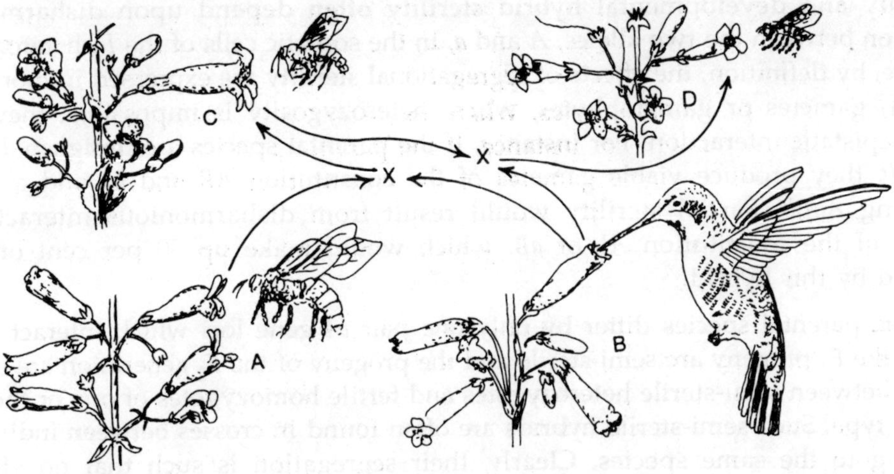

Fig. 5.2: Four species of the genus *Penstemon* found in California, together with their pollinators. Distributional evidence indicates that the two oldest species are the extreme types, mountain *Penstemon* (*P. Grinnellii*) (A) and scarlet bugler (*P. centranthifolius*) (B). The two other species, as indicated by the arrows, were derived by hybridization followed by isolation between these two, followed by isolation and stabilization of intermediate populations through the selective action of specific pollinators.

We can reconstruct a probable sequence of events which gave rise to the showy penstemon as follows. At one time in the past, the mountains of southern California were not brush covered, but were clothed with forests except for the steeper, drier slopes, which were relatively bare. Consequently only habitats suitable to the mountain penstemon and the scarlet bugler were available. At that time, hybridization between the two species may have occurred from time to time, but was of little consequence because no site was available where populations of hybrid origin could become established. As the climate in the mountains became drier, and forests were replaced by brush, such habitats became available. Some penstemons of hybrid origin moved into these habitats and in one or more places began to be visited by wasps. As a consequence, natural selection brought about stabilization of gene combinations adapting the vegetative parts of the plant to life on the dry, brush-covered slopes and the flowers to pollination by the wasps which were their commonest visitors in their new habitat. A similar course of events may have given rise to a large proportion of the diverse species of *Penstemon* found in our western states. Artificial hybridizations between many pairs of related species have usually produced fertile progeny, so that the present barriers which separate them are largely ecological or ecogeographic isolation plus floral isolation.

In respect to segregational hybrid sterility, the genetic mechanisms controlling this barrier are more complex than those controlling other isolating mechanisms, since it must depend at least in part upon unfavorable epistatic interaction between genes or chromosomal segments at different positions along the chromosome. Both hybrid inviability and developmental hybrid sterility often depend upon disharmonious interaction between the two alleles, *A* and *a,* in the somatic cells of the F_1 heterozygote. But since, by definition, the effects of segregational sterility are expressed in monoploid (haploid) gametes or gametophytes, where heterozygosity is impossible, they must involve epistatic interaction. For instance, if the parental species are designated *AABB* and *aabb,* they produce viable gametes of the constitution *AB* and *ab* and a hybrid containing *AaBb.* Partial sterility would result from disharmonious interaction in gametes of the constitution *Ab* or *aB,* which would make up 50 per cent of those produced by this hybrid.

If the parental species differ by only one pair of gene loci which interact in this fashion, the F_1 progeny are semi-sterile and the progeny of the F_2 generation are equally divided between semi-sterile heterozygotes and fertile homozygotes of one or the other parental type. Such semi-sterile hybrids are often found in crosses between individuals belonging to the same species. Clearly, their segregation is such that no effective isolating mechanism is present. If, however, the parents differ by several such pairs of complementary gene loci, the proportion of viable gametes formed by their F_1 hybrid is $\frac{1}{2}^n$, where n is the number of pairs involved. Consequently, the effectiveness of segregational sterility depends upon the number of independently segregating units of which it consists.

Because of this fact, selection for increased fertility in the progeny of a hybrid characterized by this type of sterility has a very high probability of giving rise to a stabilized type which has a different genetic constitution than that found in either parent, and which forms sterile hybrids in crosses with both of them. This fact can be illustrated by the simplest possible model, which consists of two independently acting pairs of loci (Figure 5.3). As already pointed out, the semi-sterility resulting from the cross diagrammed in this figure would not be an effective barrier to gene exchange. If, however, the parental species should differ by several pairs of complementary gene loci, fertile genotypes could be derived from them which would form much more sterile hybrids in back crosses with both parents.

If the parents differed by n pars of loci, the proportion of homozygous, fertile types in later generations, as well as that of viable gametes formed by the F_1 hybrid, would be $1/2^n$. Among this proportion, however, $1-1/2^{n-1}$ would be new combinations, forming partly sterile hybrids when crossed back to both of the parental species. If $n = 6$, this proportion becomes 31/32, or about 97 per cent. These figures show that if an effective barrier of segregational hybrid sterility is present, selection for increased fertility in later generations can occur. If it does, the resulting fertile types are almost certain to be at least partly isolated from both of the parental species and so to be potential progenitors of a new species. Since, however, selection for fertility from such hybrids would take place most easily if they were repeatedly back crossed to one of their parents, these new combinations would be likely to arise only if the species were normally self-fertilizing, or if the hybrids had migrated into a habitat not occupied by either of the parental species.

Parental Genotypes	AABBCCDD	X aabbccdd.
Gametes (All viable)	ABCD	abcd.
F_1 Genotype	AaBbCcDd	
Inviable:	AbCD, AbCd, Abcd, aBCD, aBCd, aBcd,	
Possible Gametes.	ABCd, abCd, ABcD, abcD, AbcD, aBcD, Fertile: ABCD, ABcd, abCD, abcd.	
Possible homozygous and fertile genotypes in F_2.	AABBCCDD	aabbccdd
Since hybrids between only two of these four genotypes would be partly sterile, the last two could form the starting point for new species.	AABBccdd	aabbCCDD

Fig. 5.3: Diagram showing how segregation from a partly fertile hybrid between two species which differ from each other with respect to many factors affecting fertility can give rise to stabilized, fertile derivatives which form partly sterile hybrids when crossed back to either of their original parents.

A result similar to that which would be predicted on the basis of these calculations has been obtained in progeny of an artificial hybrid between two species of grasses, the blue wild rye (*Elymus glaucus*) and the squirrel tail grass (*Sitanion jubatum*). Furthermore, the origin of a natural species of larkspur (*Delphinium gypsophilum*) occurring in central California is most easily explained on this basis (Figure 5-4). Since the few experiments which have been performed to test this hypothesis have yielded positive results and since segregational sterility based upon many genetic differences is a widespread barrier between species of higher plants, it is possible that a large number of plant species have originated in this fashion.

The Phenomenon of Introgression

There are three reasons why F_1, hybrids occurring in nature are more likely to form progeny by back crossing to one of the parental species than by mating with each other. First, the hybrids are always fewer, and usually much fewer than are individuals of the parental species. Secondly, the high fertility of parental individuals as compared to the low fertility of hybrids will make an enormously greater number of parental gametes available for fertilizing the hybrids. Finally, back cross progeny, since they will contain genes derived principally from one of the parental species, are the most likely to be well adapted to a habitat already present, which in most instances will be a modification of one of the parental habitats.

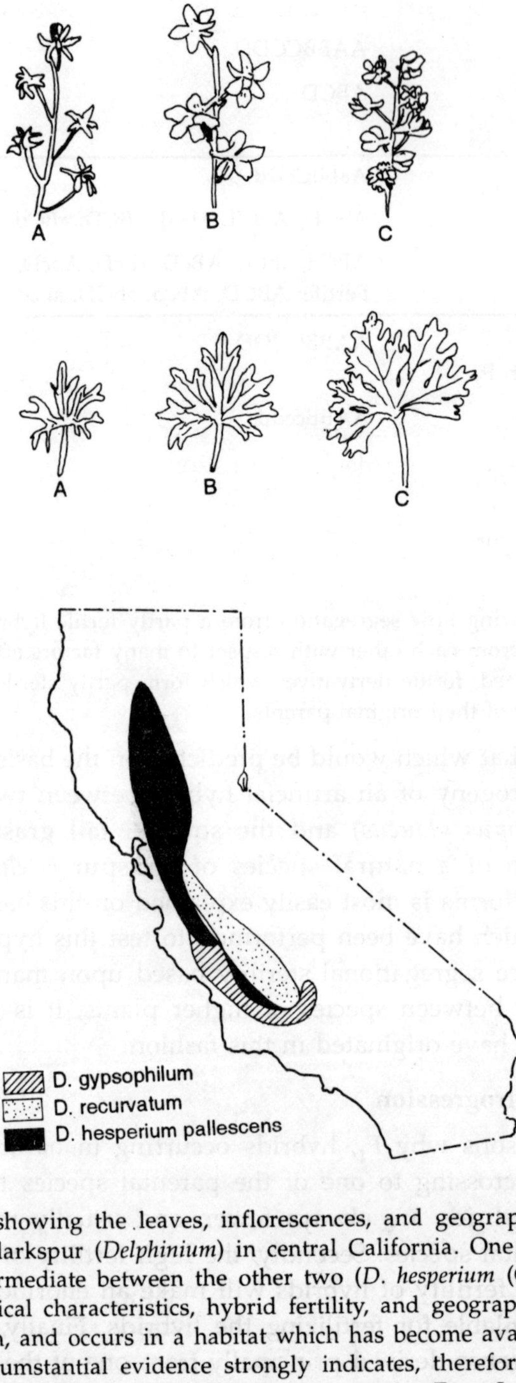

Fig. 5.4: Drawings showing the leaves, inflorescences, and geographic distribution of three species of larkspur (*Delphinium*) in central California. One of these, *D. gypsophilum* (B), is intermediate between the other two (*D. hesperium* (C), *D. recurvatum* (A)) in morphological characteristics, hybrid fertility, and geographic as well as ecological distribution, and occurs in a habitat which has become available in relatively recent times. Circumstantial evidence strongly indicates, therefore, that *D. gypsophilum* is derived from hybrids between the other two species. From Lewis and Eplin, Evolution, 13:511, 1959.

For this reason, the commonest result of hybridization is back crossing to one of the parental species, followed by selection of genotypes having a preponderance of genes derived from one species, but a few chromosomal segments introduced from the foreign species. This sequence of events—hybridization, back crossing, and stabilization of back cross types by selection—is known as INTROGRESSION. In actively evolving plant species, it is one of the commonest sources of new variability.

The best documented example of introgression is found in a cultivated species, corn. In Mexico, a weedy grass known as teosinte (*Zea mexicana*) is often found in or near fields of cultivated corn or maize (*Zea mays*). The two species are easily hybridized, and the hybrid is partly fertile. Recent archeological discoveries have shown that the ancestor of corn was a wild grass which had most of the characteristics of modern corn, but which was much smaller and had tiny ears. In a succession of layers which have been dated by the radio carbon method, corn cobs have been found which look progressively more like the modern plant. In addition, these cobs contain an increasing number of characteristics in which they resemble teosinte; a resemblance which has been known for some time to be characteristic of some modern varieties of corn. Plants resembling them have been produced artificially by crossing with teosinte the most primitive varieties of modern corn available and making suitable selections from such hybrids. By this means, the importance of introgression with teosinte as a source of some of the more valuable genes found in many modern cultivated varieties of corn has been clearly demonstrated.

Although it does not include evidence from fossils, the circumstantial evidence that introgression has played an important role in the evolution of many wild species of plants is almost as strong as in the corn-teosinte example. In the annual sunflowers, for example, the origin of new races or subspecies through introgression with other species has been repeated by artificial crossing and back crossing. Fewer examples are known in animals, but one exists in domestic cattle. In the high mountains of central Asia occurs a relative of cattle, the yak (*Bos grunniens*), which is also domesticated. Many of the herds of cattle found along the western margin of these mountains, in Soviet Central Asia, contain characteristics which show clearly the presence of genes from the yak. This has, apparently, increased their adaptability to the harsh climate of these mountain areas. Similar introgression of characters from the American bison into cattle has met with some success in the herds of western Canada.

The Origin of New Characteristics Via Hybridization

For the most part, hybridization can give rise only to populations which are intermediate between and share characteristics of the parental populations, and which occupy intermediate habitats. There are, however, three ways in which hybridization can promote the origin of new characteristics. The first is TRANSGRESSIVE SEGREGATION. Since the interaction of genes is not always additive, sometimes a segregating genotype from a wide cross can exceed both of its parents in one or more of its characteristics. A good examples is flower size in a hybrid between two species of tobacco. *Nicotiana alata,* a familiar garden ornamental with crimson flowers, has flowers which are among the largest in its genus. On the other hand, the very different looking flowers of *Nicotiana langsdorffii* are among the smallest in the genus (Figure 5.5). The hybrid between the two species is partly fertile, and a great variety of segregating types can be obtained in the F_2 and later generations. After eight generations of selection for large corolla size was practiced on progeny of this hybrid, plants were obtained which had flowers even larger than those of *N. alata,* but which were nearer in colour to those of *N. langsdorffii.* Apparently, genes from the small flowered species were able to interact with those of *N. alata* to produce larger flowers than this species ever produces on the basis of its own gene pool.

Another way in which hybridization could increase evolutionary opportunity is through the establishment of NEW GENETIC BACKGROUNDS. Placed on such new backgrounds, mutations which in the parental species were completely inadaptive could now contribute to new adaptive gene combinations. One of the principal reasons why mutations occurring in old, well-established species usually reduce the adaptability of their bearers is that nearly all of the mutations which might increase adaptability have already occurred and have been incorporated into the gene pool of the population. A population of hybrid segregates which is entering a new habitat provides, however, an entirely new environment in many ways. In such an environment, many mutations which in the parental species had occurred repeatedly and had been weeded out by selection could take place again and possess an adaptive advantage in combination with other genes already present. Although this is a strong possibility from the theoretical point of view, it is almost impossible to test. The occurrence of new mutations can be recognized only hen it takes place on a genetic background which is relatively uniform and homogeneous. In the progeny of genotypes as heterozygous as those produced by hybridization between races or species, there is no way of distinguishing between new mutations and various recombinations of pre-existing genes.

Finally, in some plant hybrids, particularly those between corn and teosinte and between species of cotton, the hybrid condition has INCREASED THE RATE OF

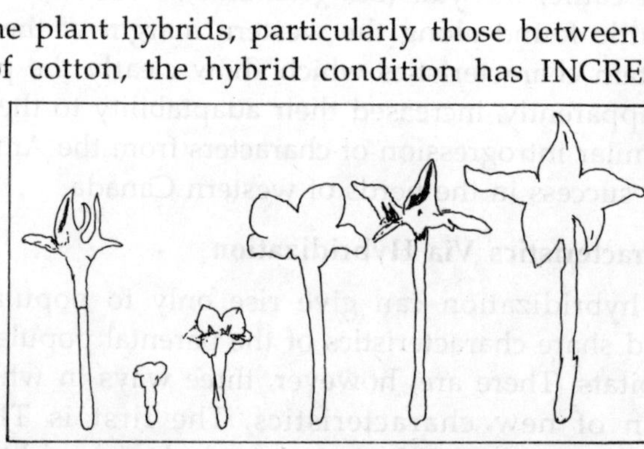

Fig. 5.5: Flowers of Nicotiana types: (I. to r.) Sanderae; N. langsdorffi; F_1 hybrids; large-flowered selecttion, F_8; backcross of large-flowered selection, F_8, to N. sanderoe; large-flowered selection, F_{10}.

MUTATION. Since as pointed out in Chapter 2, the rate of mutation is not correlated with the rate of evolution, the significance of this increase in mutation rate is hard to estimate. It should, however, help to increase the size of the gene pool from which natural selection, acting in a new environment, could sort out entirely new gene complexes adapted to this environment.

The significance of hybridization in evolution can be summarized by saying that it may provide a rapid increase in the size of the gene pool, from which natural selection for adaptation to a new habitat might quickly sort out entirely new adaptive gene complexes. This potentiality is often realized in plants. In animals, which have much more complex patterns of development, controlled by equally complex and highly integrated sequences of gene action, the proportion of successful recombinants which can segregate in hybrid progeny is much lower. Consequently, the restrictions on successful hybridization are far greater than in plants, and the adaptive recombinations which can segregate from hybrids are far fewer. Because of these facts, most zoologists tend to minimize the importance of hybridization in evolution. One must remember, however, that natural selection, particularly when it is adjusting a population to a new, poorly occupied habitat, can seize on gene combinations which occur at very low frequencies and in a few generations establish them in a population. Hence there are reasons for believing that hybridization has played a larger role in animal evolution than many zoologists think it has. This may be particularly true of hybridization between subspecies and closely related species, in which hybrid inviability and sterility may be only weakly developed.

The Role of Polyploidy in Plant Evolution

In higher plants, adaptive hybrid combinations are often stabilized by polyploidy. This process consists of doubling the chromosome number so that every chromosome present in the original diploid is exactly duplicated in the polyploid hybrid. If, therefore, the hybrid is sterile only because its parental chromosomes are too dissimilar to pair properly, this difficulty is removed in the hybrid polyploid or allopolyploid, and the plant is therefore fertile. Furthermore, since pairing in such plants is entirely or chiefly between two chromosomes derived from the same parent, segregation of the parental characteristics is eliminated, and the polyploid breeds true for its intermediate condition.

One of the best known examples is the rado-cabbage, derived from the sterile hybrids between the cultivated radish (*Raphanus sativus*) and the cultivated cabbage (*Brassica oleracea*). These two species differ radically in their fruits, (Figure 5.6) and their chromosomes are so different that they cannot pair at all in the undoubled F_1 hybrid. Consequently, although both parents from nine pairs of chromosomes at meiosis, the hybrid contains 18 unpaired chromosomes at this stage, which are distributed irregularly to the gametes. Doubling the chromosome number produces 18 pairs of similar chromosomes, so that meiosis in the allopolyploid is regular. The plant is fertile, and breeds true for the peculiar, intermediate shape of its pod.

In other examples, such as the hybrid between two species of primroses, *P. verticillata* and *P. floribunda,* the parental species are closely enough related to each other so that the chromosomes can pair in the undoubled F_1 hybrid. These chromosomes are, however, sufficiently different in structure so that crossing over and chromosomal segregation at meiosis produce gametes with disharmonious gene combinations resulting in segregational sterility. When the chromosome number is doubled, the exactly similar chromosomes derived from the same parental species pair preferentially with each other so that the doubled hybrid is fertile and does not segregate in the direction of either parental species (Fig. 5.7).

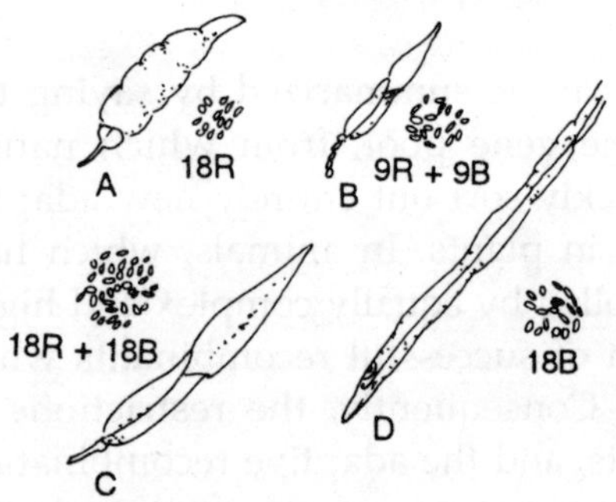

Fig. 5.6: Seed pods and somatic chromosome complements of radish, cabbage, the diploid hybrid between them, and the stabilized, true breeding new species produced by doubling the chromosome number of the sterile hybrid. From Karpechenko.

Among natural, wild polyploid species, there are many examples which resemble the rado-cabbage, others which are like the primrose example, and still others which are intermediate between the to in various ways. In addition, some natural polyploids are derived by doubling the chromosome number of a single species, usually a cross between two of its subspecies. These polyploids have four sets of similar chromosomes. This makes their segregation much more complex, and in progeny of a cross between two different autopolyploids of this type, intermediate genotypes are much more frequent relative to the parental genotypes than they are in a cross between two diploids. In this way, any hybrid vigor which may exist in interracial hybrids is buffered by the effects of polyploidy and is more easily maintained by natural selection than it is in diploid hybrids.

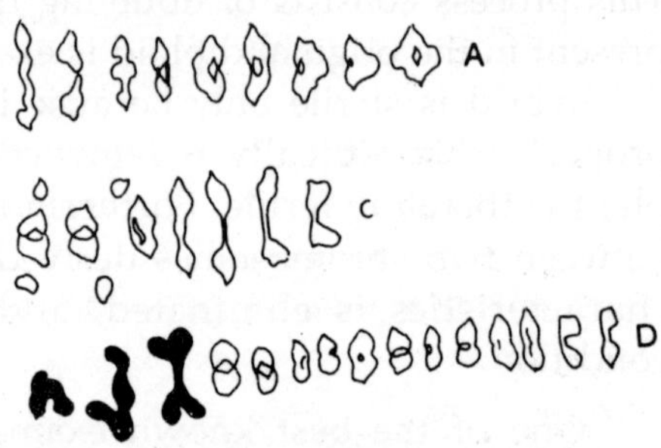

Fig. 5.7: Plants and meiotic chromosomes of two species of primrose, *P. verticillata* and *P. floribunda,* and the tetraploid species (*Primula kewensis*) produced by doubling the chromosome number of the hybrid between them. Photograph of plants supplied by the late A. B. Stout, New York Botanical Garden; chromosome drawings from M. B. Upcott in Journal of Genetics.

Once polyploidy has begun to develop in a genus of plants, it is likely to be a continuing process. A doubled hybrid derived from two original diploid species is called a tetraploid. It may cross back to either one of its parents or to a third species to form a sterile triploid. If this hybrid is doubled, a fertile hexaploid results. Two different tetraploids may similarly cross. Doubling their hybrid will produce an octoploid. In some groups of plants, this trend has continued even further, so that some species have chromosome numbers as high as ten times that of their original progenitor.

Polyploidy is a very common method of evolution in higher plants. Between one-fourth and one-third of the species of flowering plants are polyploid with reference to their nearest relatives. Familiar examples among crop plants are wheat, oats, potato, tobacco, cotton, alfalfa, and most species of pasture grasses. Familiar weeds and wild flowers which are polyploid are the eastern blue flag (*Iris versicolor*), meadow rue (*Thalictrum* spp.), some species of wild rose, chickweed (*Stellaria media*), miner's lettuce (*Montia perfoliata*), yarrow (*Achillea* spp.), and various species of violets and asters. Furthermore, there is evidence that polyploidy in the remote past has given rise to many genera and groups of genera such as the apples, olives, willows, poplars, and many genera of ferns.

Nevertheless, polyploidy has contributed little to progressive evolution. In genera which contain both diploids and polyploids, the major trends of evolution are all represented by diploid species, and the polyploids serve merely to multiply the variations on certain particular adaptive "themes." This is probably because the large 'amount of gene duplication dilutes the effects of new mutations and gene combinations to such an extent that polyploids have great difficulty evolving truly new adaptive gene complexes.

Polyploidy is rare in animals, and is confined almost entirely to forms which reproduce asexually. This is partly because the polyploid condition complicates the segregation of the sex chromosomes, so that many partly or wholly sterile intersexes are produced. In addition, many polyploid animals have disturbed developmental processes, or internal organs (kidneys, nerve centers), which function more poorly than in the related diploids. These forms, even when produced artificially, are hard to keep alive.

Polyploidy is a very striking phenomenon in higher plants, and studies of polyploids can provide much information about the past history of plant groups. This is because it is one of the very few evolutionary trends of which the direction can be clearly determined without fossil evidence. Furthermore, since polyploids can quickly be produced in a single step, many stages of polyploid trends can be reproduced under controlled experimental conditions. Nevertheless, the strong limitations on polyploid evolution render it of only secondary importance in considerations of the phenomenon of evolution as a whole.

Chapter Summan

An evolutionary definition of hybridization is crossing between members of population having different adaptive gene complexes. When such hybridization takes place, the majority of the progeny in later generations are more poorly adapted to any environment than were the parents of the hybrid. Nevertheless, a small proportion of segregates from such hybrids may be better adapted to a new environment than were any members of the parental population. Consequently, when hybridization takes place in well filled, undisturbed habitats, the hybrids are quickly eliminated and no important results occur. If, however, hybridization is accompanied by the opening up of new habitats, hybrid derivatives may enter these habitats and establish new adaptive gene combinations which contain genes derived from both of the parental species. In modern biota, examples are found most often in habitats disturbed by man, or which were strongly affected by the advance and retreat of glaciers during the ice age.

For hybrid progeny to be successful, they must become stabilized either through establishment of true breeding intermediate gene combinations, backcrossing and selection, or, in plants, doubling the chromosome number. When reproductive isolating mechanisms are determined by many genes, the possibility exists that a new gene combination will be formed which will form partly sterile hybrids when back crossed to either of the parental species populations and thus can become the beginning of a new species. This has been the origin of some species of plants.

New characteristics can arise as a result of hybridization through 1. transgressive segregation, 2. the establishment on new genetic backgrounds of mutations which were consistently rejected by selection on the backgrounds of the parental species, and 3. the effect of hybrid gene combinations which stimulate the occurrence of new mutations.

In higher plants, many new species and some genera have originated by means of hybridization followed by chromosome doubling. This is called hybrid polyploidy. It produces many species which are variations on an established theme of adaptation, but rarely gives rise to new modes of adaptation.

Questions for Thought and Discussion

1. In what way or ways will the environmental in which the process takes place affect the outcome of interspecific hybridization?
2. In what ways can the progeny of interspecific hybrids become stabilized so that they will breed true and give rise to new adaptive populations?
3. Discuss the ways in which new characteristics can arise in the progeny of hybrids.

Chapter—6

How does Evolution Differ from Development

Classification of Developmental Processes

Many different kinds of processes have been suggested which could lead to an increase in geometrical complexity. I start by offering a classification of such processes. It is not exhaustive, nor am I convinced that it is arranged in the most illuminating way. It is offered in the hope of provoking others into giving a more complete and rational classification.

We have a good idea of how cells can produce sets of specific proteins. Since Monod & Jacob (1961), we can imagine how different cytoplasmic constituents of cells can activate different sets of genes, thus amplifying initially small differences. Although such differentiation must be both more complex and more stable in eukaryotes than in prokaryotes, there are no particular conceptual difficulties in imagining how it might happen. The major problem lies in how different kinds of cells come to be arranged in space in particular ways. I suggested the following classification of the processes responsible for such spatial arrangement:

A. Molecular jigsaws.

B. Cellular mechanisms.

1. Cell division and determination coincident in time.
2. Cell sorting.
3. Determination subsequent to cell division:
    - (i) causes external to tissue (e.g. induction),
    - (ii) causes internal to tissue:
        - (*a*) positional information
        - (*b*) prepattern and competence
        - (*c*) cell shape generates tissue shape.

A. Molecular Jigsaws

We have a rather good understanding of how the primary structure of proteins is determined, and some reason to think that the primary structure determines the three-dimensional shape of the whole molecule. It is then possible that larger-scale structures arise because proteins fit together, like pieces of a jigsaw. If so, the final structure would require a specification of which proteins were to be made (i.e. which genes were to be switched on), in approximately what proportions. It could also depend on the sequence in which proteins are made, and possibly on the pre-existence of some structure analogous to the seed of a crystal. It would require that physical conditions (e.g. temperature, pH) be kept within certain bounds.

It seems certain that such processes play a role in the formation of viral structures, and probably also of subcellular organelles. I doubt if they are important on a larger scale, except insofar as the existence of such structures as ribosomes is a necessary precondition for the development of larger-scale structures. However, in those cases in which there is a large amount of cytoplasmic localization in the egg before cleavage, we cannot rule out that some of the localization arises by a jigsaw-like process.

B. Cellular Mechanisms

1. Cell Division and Determination Coincident in Time

The fate of two cells may become different at the time of cell division because of pre-existing cytoplasmic localizations. An alternative possibility is that the state of activation of genes becomes different at the time of chromosome replication. Cytoplasmic localization is known to be important in most eggs, and is probably imposed from outside by the female laying the egg. Determination occurring at the time of cell division is also involved in the differentiation of, for example, the haemoipoetic system.

If all division merely partitions, between cells, cytoplasmic localizations which already existed, the increase in geometric complexity generated would be small. Alternatively, we can suppose that new localizations are generated within daughter cells, which in turn specify new planes of division and new differentiations. If so, there could be a continued increase in complexity. The difficulty lies in imagining how successive differentiations could be coordinated on a global scale; consequently, such a process would be unlikely to be able to regulate.

2. Cell Sorting

Suppose that cells of two or more kinds are produced, and subsequently move relative to one another. If the cells have differing degrees of affinity for one another, analogous to the interlocking of jigsaw pieces, then they will arrange themselves in regular patterns, regardless of their initial arrangement. There remains the question of how appropriate numbers of different kinds of cells arise in the first place; cell sorting will merely arrange them in space once they are formed.

3. Determination Subsequent to Cell Division

I now consider cases in which a population of cells—e.g. a sheet of cells—are initially identical in their potentialities, and become different only later.

(i) Causes External to the Differentiating Tissue

In embryonic induction, a set of initially similar cells become different because some of them come into contact with another cell layer. In effect, the folding of one cell layer can induce increased spatial complexity in another. This raises the possibility of serial induction: each group of cells, once differentiated, imposing differentiation on the next group.

(ii) Causes Internal to the Differentiating Tissue

Since Turing (1952), it has become possible to imagine how an initially homogeneous region can become heterogeneous, without the need of a localized external influence. Turing imagined this happening through chemical reaction and diffusion, without flow, active transport, or electrical phenomena, and without need for cellularization. However, we can, in principle, imagine a variety of pattern-forming processes. In effect, dynamical processes in space rather readily give rise to wave-like inhomogeneities; if the wave-length is large relative to the size of the field, the resulting pattern will be gradient-like rather than wave-like.

I will not attempt to list the various physical and chemical processes which have been proposed. However, some general points are worth making. First, the pattern-generating process need not depend on cellularization. Secondly, the patterns which arise are likely to be simple—e.g. waves, gradients. Thirdly, there is no need to postulate any initial inhomogeneity, other than the presence of boundaries to the developing fields. But some models do assume the existence of some privileged region or boundary, which may, for example, act as a source or sink.

The interactions between the morphogenetic field and the cells it influences can take several forms (Fig. 6.1).

(*a*) *Positional information.* Wolpert (1969) imagines a process giving rise to a monotonic gradient (or to two orthogonal gradients). The value of some scalar quantity (or quantities) uniquely specifies each position in the field. Cells then respond to the local value of this scalar by differentiating appropriately. All the complexity of the final pattern, over and above that manifest in a monotonic gradient, is then generated by the capacity of cells to respond differently to different values of the positional scalar.

(*b*) *Prepattern and competence.* Stern (1954) and Maynard Smith & Sondhi (1961) were primarily concerned to account for patterns composed of one kind of element, or of a few kinds, arranged in repeated patterns. It was therefore natural to imagine that some field process gave rise to a standing wave of values of some scalar quantity, and that cells responded by giving rise to (e.g.) a bristle if the value rose above some threshold. Thus less of the final complexity was supposed to reside in the differential response of cells, and more in the field itself (prepattern). Logically, there is no reason why a row of bristles should not be generated by a monotonic gradient with cells responding by forming a bristle in response to (say) odd number values, and by not doing so in response to even values, but it seems implausible.

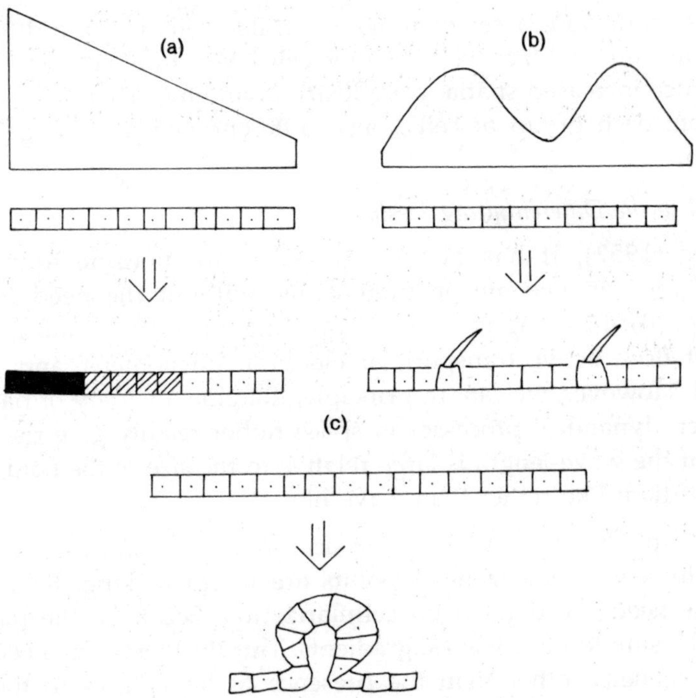

Fig. 6.1: Three kinds of field process: (*a*) positional information (Wolpert, 1969); (*b*) prepattern and competence (Stern, 1954); (*c*) elastic instability (Oster *et al.*, 1980). Shading indicates monotonic gradient.

(*c*) *Cell shape generates tissue shape.* Both positional information and prepattern models postulate the existence of some invisible spatial inhomogeneity, which elicits differentiation in cells locally, presumably by activating different sets of genes. Oster *et al.*, 1980 has proposed a model of pattern formation in which all the cells composing a sheet have the same tendency to change shape. However, since they are stuck to one another and to a basement membrane, changes of shape in one cell cause changes in others to which it is attached. A homogenous sheet of cells may be mechanically unstable, and may generate patterns such as placodes or invaginations, without further gene activation. In such a mechanism, there is no distinction between positional information and competence to respond.

The Hierarchical Nature of Development

The processes listed above are not mutually exclusive. I see no reason to think that only one basic kind of developmental process occurs, any more than the occurrence of nerve conduction rules out hormonal mechanisms in physiology. It seems to be well established that the increase in geometrical complexity occurs in a stepwise fashion. By considering the accuracy with which fixed numbers of parts can be produced, I have argued (Maynard Smith, 1960) that one reason why this must be so is that there is a limit to the amount of additional complexity that can be generated in one step; the limit is set by the accuracy with which scalar quantities can be specified.

Our picture of development, then, is one of a succession of qualitatively different kinds of process, arranged hierarchically in time, with the products of one step forming the starting point for the next. For reasons of accuracy and repeatability, only a limited increase in spatial complexity is achieved in each step. The accuracy of the process as a whole depend on the accuracy of gene replication, and the accuracy with which specific sets of genes can successively be associated.

Evolution and Development

Evolutionary biologists see evolution in a variety of ways. To a palaeontologist, evolution typically presents itself as a succession of adult forms; to a geneticist, evolution is a change in the frequencies of genes in populations, leading to the substitution of one allele by another, at a locus. These are complementary views of the same process. However, if we are interested in mechanisms rather than outcomes, I think the geneticist's view is more fundamental, for the reason stated by Weismann. If a gene in a fertilized egg is altered, that will lead to a change in the resulting adult, and to the transmission of a similarly changed gene to the next generation of eggs. In contrast, if the structure of an adult is changed, that will not lead to changes in the eggs produced which are of a kind to cause the development of a new generation of adults with the changed structure. The difference is crucial; I will discuss in a moment whether it is always true.

The issue is essentially about the nature of heredity. Heredity is a property of entities which multiply, and which vary. Thus, suppose there are entities of kinds *A*, *B*, *C* etc. If, when they multiply, with high probability *A*'s give rise to *A*'s, *B*'s to *B*'s, and so on, the entities possess heredity. Only such entities can evolve by natural selection. Hence, to an evolutionist, the interesting question about, for example, cytoplasmic localization is not whether such localization is essential for proper development, but whether, if the localizations are changed, the result will be an adult which produces eggs with similarly altered localizations.

In general, the answer to such questions is no. There are a few well-established exceptions, of which the phenomenon of 'cortical inheritance' in ciliates is perhaps the most important. Neo-Darwinists should not be allowed to forget these cases, because they constitute the only significant experimental threat to our views. However, the overwhelming majority of inherited difference between similar organisms are caused by differences between chromosomal genes. Further, most cases of 'cytoplasmic' inheritance turn out to depend on non-nuclear DNA or RNA, a fact which encourages a still more gene-centred and less organism-centred view of evolution.

Provisionally, then, I continue to hold a Weismannist and gene-centred view of evolution. It is worth asking *why* heredity should be Weismannist and not Lamarckian. The following answers are possible.

(i) Weismannism is true because the 'central dogma' of molecular biology is true; i.e. because information does not pass from proteins to nucleic acids. I think this is correct. However, it leaves open the question of why a hereditary mechanism with a one-way flow of information has been retained. After all, the hereditary mechanism itself is subject to evolutionary change, and I see no reason in principle why 'reverse translation' should be impossible; some information channels can work both ways.

(ii) Most 'acquired characters' are the results of injury, disease and old age. Therefore, a mechanism able to transmit such changes would lower the fitness of offspring. I think this is also correct, and may be the 'functional' explanation of why inheritance is rarely Lamarckian. If an organism had some way of telling which of its acquired characteristics were adaptive, a mechanism for transmitting them would be favoured. Of course, this is precisely what happens with learning and cultural inheritance, but such a mechanism could hardly be primitive.

(iii) Development is 'epigenetic' rather than 'preformationist', and this makes it hard to see how Lamarckian inheritance could work. I think this was the main reason why Weismann himself was a Weismannist, and the argument has recently been reformulated by Dawkins (1982). Thus, if development consisted merely of the growth in size of a preformed structure, gamete formation could be merely a reversal of this process; it is true that a gamete could hardly be formed merely by shrinking an adult, but some process resembling Darwin's 'pangenesis' might have the same effect. If so, when an adult structure was altered, the alteration might be transmitted. But development is not like that. It is much harder to imagine how, for example, the hypertrophied muscles of a blacksmith could cause him to produce altered sperm which, although themselves not containing anything directly corresponding to muscle size, would cause his children to have larger muscles. As Weismann remarked, if one were to come across a case of the inheritance of an acquired character, it would be as if a man were to send a telegram to China, and it arrived translated into Chinese.

This argument is, of course, not decisive. We do not understand how genetic information is translated into adult structure, but are happy to assume that it is. We cannot, therefore, rule out the possibility of reverse translation merely because we cannot see how it could work. It is best to regard the Weismannist nature of inheritance as a contingent rather than as a necessary fact about nature.

Genes, Development, and the Conservation of Form

Comparative anatomists have taught us to regard similarities of form as a better guide to relationship than similarities in habits, and the fossil record demonstrates, at least for vertebrates, that the relationships being revealed are in fact phylogenetic relationships. It follows that anatomical patterns are more conservative in evolution than are ways of life: the pentadactyl limb is more conservative than the habit of running, climbing, flying or swimming. Why should this be so?

It is easy to see how changes in genes could cause changes in the structures produced by any of the processes listed above. For example, Stern's concepts of prepattern and competence were formulated with the specific aim of asking whether genes causing particular differences in *Drosophila* acted by changing the prepattern or the cellular response. He concluded that the differences were usually of competence; Sondhi and I agreed, but argued that some of the changes we observed were more plausibly interpreted as arising from changes in the prepattern. There is, in any case, no difficulty in principle in seeing how gene changes could cause prepattern changes. If, for example, one imagines the formation of chemical waves as proposed by Turing (1952), then any gene coding for a protein affecting either the reaction or diffusion rates could alter the prepattern. For the positional information model, in its extreme form, hereditary change can arise only by changing the genes which affect the responses of cells to unchanging positional information; it is for this reason that, as an evolutionist,

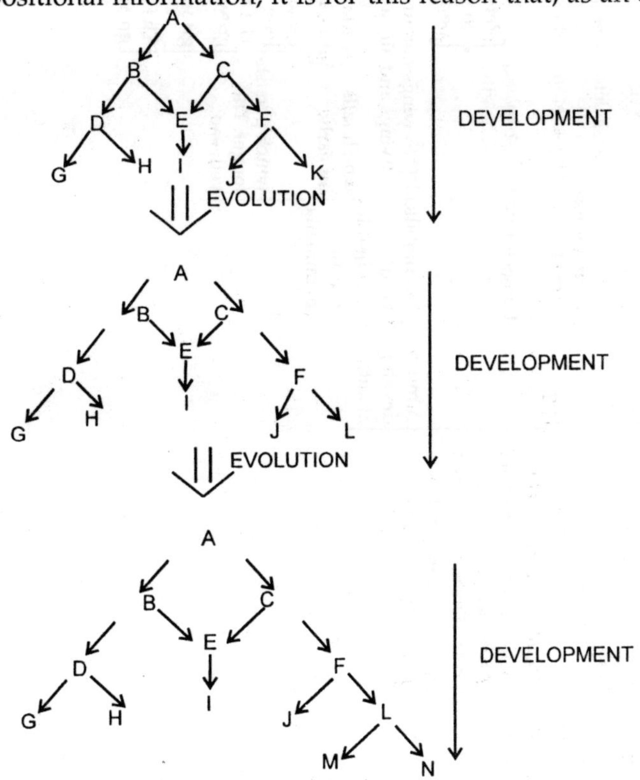

Fig. 6.2: The hierarchical nature of development. Each hierarchical tree represents the development of an individual, the three trees representing successive generations. Arrows within each hierarchy represent developmental processes, and letters represent resulting structures.

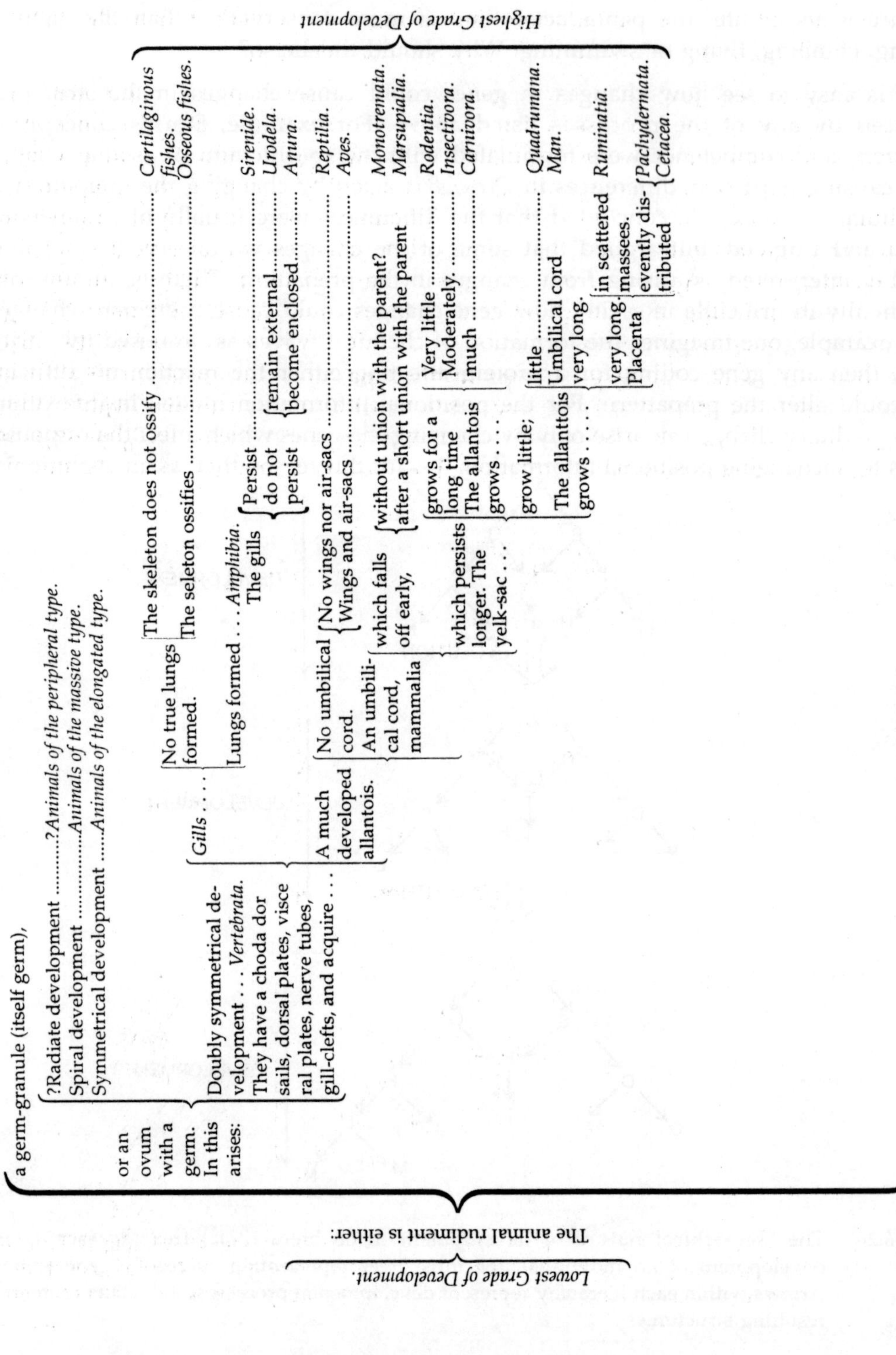

Fig. 6.3: Scheme of the progress of development in vertebrates. (From von Baer, 1828, facing p. 225, in Huxley's 1853 translation, p. 215.

I find the model less appealing, although I am ready to agree that prepatterns may often take the form of monotonic gradients.

If development is hierarchical, as most of us would think, then evolutionary changes could occur at any stage. However, in practice it is hard to see how viable alterations could occur except as terminal alterations or additions (if development passes through a free-living larval stage, neotenic changes are also a possibility). Consider the hierarchy in Fig. 6.2, in which each arrow represents a process and each letter a resulting structure. An alteration early in the hierarchy would be unlikely to be compatible with life. It is easier to imagine an alteration or addition at the end being compatible with survival. In the case of organisms with free-living larval stages, the omission of later stages (i.e. neoteny) is also a possibility.

If so, the conservation of morphological patterns becomes comprehensible. Conservation requires the accuracy of DNA replication, but by itself that is not a sufficient explanation, because it cannot explain the conservation of pattern despite change of function.

If we say that adult structure A in species 1 and structure B in species 2 are 'homologous', we may mean any one of a number of different things. For example,

(i) structures A and B have similar topological relationships with other structures,

(ii) the modes of development of A and B are similar,

(iii) structures A and B are connected by a series of small gradations to a structure C in an common ancestor of species 1 and 2.

It is rare for us to have direct evidence of C. More commonly, (iii) is a hypothesis, for which (i) and (ii) provide supporting evidence. It might seem that (i) and (ii) are circular. For example, a bone is called an articulator if it articulates with a quadrate, but to decide whether it does one must first know what is a quadrate. This objection does not seem to me particularly serious. In practice we recognize the isomorphism between two sets of objects, and regard as homologous objects occupying corresponding positions in the two sets. Similarly, we recognize the hierarchical patterns of development of two organisms as being isomorphic, and then treat as homologous structures derived from corresponding elements of the hierarchy (in Fig. 6.2, $K \equiv L \equiv M + N$).

On this view, anatomical conservatism is a consequence of the conservation of the hierarchical structure of development, which in turn is a consequence of the fact that alterations in the early stages of such a hierarchy are almost certain to be incompatible with life. It follows that there are, indeed, constraints on the kinds of structures which can evolve from any given ancestor. However, these constraints are historical and contingent in nature, arising from the existing pattern of development. The alternative to this view would be that there are absolute 'laws of form', not historically contingent, determining what kinds of organisms can and cannot exist. I can see little to recommend such a view.

Conclusions

It has never been part of Darwinism to claim that phenotypic variation is random or unconstrained. In any given taxon, some variations occur commonly and others rarely or not at all. Darwin himself noted this phenomenon, and coined for it the term 'analogous variation'. There are, however, two senses in which a contemporary Darwinist would claim that variation is unconstrained. First, at the molecular level, we know of no constraints on the kinds of changes in DNA sequence that can arise by mutation. Secondly, the phenotypic changes resulting from a mutation do not in general adapt the organism to withstand the agent which caused the mutation. There is, therefore, nothing 'un-Darwinian' about the claim that there are developmental constraints. Our difficulty is that we can rarely predict the nature of those constraints from a knowledge of development.

The question of sudden versus gradual changes in evolution is more controversial. It is known that single mutational events can and do give rise to large phenotypic changes, as a visit to any genetics laboratory will establish. What is at issue is how often such mutations are incorporated in evolution. The question is ultimately an empirical one. The most fruitful approach is through a genetic analysis of related species which differ morphologically. Such analyses do little to suggest that mutations of large effect have been important (Charlesworth *et al.*, 1982), but it would be wise to keep an open mind. However, two assertions can be defended on theoretical grounds. First, a mutation of large phenotypic effect is unlikely to be incorporated in evolution unless it increases the fitness of its carriers. Secondly, 'systemic mutation' as envisaged by Goldschmidt (1940)—i.e. complex new adaptations arising without selection from a restructuring of the genome—can be ruled out on probabilistic grounds.

I cannot see that developmental constraints can help to explain 'stasis' (Gould & Eldredge, 1977)—i.e. the morphological constancy of species over long geological periods. Thus it may be that variation occurs in some characteristics more readily than others, or that, in some cases, variation is discontinuous, permitting the existence of only a discrete subset of the conceivable phenotypes. However, it is manifestly *not* true that heritable variation is absent altogether. Indeed, many traits which show stasis in the fossil record vary in contemporary populations. One of the few generalizations on which geneticists agree is that, if a sexual population is exposed to selection for a phenotypically variable trait, then the population will change (for a rare exception, see Maynard Smith & Sondhi, 1960). Hence, if such traits remain constant for long periods, it must either be because no selection operates on them, or because selection is normalizing. Whatever one's view may be on electrophoretic variability of proteins, no naturalist would accept that morphological variations are selectively neutral.

Thus we must distinguish between the idea that developmental constraints are responsible for the conservation of major morphological patterns, which is plausible, and that such constraints are responsible for the constancy in time of traits which vary in contemporary populations, which is not.

This brings me to my final point. If we are to understand evolution, we must remember that it is a process which occurs in populations, not in individuals. Individual animals may dig, swim, climb or gallop, and they also develop, but they do not evolve. To attempt an explanation of evolution in terms of the development of individuals is to commit precisely that error of misplaced reductionism of which geneticists are sometimes accused.

How shall We Depict Phylogeny?

Gould's answer to this is: by a series of comparable or standardised stages of development, conventionally a sequence of adults. I want to look at his question from a different angle, and to consider depictions of phylogeny which employ conventions rather than organisms. There are three principal models in use, which I will call trees maps (e.g. Fig. 6.4) and spindle diagrams (e.g. Fig. 6.6). The last two examples, and other pre-Darwinian examples of all three models reproduced by Nelson & Platnick (1981), show that these models have no *necessary* phylogenetic implication.

I will concentrate first on Milne-Edward's map (Fig. 6.4) and Agassiz's spindle diagram (Fig. 6.6), both published in 1844. With his map, Milne-Edwards was arguing, like von Baer, for the importance of embryology in systematics. He used the tree metaphor to explain his views, describing ontogenies as "a multitude of. . . series which appear to branch from one another at different heights, or rather which are joined in a bundle at their base, and break up into secondary, tertiary and quaternary bundles, as in ascending towards the end of embryonic life they deviate from one another and take on distinct characters' (Milne-Edwards, 1844, my translation). Milne-Edwards compared (*ibid.*) his diagram (Fig. 6.4) to the map of 'the islands of a vast archipelago' within which 'one could distinguish secondary archipelagos' His map is readily translated into a hierarchical classification, resembling von Baer's diagram (Fig. 6.2), or into a branching diagram (Fig. 6.5; cf. Nelson & Platnick, 1981) which is a cladogram rather than a tree, since it carried no implication of descent for its author. Comparing the cladograms derived from Milne-Edwards's and von Baer's diagrams (Fig. 6.5), the two differ in the von Baer's unites birds and reptiles, whereas Milne-Edwards places those two groups in a trichotomy with mammals. However, the character von Baer used to unite birds and reptiles was 'no umbilical cord' (Fig. 6.3), that is, no placenta. Aristotle called such negative characters 'privative,' and argued that they could not logically characterise groups, since they cannot be further subdivided: absence is merely absence (Nelson & Plantick, 1981). In von Baer's scheme (Fig. 6.3), several other 'groups' are characterised in this way, by absence: fishes (no lungs), perennibranchiate amphibians (no transformation of gills), and reptiles (no wings or air-sacs). Milne-Edwards's group Anallantoidinens (Anamniota in Fig. 6.5A) has the same disadvantage, since its characters are 'ni allantoide ni amnios' (*idem* 1844, p. 89). Use of such groups in a phyletic context is discussed on p. 18, below.

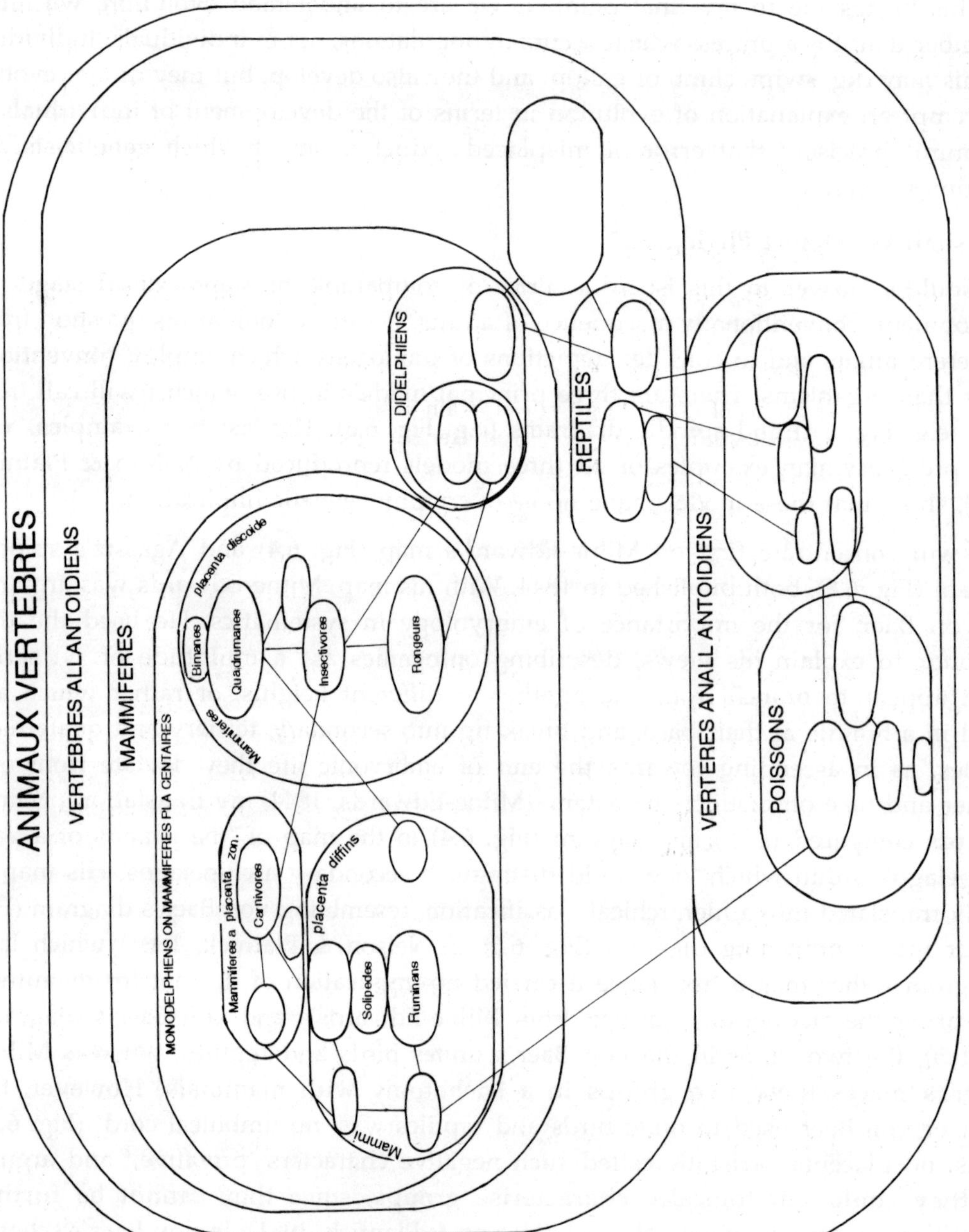

Fig. 6.4: 'Essai de distribution naturelle' of vertebrates. (From Milne-Edwards, 1844, facing p. 98.) The relative positions of the groups and the lines joining them symbolise 'the various degrees of affinity' between them.

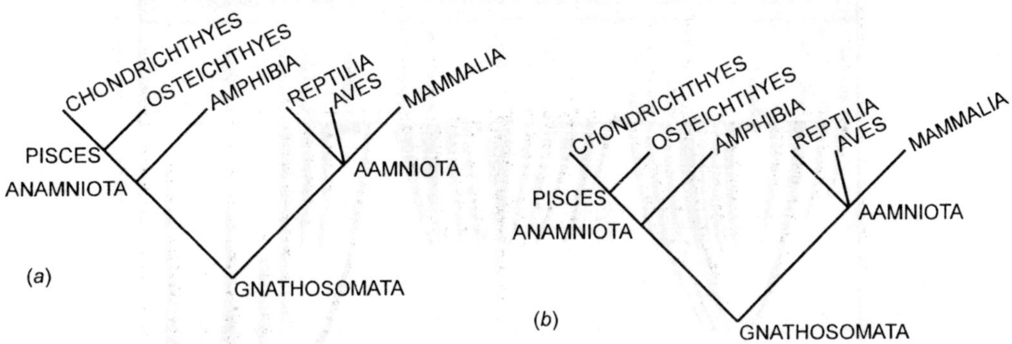

Fig. 6.5 Cladograms of the main groups of gnathostome vertebrates derived from (a) Milne-Edward's map (Fig. 6.4) and (b) von Baer's scheme (Fig. 6.3).

Agassiz's spindle diagram (Fig. 6.6) is called a 'généalogi'. Whereas Milne-Edwards used his map to illustrate the importance of von Baer's law for systematics, Agassiz presented his diagram in an essay on classification of fishes which began with a comparison between systematists and map makers (Agassiz, 1844). Agassiz goes onto discuss the correlation he observed between the sequence of features in ontogeny and the fossil record. This parallel between palaeontology and embryology was generally seen as an indication that von Baer's laws of divergence and individualisation applied to both (Ospovat, 1976; 1981). Discussing ways of depicting the relations between species, Agassiz (1844, vol. I) rejects linear series, and recommends diagrams in which clear-cut divisions are each centered about 'les types mieux connus', around which can be grouped other 'types' which may, in turn, serve as centres for groups of subsidiary rank. Palaeontological information may be added to these zoological maps ('cartes') to produce genealogical trees ('arbres'), showing the time of appearance of each group, and its importance in each geological period. Thus Agassiz's diagram can be seen as a version of Milne-Edward's map (creation actuelle' in Fig. 6.6), with the time dimension added as a summary of the fossil record. Unlike Millne-Edwards' diagram, Agassiz's cannot be readily converted into a tree, except perhaps by continuing the converging lines downwards until they meet. Agassiz avoided this, for the implication of common descent would be inevitable, and Agassiz expressly denied it. Diagrams like Agassize's are frequently modified in this way today (examples in Romer, 1966, and may other modern works).

Common descent was added to the fossil record in Haeckel's (1866, 1868) 'palaeontologisch begrundet' trees (Figs. 6.7). In these early trees the trunk and main branches are thick (Fig 6.8), as in Milne-Edward's simile of the bundle ('faisceau') at the base, and are labelled with names or numbers using five different conventions of ancestry. The first is names of higher taxa whose rank decreases as one ascends the tree (e.g. Craniota, Amphirrhina (= Gnathostomata), Amniota): the convention here is that a groups gives rise to its subgroups. Secondly, there are groups which give rise to taxa of equal or higher rank (e.g. Acrania to Craniota, Anamnia to Amniota both of

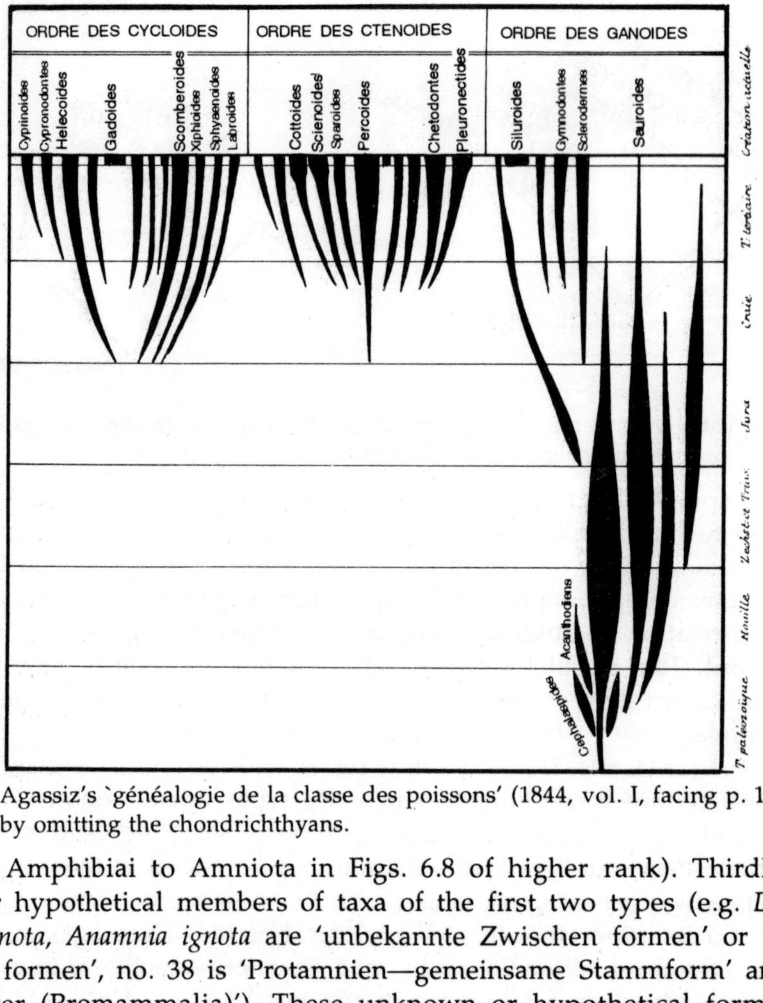

Fig. 6.6: Agassiz's `généalogie de la classe des poissons' (1844, vol. I, facing p. 170), modified by omitting the chondrichthyans.

equal rank; Amphibiai to Amniota in Figs. 6.8 of higher rank). Thirdly, there are unknown or hypothetical members of taxa of the first two types (e.g. *Dipnoi ignoti, Amphibiia ignota, Anamnia ignota* are 'unbekannte Zwischen formen' or 'unbekannte Uebergangs formen', no. 38 is 'Protamnien—gemeinsame Stammform' and no. 39 is 'Stammsauger (Promammalia)'). These unknown or hypothetical forms are really equivalent to the groups to which they belong, since 'hypothetical forms are themselves generalised classification groups' (Russell, 1916, p. 294). The fourth type includes extinct groups of rank lower than those to which they are ancestal (e.g. no. 12 in Fig. 6.7 is the genus *Onchus,* Silurian and Devonian finspines ancestral to cartilaginous fishes; no. 14 in Fig. 6.7 is the genus *Pteraspis,* ancestral to cartilaginous fishes; no. 14 in Fig. 6.7 is the genus *Pteraspis,* ancestral to bony fishes; no. 30 in Fig. 6.7 is the order Ganocephala, including the Carboniferous amphibians *Archegosaurus* and *Dendrerpeton,* ancestral to amphibians and amniotes). And the fifth type is groups inferred only from early ontogeny (e.g. Moneren, Amoeben, Synamoebien, Planaeaden, Gastraeaden in Fig. 6.8). As Gould (1977) points out, these 'groups' relied on no other evidence than early ontogeny, Monera being anucleate primordia, Amoeben the nucleate zygote, Synamoebien the morula, Planaeaden the planula and Gastraeaden the gastrula. Like unknown and hypothetical forms (type 3), these ancestral embryonic forms are

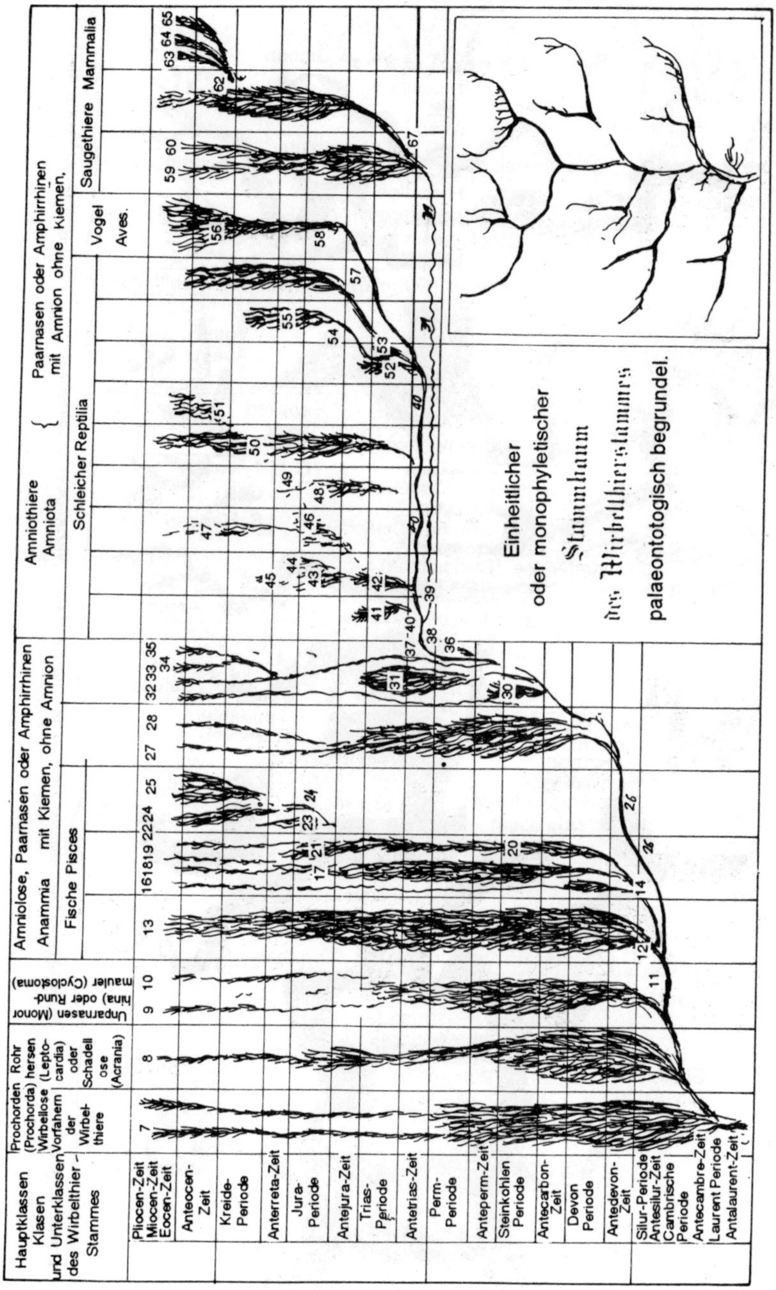

Fig. 6.7: Haeckel's monophyletic tree of vertebrates (1868, pl. 6).

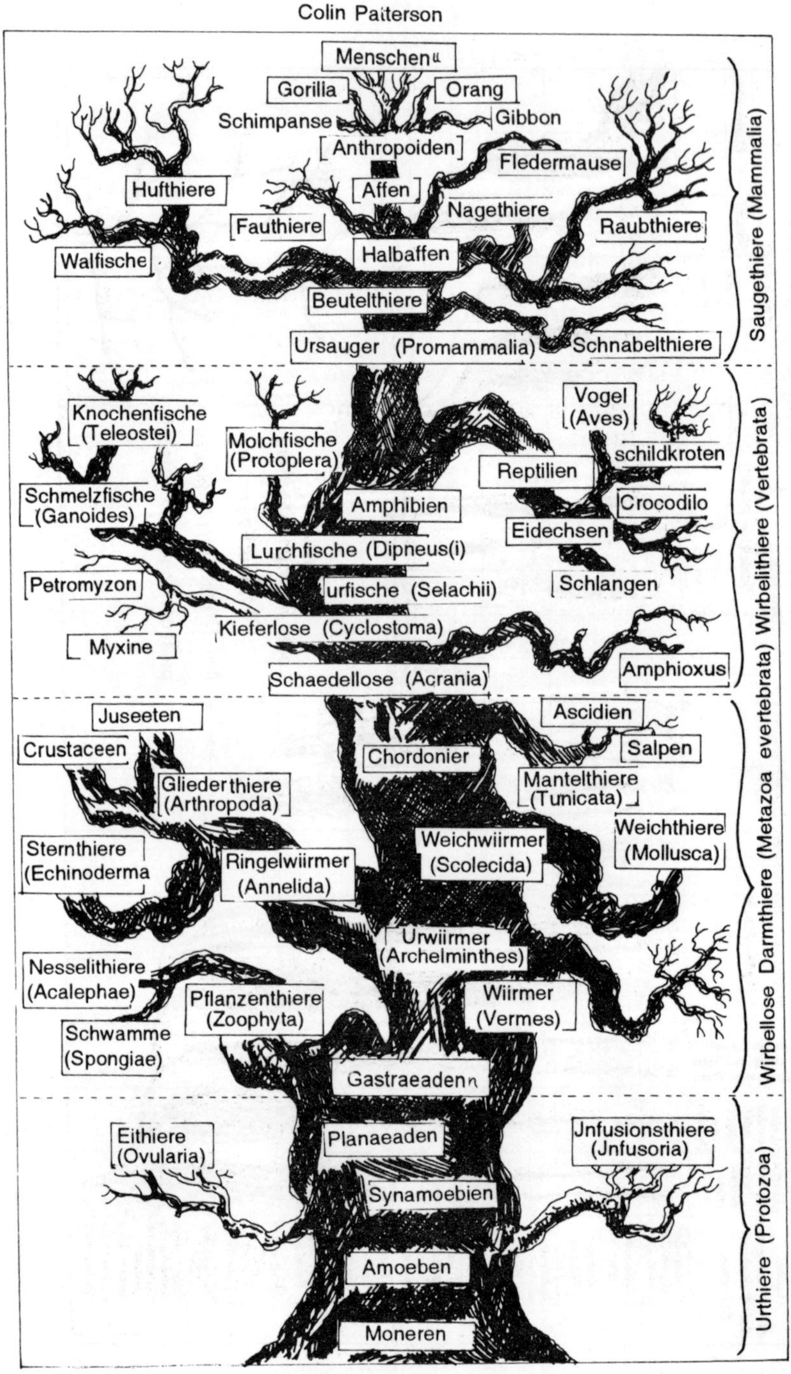

Fig. 6.8: Haeckel's phylogenetic tree of man (1874, pl. 12).

equivalent to the first two types, ancestral groups whose rank is higher than or equal to that of their descendants, since we could replace the ancestral embryos by the names of extant taxa without violating current concepts of phylogeny: Moneren = kingdom Monera (prokaryotes), Amoeben = kingdom Protista, Synamoeben + Planaeden = Metazoa, and Gastraeaden = Diploblastica (Hansen, 1977; cf. Tribe *et al.*, 1981).

The five different conventions that Haeckel used to denote ancestry do not exhaust the possibilities, for there is a sixth, ancestral species. There are no ancestral species in Haeckel's trees, yet this type of ancestry is the only one compatible with evolutionary theory, since species and groups of species are descended from species, not group (Wiley, 1979).

These six forms of ancestry may be categorised in different ways, but the distinction between phylogeny and ontogeny is demonstrated most clearly be splitting them into those which are consistent with von Baer's law of decreasing generality, or with a map like Fig. 6.4 or with a cladogram (Fig. 6.5), and those which are inconsistent with von Baer's law and with a cladogram, and can only be accommodated in a phylogenetic tree of historical descent. The first set, which may be called von Baerian conventions of ancestry, includes groups whose rank decreases as one ascends the tree and unknown or hypothetical members of such groups. The second set, which can be placed only in trees, may be called Haeckelian conventions of ancestry. They are extant or extinct taxa whose rank is equal to or lower than that of their descendants and unknown or hypotehtical members of such groups. Named ancestral species are components of trees, but they may also occur in cladograms, where an ancestral species will appear in a trichotomy or polychotomy as equal in rank to its descendant groups or species.

Von Baerian and Haeckelian ancestral groups differ in one significant way. Examples of von Baerian ancestry are the taxa named at the nodes in the cladograms in Fig. 6.5 and taxa like Craniota, Amphirrhina (=Gnathostomata), Amniota and Mammalia. The convention is that a group gives rise to its subgroups, and the pattern of character distribution is for the ancestal group to have certain characters, and the descendant groups to have certain additional characters. Examples of Haeckelian ancestry are Acrania, Anamnia and Amphibia, various extinct groups in Fig. 6.7 and Moneren (= prokaryotes) and Amoeben (= protists) in Fig. 6.8. The convention here is that a group gives rise to another of equal or higher rank, and the pattern of character distribution is that the ancestral group is always characterised only by lack of the features which distinguish the descendant group, i.e. by absent or privative characters. In terms of the hierarchical structure of the tree, rather than character distribution, Von Baerian ancestry is consistent with the hierarchical structure, which is one of decreasing generality, and Haeckelian ancestry is inconsistent with that structure.

Embryos and Ancestors

With the theory of evolution established, and the groundwork of phylogeny reconstruction laid by Haeckel's biogenetic law, the work of building genealogies forged ahead in the late nineteenth century, especially amonsts embryologists ,and then gradually fell into disrepute. By 1893, an editor could write 'three weeks at the seaside with borax-carmine and a rocker microtome, no longer result in an epoch-making paper containing a hypothetical ancestor, and a brand new pedigree' (Anon., 1893). Gould (1977) points out that one neglected factor in the decline of the biogenetic law was von Baer's demolition of recapitulation. In the nineteenth century, the only von Baerian criticism of Haeckelian recapitulation was Hurst's (1893*a-d*): 'The two views —von Baer's and the Recapitulation Theory—are irreconcilable. Von Baer's is applicable only to certain animals, but its applicability to these disproves 'recapitulation' in these' (Hurst, 1893*b*). Significantly, the only replies to Hurst came from palaeontologists, Bather (1893a, 1893b) and Buckman (1893). Baker (1893a) first demanded that 'ontogeny recapitulates phylogeny' be disproved by contrary examples 'from the actual history of extinct beings', that is, by 'direct comparison of the ontogeny with the phylogeny in a large number of cases,' as Hurst put it (1893b). Hurst went on 'that has hitherto proved impossible', though he did not say why. But what Bather demanded was phylogenies established without the help of the biogenetic law, and the only alternative method of establishing phylogenies was stratigraphic succession of fossils (e.g., Huxley, 1870; cf. Patterson, 1981). Indeed, Bather referred Hurst to 'the brilliant researches of Hyatt and Buckman on [fossil] Cephalopoda. . . which . . . have been inspired by the Recapitulation Theory' (1893a). Bather went onto accuse Hurst of the 'main fallacy' of substituting 'contemporary relations for ancestors', pointing out that 'there is all the difference in the world between filial and fraternal relationship.

This exchange illuminates the trend of phylogenetic research in the late nineteenth century, and Russell's superb survey (1916) takes the story 20 years on: 'There was a close spiritual affinity between the speculative evolutionists and the transcendentalists' (*ibid.*); 'Phylogenetic method differed in no way from transcendental—except perhaps that it had learned from von Baer and from Darwin to give more weight to embryology' (*ibid.*). Phylogenetic research fell out of favour because, under the influence of recapitulation theory, it led to the proposal of innumerable irreconcilable and uncriticisable speculations. Russell adds (*ibid.*) that 'It was a curious, but very typical characteristic of evolutionary morphology that its devotees paid very little attention to the positive evidence accumulated by the palaeontologists, but shut themselves up in their tower of ivory and went on with their work of constructing ideal genealogies.' Summarising the work of post-Darwinian palaecontologists, Russell notes first that intermediates between major groups were notably lacking (cf. Olson, 1981), but that close study of a few limited but complete series had confirmed the law of recapitulation, and had established 'the actual "phyletic series" (Russell, 1916), which, when generalised, showed 'an immense number of phyletic lines which evolve parallel to one another' (*ibid.*). And so speculative genealogies, built on embryos, fell into disrepute and were replaced by concrete phylogenies based on fossils—Haeckel''s 'infallible and indisputable records which fix the correct history of organisms' (Haeckel, 1876, vol. I).

The new view of phylogeny reconstruction that Russell was summarising—fossils as the arbiters, parallelism and polyphyly as the rule—was prevalent until the 1960s or later (e.g. Grasse, 1977). However, some biologists eventually came to regard fossils and their stratigraphic succession as no more foolproof than the biogenetic law (Hennig, 1969; Schaeffer *et al.*, 1972; reviews of particular groups in Patterson, 1977, 1981; Rosen *et al.*, 1981). The problem is that epitomised in the 1893 exchange between Hurst, who asserted that the biogenetic law is false, and Bather, who demanded that Hurst provide actual phylogenies that disprove it. How were these phylogenies to be made out? Stratigraphic succession is the only source that has even been offered for actual ancestor—descendant series. The method has recently been made more explicit as stratophenetics (Bretsky, 1979; Gingerich, 1979). But referring to the general argument from stratigraphy, Nelson & Platnick (1981) write 'This . . . is simply fallacious; stratigraphic sequence alone cannot indicate that two fossils belong to the same lineage . . . Fossils must be ordered on the basis of systematic hypotheses, and since these hypotheses may always be incorrect, fossils so ordered cannot be said to show the truth . . . of evolution.' Indeed, the possibility that several of the classic 'phyletic series' of fossils may be incorrect has come up in the recent debate over gradualism (Gould & Eldredge, 1977; Stanley, (1979).

Riedl (1979) criticises the view of fossils as arbiters from a different angle; 'construction of phylogenies without fossil evidence . . . has been condemned as methodologial sin . . . However . . . the condemnation turns the matter upside down. The sole authority is in fact morphological theory. 'Where Riedl proposes 'morphological theory' as the authority, Nelson & Platnick (above) offer 'systematic hypotheses.' But the two prescriptions are really no different. For Riedl, 'the so-called homology theorem . . . is the essential part of the principles of morphology' (1979), whereas for Nelson & Platnick 'systematics in general consists of the search for defining characters of groups' (1981). The alternative arbiters, homology and characters of groups, coincide if 'homology is the relation characterizing natural groups' (Patterson, 1982), an opinion also held by Eldredge & Cracraft (1980) and Nelson& Platnick (1981). Riedl's book, and those of Eldredge & Cracraft and Nelson & Platnick, are signs of a recent resurgence of interest in hierarchical order in nature, and in methods of discerning and representing that order (also Gould, 1980; Lauder, 1981, 1982). Aspects of this are discussed in the next three sections.

Hierarchy Versus Continuum

The change in systematic or comparative biology which accompanied general acceptance of evolution can be summed up as a shift from an attempt to discover and represent natural order, to a world view in which life was known to be a historical continuum on which biologists must impose order (Griffiths, 1974). One illustration of this change is the contrast between Milne-Edwards's (1844) and Agassiz's (1844) use of the map simile, and Haeckel's post-Darwinian use of the tree (Figs. 6.6, 6.7). Milne-Edwards and Agassiz viewed nature as imperfectly known territory which biologists were exploring and charting. But for Haeckel and his successors the *form* of the territory is known —it is a tree —and the biologist's task is to discover the unknown creatures (e.g. *'Amphibia ignota'*; 'unbekannte Uebergangs-ormen') which will complete the continuum.

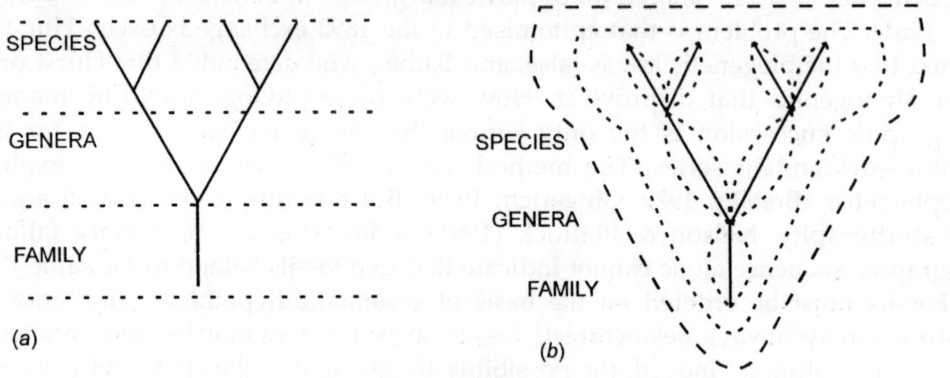

Fig. 6.9: Two views of the relationship between a tree and a classification, or between time and relationship. (After Simpson, 1961, Fig. 19.)

For pre-Darwinian biologists, species were generally thought to be real—discrete and immutable. For Darwin, they ceased to be real and became arbitrary segments of a continuum (Eldredge, 1979; Cracraft, 1979, 1981). The change is epitomised in an indirect exchange between Agassiz and Darwin: 'If species do not exist at all, as the supporters of the transmutation theory maintain, how can they vary?' (Agassiz, 1860) —'How absurd that logical quibble . . . As if anyone doubted their temporary existence' (Darwin, 1887).

Given the continuous tree composed of evanescent species, it was inevitable that the aims of systematics should change. Discovery of the natural hierarchy was replaced by the imposition of order, more or less arbitrarily, on a continuum. This change is exemplified in Fig. 6.9. which contrasts a pre-Darwinian and a neo-Darwinian view of the relation between time and hierarchy. For von Baer, the ontogeny of vertebrates is exactly congruent with the natural hierarchy of those animals: the characters of Vertebrata appear before those of vertebrate classes, and so on through the hierarchy. This ontogeny is hierarchical, just as is the natural system. Agassiz and Richard Owen extended these ideas into the fossil record: the oldest fossil vertebrates were just vertebrates, and fishes did not assume their characters until tetrapods appeared, nor did reptiles become reptiles until birds and mammals appeared (Agassiz, 1859). Owen drew similar conclusions from fossil mammals (Ospovat, 1976). So in Fig 6.9a. the time axis may be read as ontogenetic time (von Baer) or geological time (Agassiz, Owen, extending von Baer's laws into the fossil record). The neo-Darwinian view (Fig. 6.9a) is entirely different. Because the tree is a continuum, composed of species throughout, the concept of relationship becomes problematical, and some arbitrariness is necessary to form higher group (Simpson, 1961; Mayr, 1969). "The only problem was where to apply the ax. The same tree could be chopped up in many different ways, and it often was' (Nelson & Platnick, 1981). So it became orthodox to view classification as an art reflecting personal taste or discrimination rather than the hierarchical structure of the tree itself. And since tastes vary, there are many possible classifications, no one necessarily better than another.

This thesis, species and higher categories arbitrarily imposed on a continuum, has far-reaching consequences in biology, resulting in an approach named (by its critics) 'transformational' (Eldredge, 1979), characterised by concentration on process, by smooth extrapolation from population genetics to the origin of higher taxa (Gould, 1980), and so by reductionism which sees evolution and phylogeny as nothing more than changes in gene frequency. The alternative approach, christened 'taxic' by Eldredge (1979), is marked by an interest in pattern before process, or hierarchy rather than continuum, by anti-reductionism which denies that higher hierachical levels have properties reducible to or predictable from the lowest level, by an interest in testable historical hypotheses rather than those consistent with any observation (Lauder, 1981, 1982), and by the belief that taxa are real and may be discovered (Ball, 1976, p. 426). If taxa are real, it follows that there is only one real or natural hierarchy, one correct or true classification. In systematics, these ideas originated with a new definition of relationship, which led to a new conception of types of group.

Ancestry and Types of Group

Haeckel used five different conventions to denote ancestry in his trees, and there is a sixth possibility, ancestral species. I suggested above that these could be divided into two; first, those which invoke only the decreasing generality of von Baer's law (cf. Fig. 6.9a); and secondly, those which concern actual historical descent (cf. Fig. 6.9). As phylogeny, documented by fossils, and the historical viewpoint came to dominate higher classification (Simpson, 1945), and since ancestral species remained elusive, groups of the second type occupied the trunks of trees and played a major part in phylogenies. These ancestral groups, or Haeckelian conventions of ancestry, may be extinct, containing fossils only, or extant. For example, Romer's (1966) spindle diagram of placental mammals shows an extant order, Insectivora or a, as ancestral to all other orders, and an extinct order, and an extinct order, Condylarthra, as ancestral to all the ungulate orders.

With this view of phylogeny, groups rise to groups, the concepts of monophyly and polyphyly, which Haeckel and introduced (1866), were much discussed, and 'minimal monophyly' was proposed (Simpson, 1945, 1961): a group may be considered monophyletic if it is descended from a group of equal or (perferably) lower rank. This criterion allowed descendant groups to be polyphyletic, in the sense that they were derived by more than one lineage from the ancestral group, and during the 1950s and 60s much paper was used in discussions of the mode of origin of various groups. These discussions ceased when the consequences of Hening's (1966) definition of relationship became widely diffused. Hennig defined realtionship became widely diffused. Hennig defined relationship by recency of common ancestry: given three species or groups, those two are more closely related which share an ancestral species not shared by the third. One consequence of this definition is that groups of related species exactly match the hierarchical structure of the tree. In other words, Hennig used evolutionary theory to rediscover a congruence between groups and the hierarchy like that used by pre-Darwinians such as von Baer, Milne-Edwards and Agassiz (Figs. 6.3–6.6).

A second consquence of Hennig's definition of realtionship is that there are not two types of group, monophyletic and polyphyletic, as evolutionists from Haeckel to Simpson had supposed, but three. Hennig named the third type paraphyletic: a monophyletic group includes an ancestor and all its descendants, a paraphyletic group includes an ancestor and some but not all of its descendants, and a polyphyletic group does not include a common ancestor (those statements paraphrase later discussions of Hennig's original definitions—Holmes, 1980). Fig. 6.9b illustrates the three concepts as applied to a theory of relationships: the whole family would be monophyletic, since it includes the common ancestral species and all its descendants; the extinct genus would be paraphyletic, since it includes the common ancestral species and some but not all of its descendants; and the two extant genera would be polyphyletic, since neither includes the species ancestal to its members. The three concepts may also be defined in practical rather than theoretical terms: monophyletic groups are characterised by shared derived characters (Hennig's synapomorphies), paraphyletic groups by shared primitive characters (Hennig's symplesiomorphies), and polyphyletic groups by convergent characters. Again, these definitions are illustrated by Fig. 6.9b. For example, the two extant genera can only be characterised by independently evolved (convergent or non-homologous) characters, since whatever they uniquely share was absent in their common ancestor.

The concepts of monophyly, paraphyly and polyphyly have since been much discussed (see Holmes, 1980; Wiley, 1981), but there is general agreement that Haennig's discovery of paraphyly was new and important. This is evident from Fig. 6.9b, taken from Simpson (1961), who regarded all the groupings as valid monophyletic taxa. Hennig recommended (1966) that paraphyletic groups should be avoided, since they have no independent history, reality, individuality, or exclusive common ancestry and point of origin; 'The concept of "extinction" has thus a different meaning for them than it does for monophyletic groups.' For a paraphyletic group can become extinct only by 'transformation'—this is true of all ancestral groups. Hennig's references to reality and individuality reflect his belief that higher taxa are real, and 'exist whether or not there are any systematists around to perceive or name them' (Wiley, 1981). This idea is diametrically opposed to the prevailing neo-Darwinian view that taxa are artiffacts, invented or imposed on nature according to the skill and taste of systematists. The theme of the reality of monophyletic taxa, in Hennig's sense, has come to be one of the central tenets of cladistics, the school of systematics that he founded. And the equation of monophyletic groups with natural group (e.g. Farris, 1980; Nelson & Platnick, 1981; Wiley, 1981) is another sign of convergence between modern systematics and the pre-Darwinian systematists' attempt to map or mirror a discoverable natural hirarchy.

Hennig's objections to paraphyletic (= ancestral) groups were theoretical. An additional theoretical objection to such groups is that they are inconsistent with evolutionary theory: species, not groups of species, are the ancestral units in evolution (Wiley, 1979). Practical objections are that paraphyletic groups can only be defined arbitrarily or by authority, since an arbitrary line must be imposed to separate the ancestral group from the group or groups descended or detached from it, and this residual cluster has no characters of its own (Patterson, 1982)—it is an 'almost meaningless set' (Ball, 1976). To be consistent with evolutionary theory, evolutionary trees must consist only of species. Those species may be hyppothetical or named, but in neither case is it legitimate to combine them in ancestral groups.

Once paraphyletic groups are recognised as systematists' inventions, the trunks of phylogenetic trees melt away into hypothesis. Nevertheless, Hennig's concepts of monophyly and paraphyly were firmly tied to phylogeny, both in theory, by common ancestry, and, in practice, by derived (= advanced) versus primitive (= ancestral) characters, which necessitate determination of character phylogenies, morphoclines, or 'transformation series', as Hennig called them.

Character Phylogeny

Methods of distinguishing primitive and derived characters, or determining the polarity of morphoclines, have recently been reviewed from several angles. De Jong (1980) distinguished no less than 21 different arguments, and recommended seven, particularly out group analysis. Stevens (1980) discussed seven groups of criteria (ten criteria in all), and also recommended outgroup analysis. Wiley (1981) reviewed six methods and recommended two, out group comparison'. Rieppel (1979) discussed, and rejected, the ontogenetic criterion. Watrous & Wheeler (1981) analysed out group comparison, and commanded it above all other criteria; and Bishop (1982) reviewed five criteria and rejected all save functional morphology. This most authors choose outgroup comparison as the decisive criterion, and reject ontogenetic sequence. The reasons given for rejecting ontogeny are varied. De Jong (1980) writes 'the ontogenetic argument is of little use for the distinction between apomorphous and plesomorphous character states' because 'parallels between ontogeny and phylogeny . . . an only be observed if the phylogeny is known . . . Observation of gill arches in adult fishes and embryonic mammals does not *per se* tell us that the mammals evolved from a fish-like ancestor. Without the help of fossils and other evidence (i.e. without a presupposed phylogeny) the fishes could have ben regarded as derived from mammals.' In short, the argument from ontogeny is 'based on tendencies observed in established phylogenies' (*ibid*). Like Wiley (1981), Stevens (1980) argued that the ontogenetic criterion must be used in a comparative context, 'i.e., in some from of out-group analysis,' and that it depends on 'the implicit assumption that evolution leads to an increase of complexity.' Both de Jong and Stevens comment that the ontogenetic argument is invalid whenever neoteny or paedomorphosis occurs. Rieppel (1979) criticised the ontogenetic criterion by giving examples from reptiles (vertebrae, palatobasal articulation, trabeculate) to show that ontogenetic criteria fail to match phylogenies established by outgroup comparison. Rieppel accounts for that failure by heterochrony, deletion, deviation or paedomorphosis.

All those criticism seem to miss the point. Hennig (1966) introduced the terms 'apomorphous' and 'plesiomorphous' with a discussion of evolution as transformation, with the equation of 'derived' (as in 'shared derived characters') and 'transformed,' and with the expression 'transformation series' to describe the difference between symplesiomorphy and synapomorphy. The notion of character phylogeny is therefore one of transformation, or of direction of transformation.

Phylogeny is generalised transformation, but we have no empirical experience of phylogeny; the only transformations of which we have empirical evidence are those of ontogeny. When de Jong suggests that, on the evidence of gill arches in the embryo, we might as well derive fishes from mammals as the reverse, ontogenetic transformation shows him to be wrong. For we know organisms (anurans, urodeles) in which we observe transformation of serially similar, gill-bearing gill arches, separating open clefts (the fish condition), into serially differentiated, cleftless, gill-less arches (the mammal condition), and we know no examples of the reverse transformation. This does not show that mammals are descended from fishes, for the same observations were available to von Baer and Agassiz. Rather it shows, as it showed von Baer, that there is a general condition, seen in embryos of fishes, amphibians and mammals, and a special condition in adults of each of these types; and that the transformations in amphibians are more extensive than in fishes, and those in mammals are more extensive than in amphibians. Empirical transformation is, I believe, what Nelson (1973, 1978) meant when he differentiated the ontogenetic criterion of polarity from all others as a direct argument 'not involving other species or groups' (1973), or not requiring 'a prior assumption of relationship (or a higher level phylogeny)' (1978). All other criteria, including out group comparison, are indirect, and require prior assumptions.

The mistake made by those who criticise or neglect the ontogenetic criterion is, in my view, that their training forces them to see it only in a phylogenetic, or evolutionary, or Haeckelan context. For example, Wiley (1981) writes, 'The ontogenetic criterion assumes that ontogenetic transformation toward a particular character reflects the phylogenetic development of that ontogeny.' But no such assumption is necessary. All one need assume is that, since ontogenetic transformation is consistently observed to be in one direction, and never in the reverse, we have direct evidence of transformation, and may rate the untransformed state as more general ('primitive') than the transformed state. Hence, Nelson's (1978) re-statement of Haeckel's biogenetic law in von Baerian terms:'given an ontogenetic character transformation, from a character observed to be more general to a character observe to be less general, the more general character is primitive and the less general advanced.' Just as the transformation is empirical, so is the generality, in terms of numbers of species. Wiley (1981) rephrased the law, replacing 'observed to be more general' by 'found in the out group', and replacing 'observed to be less general' by 'found only in the group considered,' and so argued that the ontogenetic criterion is complementary to out group comparison. But Wiley's modification brings in the problem with outgroup comparison, and all other indirect methods; what criterion is to be used to establish the in group and the outgroup? (Patterson, 1982). In other words, without some prior knowledge of character phylogeny, how can we investigate phylogeny? More to the point is that Wiley's reformulation obscures the fact that the general condition is not 'found in the outgroup', but found both in the outgroup *and* the in group (in early ontogeny), so that no prior establishment of in group and outgroup is necessary. Without the transformations of ontogeny, which by themselves define nested sets (outgroup + in group → in group), systematics would be impossible, as Cartmill (1981), who neglects this criterion, suggests it is.

Von Baerian and Haeckelian Recapitulation

Lovtrup (1978) distinguished these two types of recapitulation. He epitomised von Baer's laws as 'Ontogeny recapitulates the taxonomic course and the mechanism of phylogeny' (*ibid.*), and contrasted that with Haeckel's biogenetic law. Lovtrup differentiated the two theories by the Haeckelian demand that all phylogenetic novelties be terminal, so that terminal features of ancestors are shunted back indescendants to become non-terminal, and the von Baerian tolerance of non-terminal novelties. Lovtrup concluded (*ibid.*) 'All instances of ontogeny are accounted for by von Baerian recapitulation', whereas Haeckelian recapitulation is excluded during morphogenesis (the creation of form, as opposed to growth, the modification of form), and can concern only terminal or subterminal features.

Leaving aside the propriety of citing von Baer as an advocate of recapitulation (Gould, 1979), Von Baerian and Haeckelian recapitulation can also be sharply differentiated in phylogenetic rather than ontogenetic terms: von Baerian recapitulation is a guide to *common* ancestry, and Haeckelian recapitulation is a guide to *direct* ancestry. Since direct ancestry is a special case of common ancestry (Nelson, 1974), this distinction agrees with Lovtrup's (1978) proposition that Haeckelian recapitulation is a special case of von Baerian recapitulation. The distinction between the two types of recapitulation in terms of ancestry recalls the differentiation between von Baerian conventions of ancestry (a group gives rise to its subgroups) and Haeckelian conventions—ancestral group of rank equal to or lower than that of their descendants.

There is not space here to summarise Riedl's (1979) analysis of terminal and non-terminal features (which he calls metaphenes and interphenes). Riedl does not distinguish between Haeckelian and Van Baerian recapitulation, regarding the biogenetic law as true and all recapitulation as Haeckelian, but his discussion of recapitulatory and caenogenetic features in terms of his concept of burden ('the responsibility carried by a feature or decision', quantified in numbers of dependent homologues) is relevant and illuminating.

Does Phylogeny Differ from Ontogeny?

In seeking answer to this question, we must bear in mind Haeckel's distinction between the two aspects of phylogeny, the chain of direct ancestors which he saw as recapitulated in the individuals's ontogeny, and the branching phylogenetic tree, which he saw in the natural hierarchy of systematics, and which is expressed in ontogenetic terms in Von Baer's laws. The difference between the two aspects is shown in the contrast between Haeckel's biogenetic law of serial transformation and von Baer's laws of hierarchical individualisation. Further, in order to specify where or how phylogeny differs from ontogeny, we need some knowledge of phylogeny. Regarding the first aspect, the chain of ancestors in any particular lineage, we are hardly better informed than in Haeckel's time, for no general method of identifying ancestors has yet been found. Instead the chain of ancestors has been approximated by what I have called Haeckelian conventions of ancestry, chopping the systematic hierarchy horizontally into extinct ancestral groups, or phenetic envelopes surrounding certain known fossils. Such groups are paraphyletic, or unreal, or invented, and tell us nothing more about phylogeny than that monophyletic groups possessing certain characters have evolved from ancestors which lacked those characters. Since this is no more than a rephrasing of von Baer's laws, it can hardly be inconsistent with ontogeny, in which adults possessing certain characters evolve from embryos which lack those characters.

With the branching-tree aspect we seem to be on firmer ground, for the systematic hierarchy has been refined since Haeckel's time. Fig. 6.16. compares the cladogram of gnathostomes derived from von Baer's 1828 scheme (Fig. 6.5b) with one modern version. The significance of the differences between the two is simple. Just as Milne-Edward's diagram (Fig 6.5a) differed from von Baer's in lacking one 'group' (reptiles + birds) whose only character in Von Baer's scheme was a lack (of the placenta), so the modern cladogram of gnathostomes differs from von Baer's in having fragmented other 'groups' which were uncharacterised. Von Baer's Pisces, characterised by lack of lungs (Fig. 6.3) is broken up into four groups, each the sister-group of everything to its right: chondrichthyans, actinopterygians, and two groups undiscovered in von Baer's time, coelacanths and lungfishes. Von Baer's Anamniota, uniting fishes and amphibians and characterised by gills, has also disappeared; it comprises the four successive groups of fishes and the amphibians. And Von Baer's Reptilia, characterised by lack of wings or air-sacs, is broken up into three groups, lepidosaurs, chelonians and crocodiles, each the sister of everything to its right. The change comprises the elimination by subdivision of all the uncharacterised, or paraphyletic, or potentially ancestral groups.

Von Baer's laws can be tested against the cladogram in Fig. 6.10a. His first law—more general characters appear earlier in the embryo than more special —predicts that the characters of the groups named up the axis of the cladogram should develop in sequence. They do not: the amnion and allantois, characterising Amniota, a subgroup of Tetrapoda, appear before any feature of the latter group, and even before features of Gnathostomata such as jaws and limb girdles. Von Baer's first law therefore fails as a sequential guide to hierarchical ranking of features (cf. Muller, 1864, on Crustacea).

Von Baer's second law—less general characters are developed from the most general, and so on—is not presented as a guide to sequence in the whole organism, but to sequence in structural relations. The embryonic membranes of amniotes, as developments (outgrowths) or more widely distributed structures, are consistent with this law, as is, so far as I know every other observation invertebrate morphology. Von Baer's third and fourth laws are denials of recapitulation of adult stages. He summed up his work—'the most general result'—as 'The history of the development of the individual is the history of its increasing individuality in all respects' (Huxley's (1853) translation). That statement seems to be applicable to Fig. 6.10a and to any more detailed analysis of vertebrate relationships.

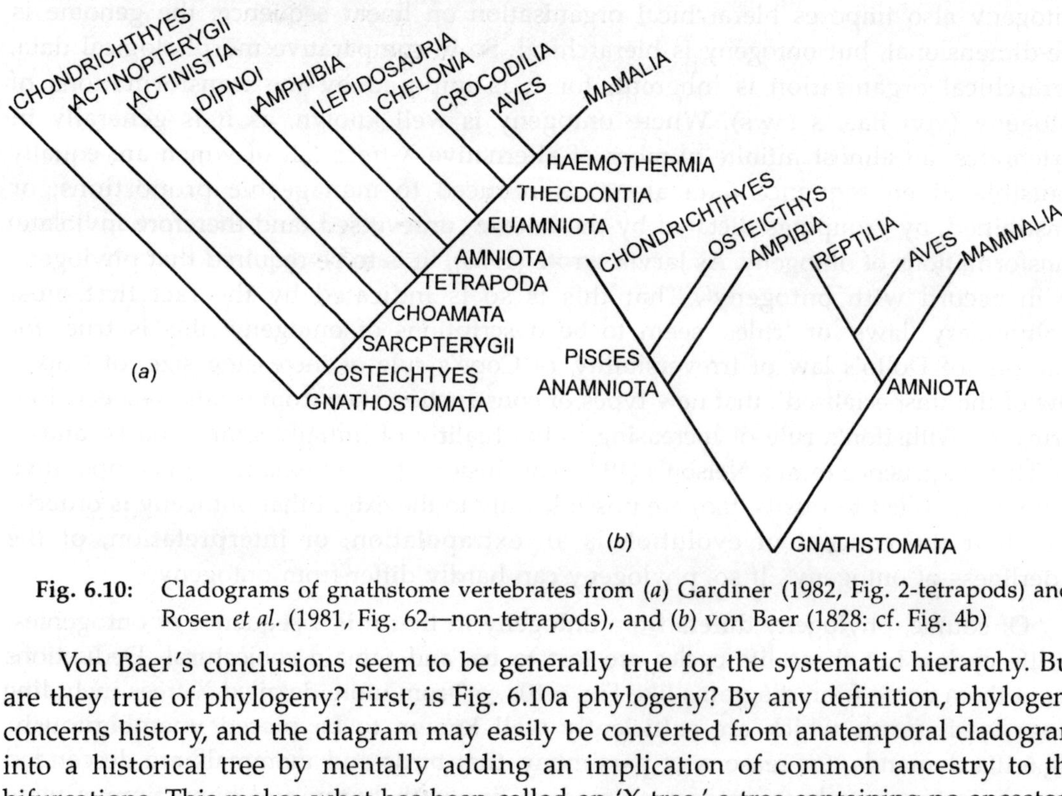

Fig. 6.10: Cladograms of gnathstome vertebrates from (*a*) Gardiner (1982, Fig. 2-tetrapods) and Rosen *et al.* (1981, Fig. 62—non-tetrapods), and (*b*) von Baer (1828: cf. Fig. 4b)

Von Baer's conclusions seem to be generally true for the systematic hierarchy. But are they true of phylogeny? First, is Fig. 6.10a phylogeny? By any definition, phylogeny concerns history, and the diagram may easily be converted from anatemporal cladogram into a historical tree by mentally adding an implication of common ancestry to the bifurcations. This makes what has been called an 'X-tree,' a tree containing no ancestors (each ancestor is an unknown 'X') but carrying the implication of a time-scale and of descent and speciation. Von Baer's second law and his general conclusion must apply to this tree just as they applied to the cladogram, for the mental operation of converting it from tree to cladogram cannot have modified its content. How then does phylogeny differ from ontogeny?

There is one class of phylogenies, or X-trees, that does differ from ontogeny: molecular phylogenies, derived from nucleotide or protein sequence data. The transformations postulated in molecular phylogenies are gene duplications and changes in nucleotides that are nowhere matched in ontogeny. Yet, paradoxically, this difference brings out one of the similarities between phylogeny and ontogeny. Molecular sequences are linear, or one-dimensional, and do not themselves display hierarchic organisation. In building phylogenies from homologous sequences, methods such as parsimony (minimum evolution), compatibility or likelihood (Felsenstein, 1979) must be used to link the sequences by hypothetical ancestral sequences. Different methods (assumptions) may give different phylogenieis from the same data (e.g. Romero-Herrera *et al.*, 1978); that is, the tree topology and the ancestral sequences hypothesised are assumption dependent. This means that hierarchical structure is not inherent in comparative sequences data (except with very limited samples), but is imposed by the investigator. Ontogeny also imposes hierarchical organisation on linear sequence: the genome is one-dimensional, but ontogeny is hierarchical. So in comparative morphological data, hierarchical organisation is inherent, for it is imposed by the transformations of ontogeny (von Baer's laws). Where ontogeny is well known, as it is generally in vertebrates, an almost infinite number of alternative X-trees (all of which are equally plausible given sequence data alone) is reduced to manageable proportions, or constrained, by groupings dictated by the known, unreversed (and therefore inviolate) transformations of ontogeny. As Jarvik wrote (1980), 'it is to be required that phylogeny be in accord with ontogeny'. That this is so is indicated by the fact that most evolutionary 'laws' or 'rules' seem to be descriptions of ontogeny: this is true, for example, of Dollo's law of irreversibility, of Cope's rule of increasing size, of Cope's 'law of the unspecialised', that new types of construction start from small, unspecialised forms, of Willistion's rule of increasing individualitiy of initially similar parts, and so on. This congruence recalls Nelson's (1978) conclusion' that systematics and comparative anatomy (applied to fossils, too) are possible only to the extent that ontogeny is orderly' and 'that the concept of evolution is an extrapolation, or interpretation, of the orderliness of ontogeny'. If so, phylogeny can hardly differ from ontogeny.

Of course, phylogeny differs from ontogeny in that it is a sequence of ontogenies, or life cycles, but those life cycles are lost to us, and remain conjectural. Deductions about them come from diagrams like Fig. 6.10a, or from more detailed X-trees including those fossil forms which are sufficiently well known to be placed unambiguously. Fig.6.10a demands, for reasons of parsimony, that birds and mammals are descended from ancestors in whose ontogenies median fins with horny and bony finrays were developed; several generations of teeth developed on the jaw bones and on all the dermal bones of the buccal cavity and pharynx; enamel and dentine developed on the entire body surface rather than on the teeth alone; dermal bones developed all over the body, and were more numerous in the skull; enclosed sensory canals with neuromasts developed on the head and trunk; and so on. Transformations from these conditions to those found in adult birds and mammals are not seen in ontogeny, and so phylogeny differs from ontogeny. Yet if the transformations required are generalised, they comprise only loss or suppression (e.g. of median fins and finrays, of teeth in birds), and restriction or individualisation (e.g. of dentine and enamel formation, of dentition, of dermal bones). And of course, loss or suppression, restriction and individualisation are well-known in ontogeny.

Finally, phylogeny also differs from ontogeny in that it is conceived as a random walk, whereas ontogeny is the epitome of economy, or purposefulness. However, Riedl asserts (1979) 'that evolution is excluded from accident to a much greater degree than hitherto supposed and that this shielding is the necessary result of a selection dictated . . . by the functional systemic conditions in the organization of organisms themselves'. Which again sounds like a description of ontogeny.

REFERENCES

Agassiz, J.L.R. (1833-44). *Recherches sur less poissons fossiles.* 5 vols. Petitpierre: Neuchatel. Dates of publication of parts are given by W. H. Brown in Woodward, A. S. & Sherborn, C. D (1890) *A catalogue of British fossil Vertebrata,* pp. XXV-XXIX, Dulau: London.

Agassiz, J. L. R. (1845). *Monographie des poissons fossiles du vieux gres rouge.* Petitpierre: Neuchatel.

Agassiz, J. L. R. (1859). *An essay on classification.* Longman, Brown, Green, Longmans & Roberts, and Trubner & Co.: London.

Agassiz, J. L. R. (1860). Professor Agassiz on the origin of species. Individuality and specific differences among Acalephs. *Am. J. Sc.* (2) 30, 142–55.

Anon. (1893). Vitalism. *Nat. Sci. Lond.* 2, 82–3.

Ball, I. R. (1976). Nature and formulation of biogeographical hypotheses. *Syst. Zool.* 24, 407–30.

Bather, F. A. (1893a). The recapitulation theory in paleontology. *Nat. Sci. Lond.* 2, 275–81.

Bather, F. A. (1893b). Phylogeny *v.* ontogeny. Nat. Sci. Lond. 3, 238–9.

Bishop, M. J. (1982). Criteria for the determination of character state changes. *Zool. J. Linn. Soc.* 74, 197–206.

Bowler, P. J. (1975). The changing meaning of 'Evolution' *J. Hist. Ideas,* 36, 95–114.

Bretsky, S. S. (1979). Recognition of ancestor-descendant relationships in invertebrate paleontology. *In Phylogenetic analysis and paleontology, eds. J. Cracraft* and N. Eldredge, pp. 113–63. Columbia University Press: New York.

Buckman, S. S. (1893). The recapitulation theory in biology. *Nat. Sci. Lond.* 3, 138–9.

Cartmill, M. (1981). Hypothesis testing and phylogenetic reconstruction. *Z. zool. Syst. Evolutionsforsch.* 19, 73–96.

Charlesworth, B., Lande, R. & Slatkin, M. (1982). A Neo-Darwinian commentary on macro evolution. Evolution *(in press).*

Cracraft, J. (1979). Phylogenetic analysis, evolutionary models, and paleontology. In *Phylogenetic analysis and paleontology, eds.* J. Cracraft and N. Eldredge, pp. 7–39. Columbia University Press: New York.

Cracraft, J. (1981). Pattern and process in paleobiology: the role of cladistic analysis in systematic paleontology. *Paleobiology,* 7, 456–68.

Darwin, F. (1887). *The life and letters of Charles Darwin.* 3 vols. John Murray: London.

Dawkins, R. (1982). *The extended phenotype.* W. H. Freeman: Oxford.

Eldredge, N. (1979). Alternative approaches to evolutionary theory. *Bull, Carnegie Mus. Nat. Hist.* 13, 7–19.

Eldredge, N & Cracraft, J. (1980). *Phylogenetic patterns and the evolutionary process.* Columbia University Press: New York.

Farris, J. S. (1980). The information content of the phylogenetic system. *Syst.* Zool. 28, 483-519.

Felsenstein, J. (1979). Alternative methods of phylogenetic inference and their interrelationship. *Syst. Zool.* 28, 49–62.

Gardiner, B. G. (1982). Tetrapod classification, *Zool. J. Linn. Soc.* 74, 207–32.

Gingerich, P. D. (1979). The stratophenetic approach to phylogeny reconstruction in vertebrate paleontology. In *Phylogenetic analysis and paleontology, eds.* J. Cracraft and N. Eldredge. pp. 41–77. Columbia University Press: New York.

Goldschmidt, R. B. (1940). *The material basis of evolution.* Yale University Press: New Haven, Conn.

Gould, S. J. & Eldredge, N. (1977). Punctuated equilibria: the tempo and mode of evolution reconsidered, *Palaeobiology* 3, 115–51.

Gould, S. J. (1977). *Ontogeny and phylogeny.* Belknap Press: Cambridge, Mass.

Gould, S. J. (1979). On the importance of heterochrony for evolutionary biology. *Syst. Zool.* 28, 224–6.

Gould, S. J. (1980). Is a new and general theory of evolution emerging? *Paleobiology,* 6, 119–30.

Gould, S. J. & Eldredge, N. (1977). Punctuated equilibria: the tempo and mode of evolution reconsidered. *Paleobiology,* 3, 115–51.

Grasse, P. P. (1977). *Evolution of living organisms.* Academic Press: New York.

Griffiths, G. C. D. (1974). On the foundations of biological systematics. *Acta biotheor.* 23, 85–131.

Haeckel, E. (1866). *Generelle Morphologie der Organismen.* 2 vols. G. Reimer: Berlin.

Haeckel, E. (1868). *Naturliche Schopfungsgeschichte.* G. Reimer: Berlin.

Haeckel, E. (1874). *Anthropogene oder Entwickelungsgeschichte des Menschen.* W. Engelmann: Leipzig.

Haeckel, E. (1876). *The history of creation.* 2 vols. H. S. King. London.

Hansen, E. D. (1977). *The origin and early evolution of animals.* Wesleyan University Press: Middletown, Conn.; and Pitman: London.

Hennig, W. (1966). *Phylogenetic systematics.* University of Illinois Press: Urbana, III.

Hennig, W. (1969). *Die Stammesgeschichte der Insekten.* W. Kramer: Frankfurt am Main.

Holmes, E. B. (1980). Reconsideration of some systematic concepts and terms. *Evolut. Theory,* 5, 35–87.

Hurst, C. H. (1893a). The recapitulation theory. *Nat. Sci. Lond.* 2, 195–200.

Hurst, C. H. (1893b). The recapitulation theory: a rejoinder. *Nat. Sci. Lond.* 2, 364–9.

Hurst, C. H. (1893c). 'Recapitulation' and 'earlier inheritance'. *Nat. Sci. Lond.* 3, 239.

Hurst, C. H. (1893d). Phylogeny and ontogeny. *Nat. Sci. Lond.* 3, 316.

Huxley, T. H. (1870). The anniversary address of the President. *Q. J. geol Soc. Lond.* 26, xlii-lxiv.

Maynard Smith, J. (1960). Continuous, quanitzed and model variation. *Proc. R. Soc. Land. B* 152, 397-499.

Maynard Smith, J. & Sondhi, K. C. (1960). The genetics of a pattern. *Genetics* 45, 1939–59.

Maynard Smith, J. & Sondhi, K. C. (1961). The arrangement of bristles in *Drosophila*. *J. Embryol. exp. Morph.* 9, 661–72.

Monod, J. & Jacob, F. (1961). Teleonomic mechanism in cellular metabolism, growth, and differentiation. *Cold Spr. Harb.* Symp. quanti. Biol. 26, 389–401.

Oster, G., Odell, G. & Alberch, p. (1980). Mechanics, morphogenesis and evolution. *Lect. Math. Life Sci.* 13, 165–255.

Stern, C. (1954). Two or three bristles? *Am. Sci.* 42, 231–47.

Turing, A. M. (1952). The chemical basis of morphogenesis. *Phil, Trans. R. Soc. Ser. B*, 237, 37–72.

Wolpert, L. (1969). Positional information and the spatial pattern of cellular differentiation. *J. theor, Biol.* 25, I–47.

Chapter—7

Homeostasis in Evolution

1. Growth as the Guarantee of Homeostasis in Spite of Wear

The tendency to increase in size is, in a sense, the most fundamental feature of living things. We are all familiar with it, and take it for granted as a 'natural' process. Yet even slight study will show that growth is very far from simple and that we do not properly understand it. Moreover, it is very 'unnatural', since it is unusual among physical phenomena. It involves incorporation of less-ordered material into the highly ordered living systems, contrary to the general tendency for increase of entropy. It is indeed worth while to study growth closely, and we may expect to have surprises when we do so.

There are many ambiguities even in attempting to define growth. It has been called 'the net balance of mass produced and retained over mass destroyed and otherwise lost' (Weiss and Kavanau 1957). But is all increase of mass to be called growth? Is addition of storage fat or carbohydrate as much 'growth' as addition of protein? And what about the addition of water? How are we to define the mass of any organism? It fluctuates from moment to moment.

Features other than mass may serve to define growth. Height would be regarded as a good criterion by many children. Increase in cell number is obviously linked in some way with the concept of growth, but individual cells may grow, so the increase of number of cells alone cannot define the process of growing. From this point of view we might distinguish between (a) the addition of material to cells, (b) increase in their number, and (c) increase of intercellular material.

It would be satisfactory to be able to put forward some concept that would unify these various aspects of growth. The differences between them are semantically puzzling but trivial. A true measure of growth might be related to the rate of increase of order that is implied by the incorporation of non-living into living matter. The increase of DNA in a population might serve as the basis for an assessment of the degree of order, which would evade these difficulties about fat, water, etc. This statistic (amount of DNA) is indeed very useful for some purposes, but we may also have to be content to measure growth by whichever of the paprameters discussed above is most readily available to us. If a general phrase is wanted we can perhaps say that growth is 'the addition of material to that which is already organized into a living pattern' (Young 1950).

We shall be concerned here not so much with definitions or even with finding quantitative expressions for growth (though this will follow). It is more important to try to see the relation of growth to the processes of self maintenance. In fact we begin by asking why should there be growth in order to ensure homeostasis? This may seem a very large question. And indeed so it is, for it touches the very nature of living organization. A system of steady-state interchange is able to continue only because it is provided with 'instructions' that ensure correct responses in the situations that are likely to arise. To put it in another way, an organism an continue to live only because it has been provided with the necessary organs. The instructions thus constitute a system for forecasting the situations and risks that the creature is likely to meet. But by the very nature of the randomness of events, something will happen sooner or later that destroys part of, and eventually the whole, organism. Continued survival in a steady state by any one given set of materials is therefore impossible. Survival of the organization can, however, be ensured by a suitably arranged system of replacement. This is, therefore, the 'function' of growth. It operates at a whole series of levels, from the continuous replacement by turnover within cells, to continuous replacement of cells, to replacement of whole individuals by others, in the processes we ordinarily call reproduction, and in the replacement of one species by another in evolution. All these are a single series of phenomena by which the power of replication and of incorporation into the living system serves to ensure the maintenance of life.

2. Exponential Growth

One of the most unexpected findings for the layman is that growth, if it is not restricted, continues without limit. What grows is itself capable of growing. This is well illustrated by bacteria. When bacterial cells are isolated in a culture medium there is first a lag period and then exponential growth begins. (Fig. 7.1) Each bacterium divides about once every 30 to 40 minutes and the products then 'grow' until they reach the size of the original. This process continues until the supply of nutrients is exhausted, after which special conditions set in, which do not interest us here. If the culture medium is renewed, growth continues indefinitely. Use is made of this for continuous cultivation methods. If all the growing organisms were to be supplied with the required nutrients there would have to be an exponential increase of the size of the vessel and the amount of inflow. This is obviously not possible, but if part of the volume of multiplied micro-organisms is withdrawn at the same rate as nutrient medium is supplied, then a steady state can be maintained in which all the materials are utilized (see Malek and Fencl 1966).

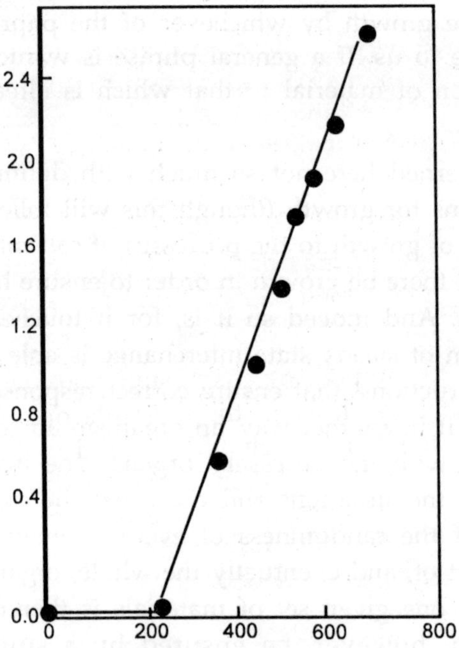

Fig. 7.1: Growth of bacterial cells in culture after explanation. (After Dean and Hinshelwood 1966.)

More penetrating observations have shown that we must think more carefully about continuous growth. Does each individual grow at an increasing rate per unit of mass, as would be required if what grows is capable of growing? Mitchison studied the growth of single cells of yeast, using interference microscopy to determine the 'dry weight' of single cells continuously. He kept the yeast in a sterile medium so that there was no question of interference by the growth processes of other organisms. Each yeast cell grew at a constant rate independent of mass, until late in the process of division, which occurred every four hours, and then the rate doubled and remained the same throughout the next generation (Mitchison 1957). The rate of synthesis therefore depends upon a group of components, perhaps the ribosomes, which may double their numbers before cell division.

The whole process of division must obviously be related to the replication of DNA and this, in many organisms, occurs at about the middle of the period between divisions. The determination of the DNA cycle probably depends upon a complex relation between the replicating material and the associated polymerase enzymes, but the system is not fully understood.

The whole question of the basic cycles that regualte cell division is still obscure. The cell does not divide because it has reached a certain size. By various means the growth can be stopped prematurely, but division will take place, though delayed. Addition of material is therefore one factor influencing division, but not the only one.

Probably replication of the DNA is a second, and perhaps cycles of the RNA and ribosomes a third. A fourth may be a rhythmic change in the energy-yielding mechanisms and other enzymatic process of the cell, perhaps related to the cycle of reducing sulphur-containing compounds (e.g. glutathione).

In examining the fundamental process of control of increase we find, therefore, many facts that are interesting and sometimes paradoxical, but difficult to reduce to any single system of control. Life may perhaps once have been simple, but today even the lower organisms maintain their balance through a web of interacting control mechanisms.

We have, however, the important conclusion that the general condition is to continue growth indefinitely, at least in these simple organisms. The situation with which we as humans are familiar, namely, decline and cessation of growth, followed later by death, thus appears as a limitation of the general condition. Our thesis will revolve around the concept that our life cycle indeed includes special sets of limitations upon continuous growth, serving to ensure that the population as a whole can continue to survive by means of a particular way of life.

3. Differentiation and Growth Control

Perhaps it should not surprise us to find that continuous growth is the general condition. Since continuous replacement is the absolutely necessary condition of self-maintenance we should expect to find it everywhere in living processes. The growth must of course be regulated, especially in the higher organisms, where survival depends upon the presence of a most complicated system of tissues, differentiated for particular functions. Differentiation is in this sense the inverse of growth; indeed, it often follows growth. We shall expect to find, then, that the study of growth consists largely in the study of the control and limitation of growth, especially in the higher organisms.

4. Intermittent Growth

When we look more closely, it thus appears that even in bacteria growth is not continuous. Addition to the mass of a whole culture masks the cycles in the growth of each individual. Replication of the DNA occurs at set intervals, and something is known of the means for regulating this (see Mandelstam and McQuillen 1968). In the absence of limitation by lack of raw materials, it seems that the single strand of DNA in each bacterium is replicated by the attendant polymerases at intervals that are not closely related to the total increase in mass. The factor that regulates this cycle thus serves to control the whole growth potential of the species. Presumably, however, this cycle is itself genetically controlled.

This question of the power of increase of a set of cells comes very close to the heart of all biological problems. Unfortunately, we have only a glimpse of it in bacteria and viruses and known almost nothing of it in higher animals and plants. It is difficult to conceive of the process of growth as a wholly continuous phenomenon, for the simple reason that organisms themselves are essentially heterogeneous. We can think about the continuous synthesis, say, of a given protein. But the more interesting proteins do not simply accumulate, they associate with other substances to form membranes, organelles, fibres, and so on. When any one of these has been formed the continuity is in a sense broken as the building of another organelle is begun.

The point is not trivial, for organisms are assemblies of such units. Higher organism are assemblies of assemblies of them (cells, tissues, colonies, populations). Since the components of such systems do not increase together and continuously we have to think of them separately and individually. The origin and fate of each mitochondrion or man is therefore inevitably a large part of the study of biology. Investigation of the cycles of their multiplication and growth may reveal some general principles, but is also likely to show that there is a *great diversity of rules of growth.*

5. Turnover and Growth

A further complication is that addition to living systems is not a one-way traffic. We must also include the subtraction that is involved in breakdown, and then compute the growth as the net gain. It seems so obvious that growth is the resultant of anabolism and catabolism that we do not pause to ask 'Why should there be breakdown at all?' More practically, we may seek an answer by looking at lower organisms to see whether turnover is always present. Perhaps it is necessary feature for the maintenance of a steady state.

Viruses cannot be said to show turnover, but they are not capable of independent self-maintenance. In bacteria there is turnover, btu at a simple level than can hardly be distinguished from reproduction.

In the more familiar situation of multicellular organisms, it has been clear since the time of Bizzozero (1894) that some parts of the body become changed much faster than others. Whether we are considering cells alone or their components (fibres organelles, etc.) it is convenient to think of three types.

1. Labile populations, multiplying throughout life (vegetative intermitotic cell of Cowdry (1952)).
2. Stable populations, usually multiplying only during development, but capable of doing so in the adult after damage (reversible postmitotics).
3. Permanent populations, never multiplying once formed (fixed postmitotics)

As applied to cell populations in mammals this would produce the following classification:

Labile (renewing) populations

Epidermis and its derivatives

Endodermal epithelium

Transitional epithelium of ducts

Blood-forming tissues (in part)

Endometrium of uterus

Germ cells of the testis

Stable populations (able to expand when necessary)

Liver

Kidney

Exocrine glands

Lens

Connective tissues

Skeletal tissues (in part)

Permanent populations

Striated muscle

Neurons

Germs cells of the ovary

6. Turnover and the Anticipation of Risks

In spite of our considerable understanding of the functions of the various tissues in the economy of the body it is not always clear why some tissues should remain intact and others be regularly replaced. Moreover, little is known about the important question of what determines the rate of production of the labile tissues or stops the stable ones from further expansion.

The products of the epidermal line are ultimately keratinizeid. Thus removed from metabolic sustenance, they are subject to the cruder forms of wear and must be replaced. This simple situation may serve as an example of the principles that we may search for. The body as a self-maintaining system inevitably suffers the physical hazards imposed by a more or less tempestuous environment. Some of these risks are of simple physical nature (like the wearing away of the skin), others are chemical changes or fluctuations of temperature. No portion of the body can be regarded as permanently immune from risks. However strong the bones, in some individuals one or another of them will ultimately be broken. To make them still stronger would involve not only collecting much material but, worse, carrying it around. The tortoise provides himself with a house of inert material that is hard to break. Within it he lives longer than, perhaps, any other creature, but tortoises are not among the most widespread animals and have not been able to colonize may habitats, for example the trees or the air. Such massive over-insurance has not proved to be the ideal policy for life.

In general, the best protection seems to be provided, not by excessively strong or inert materials, but by a system of turnover that allows for replacement. Cell replacement may proceed continuously, as in the skin, or be called into play when needed, as in the repair processes of the connective tissues or skeleton.

In this sense the continuous change of the renewing populations is an aspect of homeostasis. It provides a permanent insurance against the inevitable damage to the components that is imposed by the environment. That this is likely to be an efficient policy is clear, for example, in the skin, whose material is directly subject to wear. It is also easy to understand for the lining of the alimentary canal, which must deal with all the chemical influences that are introduced with the food, including the products of the bacteria that thrive there. In other tissues it is less clear why continuous cell-replacement is needed. There is no obvious need for rapid renewal of the red corpuscles of the blood. Possibly the limiting factor here is the energy needed to pump them round the body. It is cheaper to pump sacks of almost pure haemoglobin then to destroy them after a while, and replace them from a factory, than to pump also the instructions, machinery, and materials for their renewal. Incidentally, birds find the latter course better. Their red cells, though they are still end cells, have nuclei, and this enables them to manage with a small factory and so to lighten their bones with air sacks, while haemopoiesis on a much reduced scale proceeds in the liver, spleen, and elsewhere.

Cell replacement is obviously not the only possible means of renewal. In some tissues replacement is probably undesirable for functional reasons. Thus the activities of the nervous system depend upon an elaborate system of connectivity between the cells. To renew this would need complicated instructions and in fact regeneration can be achieved only in simpler animals and simpler parts of the nervous system. No effective regeneration takes place in the brains of mammals (or of cephalopods). Where the brains contain memories, capable of holding records of the experiences of the individual, it may obviously be difficult to provide also instructions to replicate these records if cells are destroyed. Correspondingly the neurons lose all power of cell division. They do not, however, lose their powers of synthesis. On the contrary, protein turnover is especially active in the nervous system. Nerve cells are characterized by a large amount of ribonucleic acid in their cytoplasm—the basophilic Nissl bodies. Probably there is a continuous flow of material down the nerve fibres from the cell bodies (see Ochs 1966). What happens to it at the peripheral end is uncertain.

It is clear, therefore, that these static post-mitotic' cells are not really different from the rest. Indeed it may be that they are provided with especially active mechanisms for self-renewal. For, unlike the other populations, they cannot be replaced if damaged, and must therefore continuously repair themselves. This presumably applies to striated muscle as well as to nerve. The very large number of nuclei in each syncytial muscle fibre suggest the operation of continuous processes of replacement (though myosin itself turns over little it at all).

It becomes clear, therefore, that the phenomenon of growth, that is to say, synthesis of new material, is much more than the production of a new individual, as we commonly regard it. Growth is one of the fundamental features of homeostasis; growth provides the means of replacement of what will inevitably sooner or later be damaged—a guarantee, so far as possible ,against the risks of destruction. The growth with which we rare familiar, that of the child, after the act of reproduction, we shall find to be the special from of this guarantee, since it allows survival *in face of almost all risks,* a promise for the future that would seem impossible to give, but which has so far served to ensure life for 3000 million years or more.

For the purpose of a general study of growth, then, our conclusion is that addition to the existing mass of living material is a fundamental property, presumably existing since life began. The numerous special forms of life that exist today have been arrived at partly by limiting this general power of growth. Thus all the higher organisms are colonies of different types of cells, serving different functions in the homeostatic system. the number and position of each type is regulated to produce a smoothly working whole. The study of growth thus becomes, in effect, the study of what limits growth in each tissue, during development and in the adult.

REPAIR OF THE INDIVIDUAL

1. Maintenance, Repair, Replacement, and Regeneration

These are a series of similar processes, linked both in their significance for continued maintenance and through the methods by which they operate. All are processes by which the individual life is enabled to persist in spite of actual wear or damage. As we have seen in the last chapter, in most tissues there is continual turnover and in some the cells are regularly replaced. These maintenance operations anticipate the damage or destruction of the parts in question, and they are found particularly in such vulnerable areas as the outer skin and lining of the gut.

Other parts of the body, though liable to damage, may suffer it only 'accidentally', that is, occasionally or not at all. For these the genetic system provides processes of healing, repair and replacement that are set into action only when called for as a result of traum. 'Regeneration' is usually kept as the name of a similar process but on a larger scale, leading to the production of a whole new part such as a limb.

These powers are therefore, as it were, a second line of defence when the forecasting mechanisms that avoid damage have failed. For example, the mechanisms of heredity and development provide a mammal with bones of strength adequate to meet most of the stresses likely to fall upon them. If greater stresses are met and a bone breaks, then a specific mechanism of regeneration is available to mend it.

All these processes involve growth, and are linked with the daily turnover within the tissues. They also involve cell migration and differentiation, and in this are linked with normal development from the egg. All of these repair operations that we usually separate conceptually for convenience under the headings given above form a series running from those more closely related to turnover to those that partake more nearly of reproduction and development. Each uses specific mechanisms to contribute to the homeostasis of the system. In this sense all are 'normal' processes. They differ in the regularity and timing of the wear or accidental events that they are designed to correct. Processes of 'maintenance' are designed to meet the damage that occurs in certain tissues regularly every day, for example in the skin of man, or the incisor teeth of a rodent. 'Repair', 'replacement,' and 'regeneration' are processes that come into operation only intermitently, when the organism has suffered one of the various hazards that are likely to befall it. In this sense they approach nearer to the mechanism of reproduction, which is the act that initiates the replacement of the whole creature before eventually something happens by which it is destroyed.

The earlier categories, maintenance and repair, are thus more directly concerned with minor, recurrent, and temporary damages; the last two, especially regeneration, constitute major acts of differentiation, approaching reproduction. Indeed, in some organisms it is hard to distinguish between regeneration and reproduction by budding (e.g. sea squirts (tunicates) or tapeworms).

Since the processes of repair, replacement, and regeneration are characteristically brought into operation after accidents or unusual events in the life cycle, they are obviously of particular concern for the medical man. He is, for many people, the specialist who is called in to help with the results of such untoward stresses, rather than the guardian of longer-term homeostasis (though he may well play a great part in this to, and will perhaps do so even more in the future). In order to be able to assist the repair processes of the body effectively, the doctor must obviously know about their means of operation. It is to some extent the function of pathology to provide this understanding. Here, however, we meet with the complication that, as a prelude to the study of repair, the pathologist also investigates the processes of damage themselves, especially as inflicted by invasive agents such as bacteria and viruses. We shall not attempt to study pathological conditions here but shall try to investigate the nature of the healing processes as exemplified in a few examples in which they have been set into action by controlled surgical experiment. Elementary principles can be illustrated in this way, but while the reactions to particular pathogens are alike in the early stages, later the picture may change specially for each disease condition.

2. Maintenance

Of the four categories that we are considering this approaches most closely to the normal turnover processes. The concept of maintenance, as borrowed from engineering and household management, implies the repair of wear and tear that is normal in the sense perhaps of 'expected'. We have to paint our houses and renew electric light bulbs or motor-car tyres or brake-linings, because the designers of these items were not able to make them more permanent. A similar distinction can be made for the body, though it is not perhaps altogether appropriate. The body has no permanent components (unless it be the instruction in the DNA), but it has some parts that turn over very little (e.g. the collagen of tendons, actomyosin of muscles, or antibodies), while others turn over so efficiently that we do not ordinarily think of them as suffering wear (e.g. the continual replacement of red cells or intestinal epithelia). Perhaps, however, this is indeed a wrong attitude and we should recognize that it is exactly because of the inevitable molecular wear and degradation that the turnover of these tissues is so active. In addition, of course, it allows the continual adjustment of the level of their functioning by adaptation.

The conceptions of maintenance and repair should be extended for living organisms to the molecular level. Some enzyme molecules persist only for a few minutes (Haldane 1954). It is not clear in what sense that is due to wear, but in muscle there is an increased turnover after activity.

The replacement of the surface of the skin or nails from deeper layers is more obviously an act of maintenance. We find it easy to conceptualize this process because we can literally *see* it at work on our own bodies. There is no reason to doubt that with sufficient insight we should equally recognize that the replacement of our red corpuscles and all the other replacing tissues shows the operation of specific mechanisms under the control of the DNA, ensuring maintenance of the integrity of the system in spite of wear.

As a steady-state open system, the organism is continually subject to stresses resulting from the tendency to merge with its surroundings. These knocks provide the stimuli for the operation of the synthetic growth processes, which are made possible by the replicative powers of the instructional molecules within. Each tissue is thus maintained at the appropriate level of functioning by a *double dependence,* on stresses from without and synthetic forces from within (Young 1946). The problem for the student of growth is to learn to identify and measure these forces so that he can estimate the probability of successful continuance and show how to increase it by appropriate action. To take a simple example, in managing the healing of bone the physician needs to know how the effect of mechanical stress will influence the process at each stage (specifically, when and how much to let the patient with a broken leg walk.). On the other hand, he needs to know whether the bone-growth potential needs the assistance of particular dietary factors; say, vitamin C (for collagen), vitamin D, or extra calcium. It would also, of course, be of great importance to know of any specific agents that stimulate or retard healing, a question to which we shall return.

3. Nerve Repair as an Example of Healing

Healing, repair, replacement, and regeneration are so closely related that there is no sense in trying to distinguish sharply between them. All are distinguished from maintenance in that they are processes initiated by the occurrence of some incident that is likely to happen but does not recur regularly. The instructions of the organism nevertheless contain provisions to meet such accidents, the occurrence of which is in this sense predicted and expected, although arhythmic. We may use the name as headings under which to describe the repair of damage of increasing severity, by operations ranging from simple clotting of the blood to replacement of a whole limb. Healing has many meanings, but refers especially to such processes as the closure of wounds in the skin or of other organs, where the damge is due not so much to the loss of substance, which may be minimal, as to the interruption of the continuity of the tissues. The effect of severing a nerve provides an interesting example. The physical break in the nerve trunk damages simply when it interrupts, not because of loss of tissue, but because the contained nerve fibres are dependent for their trophic maintenance on continuity with the nerve-cell body. After severance, therefore, only the central part remains alive. The peripheral axons undergo degeneration and if the two stumps are close together the old axons are then replaced by outgrowth from the central end. We can therefore regard the activities that reunite the stumps as an example of healing and the function of the new fibres as replacement or regeneration (Fig. 7.3).

After any wound (in a mammal) the repair can be considered to proceed in three phases (Ross 1968): (1) inflammatory, the first-aid operations of closure and removal of debris, (2) proliferative, the advent of fibroblasts, able to lay down new structural materials, and perhaps of specific types of cell; (3) remodelling, such as the maturation of scar tissue or, in nerve repair ,the growth to maturity of some nerve fibres and the atrophy and disappearance of others.

The first aid is provided by clotting of the blood. This is produced by a special set of enzymes and other proteins provided for the purpose. The inactive enzyme prothrombin of the blood is activated by an enzyme thrombokinase, liberated by the damaged tissue, which, in the presence of calcium, converts prothrombin to the active enzyme thrombin. The latter then turns the protein fibrinogen, circulating in the plasma, into threads of fibrin. These help to close the openings of the vessels, stopping the loss of blood. Contraction of the vessels themselves assist this. The fibrin clot serves also immediately to meet the first need of repair, which is to fill up the space left by the wound. But the clot is not itself strong enough to resist any great compression or tension and must be replaced by stronger tissues. Almost at once, therefore, it becomes infiltrated by white blood corpuscles. The first are small microphages (polymorphs), able to remove debrisi and infective agents; later, larger macrophages move in. together these cells remove the fibrin clot, probably by fibrinolytic enzymes, made available by breakdown of the neutrophil white blood corpuscles and release of their lysosomes. These enzymes work differentially on the fibrin molecules, attacking first those that are not under tension. The clot thus comes to present orientated strands. Along these migrate the fibroblasts, which then proceed to produce collagen to unite the severed surfaces more firmly.

When a nerve trunk has been cleanly severed, without other damage, the process of healing varies according to whether the two severe ends remain close together. Let us assume that they do so, or that, having been separated, they have been sewn together by a surgeon. Such clean interruptions may occur when a nerve is cut by glass, say in the wrist when a hand is pushed through a window. There may often be complications from damage to tendons, and nerves may be equally or more seriously damaged by being pulled, as when a humerus is broken and the radial nerve is bruised by contusion. However, if the severed ends of the nerve are together, a blood clot forms between them. After three or four days the fibrin will be dissolving away and fibroblasts will move in from both the central and distal stumps. Peripheral nerve normally contains much connective tissue, and the fibroblasts presumably come from this, though what activates them will have to be discussed below (Figs. 7.3 and 7.4).

The fibroblasts, moving along the strands of fibrin, soon begin to lay down their characteristic protein, collagen, and mucopolysaccharides. If the two joined ends of the nerve are under light tension, then the new collagen fibres will run longitudinally between the ends, in the direction that is most calculated to join them firmly. It is probably desirable, therefore, hat nerve stumps should be left under slight tension, rather than pressed together, which would produce a scar of collagen fibres not all orientated along the line of the nerve. A similar argument applies even more strongly to the union of the severed ends of a tendon.

The strands of fibrin and new strands of collagen serve a further function besides this resistance to stretch. They also guide the course of the outgrowing nerve fibres, or rather of the satellite Schwann cells, which precede the latter. Vertebrate nerve fibres are surrounded by these Schwann cells, which in some make an insulating myelin sheath (Fig. 7.1) that greatly increases the speed of nervous conduction. Even those peripheral nerve fibres that have no myelin all have some from of Schwann cell. In regeneration these satellites have a special function. They cease to be ensheathing cells and move out into the site of the wound by active amoeboid movement (Fig. 7.2). They creep along the threads of fibrin and collagen, froming a web of nucleated tissue between the two stumps. They emerge mainly (perhaps wholly) from the lower (distal) stump, where they are released from their enshething function by the gradual breakdown and dissolution of the axons, which have been separated from their cell bodies and are therefore degenerating (Fig. 7.3). A piece of protoplasm separated from its nucleus sooner or later dies. Nerve fibres, being very elongated, are especially vulnerable to this hazard. The nuclei of the nerve fibres that run to the toes are in the spinal cord and must therefore sustain the protoplasm of threads about a metre long, but only to µm in diameter. It is not known what influence passes along to do this, but there is evidence of an actual transport of material down the fibre.

Therefore, 2-3 days after a nerve fibre in a mammal has been interrupted, the isolated part breaks up, first at a few widely separated points and then into a string of avoids and ultimately spheres. The process is not complete for many days; indeed, some pieces of axon remain for several weeks (or longer in cold-blooded animal).

Meanwhile the Schwann cells of the peripheral stump divide mitotically, forming strands within the tubes of the supporting tissues of the distal stump (Fig. 7.3). The strands emerge from the cut end and grow up to join the central stump. Here they may make contact with the severed central ends of the nerve fibres. These have formed swollen amoeboid tips, putting out 'feelers' into the tissues between the stumps (Fig. 7.4). When one such feeler touches a strand of Schwann cells it makes some

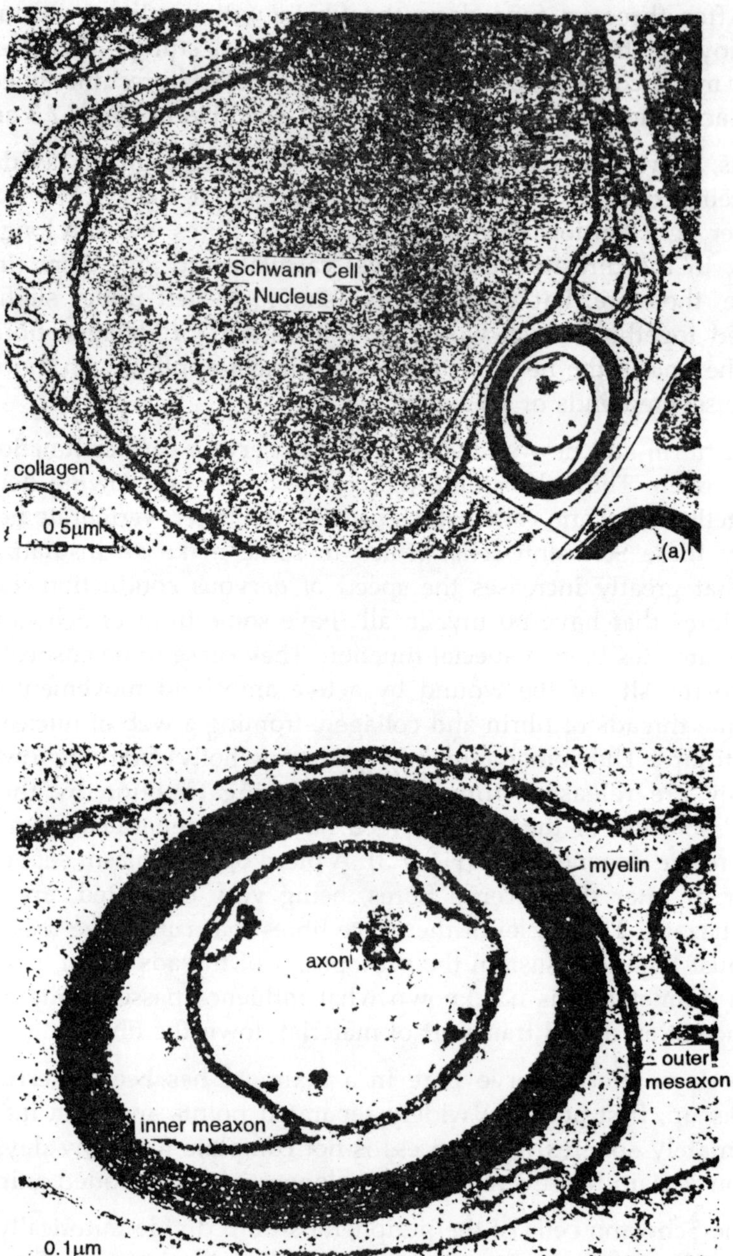

Fig. 7.2: Electron micrographs to show the Schwann cell nucleus (a) and the myelin sheath around the axon (b). (J. D. Robertson in D. Fawcett 1966.)

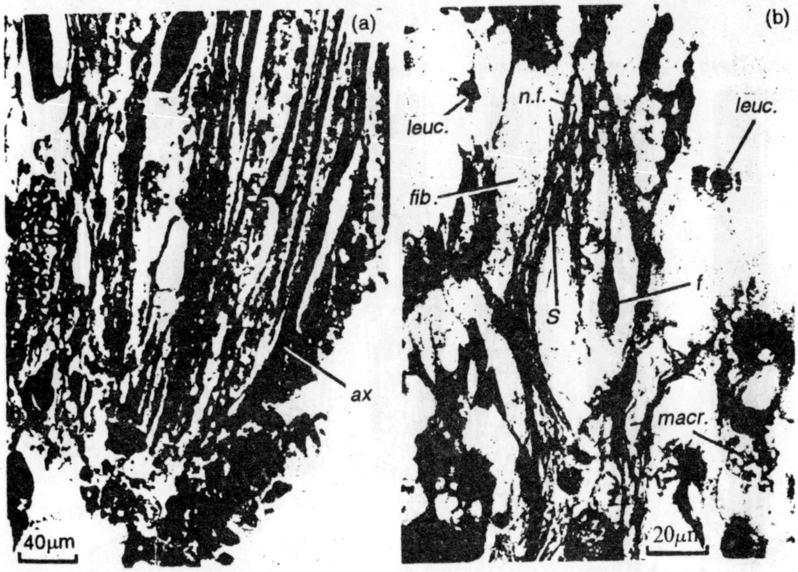

Fig. 7.3: (a) Material emerging at the central end of an axon (*ax*) of a rabbit's nerve served 2 days previously. (b) Portion of the tissue between the two ends of a nerve severed 5 days previously. The central end is above. From it there proceeds a nerve fibre (*n.f.*) which has attached itself to the surface of a column of Schwann cells (*S*) that have emerged from the peripheral stump (below). The tissue contains the remains of the fibrin clot (*fib.*) and several types of cell including white corpuscles (*leuc.*) and fibroblasts (*f*). Large phagocytic cells (macrophages) are seen in the peripheral stump (*macro.*).

specific from of union and the tip moves away down into the distal stump, spinning a fine thread, the new axon, behind it. We shall discuss its further course later on as an example of regeneration.

4. The Nerve-Growth Factor

Some information is available about chemical factors that stimulate the growth of nerve fibres in the embryo, and these may be related to those of the adult. If tumour cells of mice are implanted in place of the limb bud of a 3-day-old chick the nerve fibres of the latter wander around among the cancer cells. The sensory or sympathetic cells from which the fibres originate are stimulated to increase in number several times so that their nerve fibres invade all the tissues of the host embryo. They even enter the blood-vessels in such numbers as to interrupt the circulation (Levi-Montalcini 1964, Levi-Montalcini and Angeletti 1968). In the search for the chemical nature of the 'nerve growth factor' (NGF) responsible for the increase it was found that snake venom has a great power to produce the phenomenon, and that an even more potent source of NGF is the salivary glands of a male mouse. The substances are both proteins but no identical. That from snake venom has a molecular weight of about 40, 000 (Vernon *et al.* 1969). The effect of the factor isolated from the male mouse salivary glands and injected into young mice is to produce a specific hypertrophy only of the sympathetic

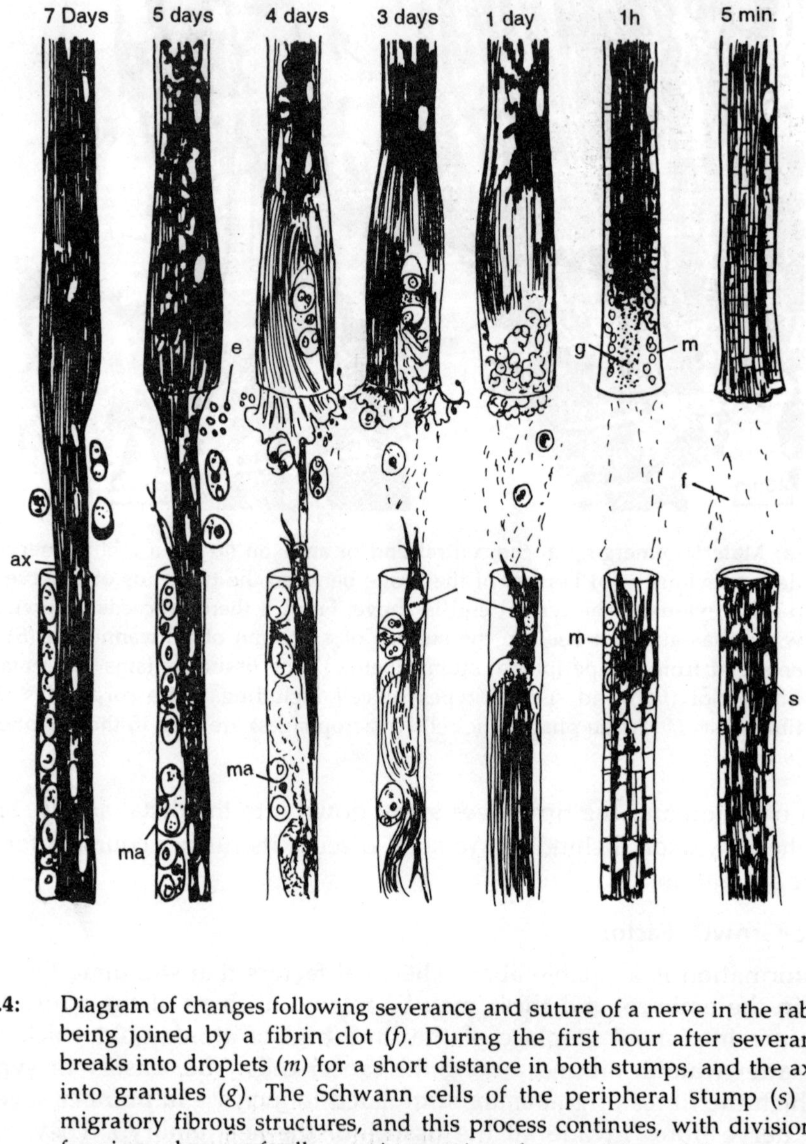

Fig. 7.4: Diagram of changes following severance and suture of a nerve in the rabbit, the stumps being joined by a fibrin clot (*f*). During the first hour after severance the myelin breaks into droplets (*m*) for a short distance in both stumps, and the axon also breaks into granules (*g*). The Schwann cells of the peripheral stump (*s*) begin to form migratory fibrous structures, and this process continues, with division of the nuclei, forming columns of cells growing out into the scar. Meanwhile, the axons and myelin of the distal stump break up and the tubes are invaded by macrophages (*ma*). The central stump axon swells, and the material pours from the end of the tube until some of it makes contact with the Schwann cell columns and is led back as a new axon (*ax*) into the peripheral stump. Unconnected portions break off to form spheres and loops (*e*). The Schwann cells of the central stump first draw in their processes and then advance again over the outflowing axoplasm. The migratory Schwann cells of the peripheral stump also put out processes over the surfaces of the new fibre. (From Young 1949.)

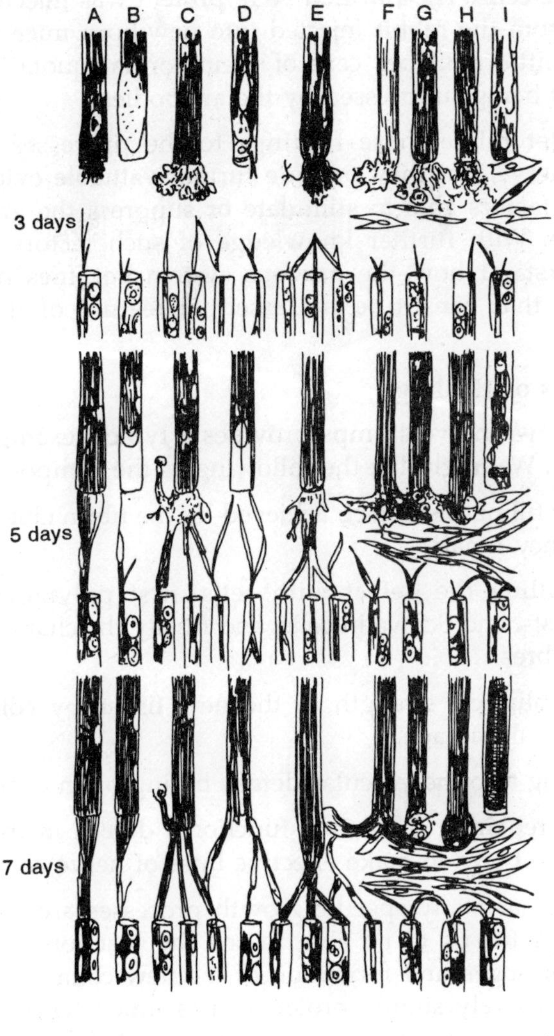

Fig. 7.5: Diagram of several nerve stumps of the rabbit on the third, fifth, and seventh days after union, to show some of the variations that occur. A, a direct union, as Fig. B, some retrograde degeneration in the central stump. C, the axoplasm divides into various streams in the scar and these make union with several distal tubes. The same result is achieved in D by branching within the central tube, and in E by branching at entrance to the distal stump. All these unions are satisfactory, but in those shown in F-I there is a lateral invasion of fibrocytes, preventing union of the stump, with resultant formation of giant clubs (H), or spirals (I). (From Young 1949.)

system. All the rest of the nervous system remains unaffected, even the parasympathetic ganglia.

A further development was the preparation of an antiserum that *suppressed* growth of the sympathetic cells. The purified NGF protein was injected into other foot-pad of a rabbit. Serum from the rabbit injected into new-born mice then caused the latter to develop with as little as 3 per cent of their normal quota of sympathetic cells, the growth of the rest being suppressed by the antibodies.

We cannot yet relate these findings to the processes of normal growth and development of nerve, but they provide further valuable evidence of the existence of specific chemical factors able to stimulate or suppress the growth of particular types of nervous tissue. With further knowledge of such factors it should eventually be possible to understand how the nervous system becomes provided with the many types of channel that constitute the essential feature of its system of coding and signalling.

5. The Processes of Healing

The union of two nerve stumps provides a typical example of the basic processes of wound healing. We might list the following as the components.

1. Quickly filling the defect, achieved by the fibrin clot—as it were, a provisional emergency repair.
2. Next, filling the defect with cells, first polymorphs, then macrophages, fibroblasts, and Schwann cells, and finally the characteristic cells of the tissue, nerve fibres.
3. Giving physical strength to the new tissue by collagen, making good the mechanical damage.
4. Repairing also the vascular defect, by ingrowth of new vessels.
5. Finally, repairing the actual functional defect, in this case by maturation of the nerve fibres to make effective lines of nerve conduction.

It is obvious that many specific growth processes are concerned. A few of them are simple enough for us to be able to see the reactions in some detail. Thus, the clotting of the fibrinogen and formation of the new collagen are the results of change or synthesis of relatively simple proteins. It is much harder to understand how the orderly sequence of so many processes is achieved. In particular, we should like to know what activates the cells to divide and to migrate after wounding and what stops them when their aim has been achieved.

6. Theories of Wound Healing

The various theories of the stimulus to wound healing offer different levels of explanation.

1. The simplest is that there are wound hormones produced by the damage to the cells. Thee are two versions of this,

(a) that the substances stimulate the cells to divide and migrate,

(b) that they remove an inhibition on these activities.

2. A second group of theories suggests that as the repair process serves to fill up the defect left by the injury, it is this deficiency or lack that somehow sets in motion the actions of repair.

3. The third theory is that it is the functional lack that is sensed by a suitable system and thus initiates repair.

   The second and third approaches seem at first sight much vaguer than the first, but may have more value than appears. In particular, we have to study not only what initiates repair but also what ends it at the correct point. The fact that the defect has been filled or the function restored may provide the relevant information that stops the tissue growing.

It is unfortunately true that at present we do not know what initiates or controls repair mechanisms. This is indeed a great weakness, but biological and medical science are faced here with a fundamental difficulty of method. One cannot distinguish between theories of the three types outlined above because one can not produce a loss of tissue without producing traumatized cells. Conversely, one can not do damage without causing some from of deficit (see Abercrombie 1964).

In plants it has been established that injured cells release a substance, traumatic acid ($HOOC(CH_2)_8CH{=}{=}COOH$), which stimulates cell division near by. No similar relatively simple wound hormones have been identified in animals. Indeed, the whole question of growth-promoting substances is highly complicated. From the earliest days of tissue culture it was found that growth is stimulated by extracts, especially by extracts of embryos, but also of other tissues. But such extracts are very complex mixtures and it is still not clear to what extent their effects are due to the nutrients they supply or to specific growth stimulants. There is, however, a considerable body of scattered evidence that the growth of each tissue can be stimulated or inhibited by specific substances produced within that tissue. Examples of these substances are the epidermal growth factor in the skin, nerve growth factor, and erythropoietin, which stimulates the bone marrow. Extracts of various adult and embryonic tissues have been shown to stimulate growth in tissue culture under carefully controlled conditions (Kutsky 1959, Abercrombie 1964). The stimulating power is associated with protein.

7. Wound-Healing in the Skin

Repair of damage to the surface has perhaps a special human interest since we all observe it so often in ourselfes (see Schilling 1968). Moreover, its superficial position makes the skin especially easy to study. After damage to the epidermis and underlying dermis there is first the formation of a fibrin clot and then its dissolution after a few days. Cells of the neighbouring epidermis are stimulated to division. They are normally, of course, in a cycle of continuous division, and presumably the mechanism for ensuring the increase is related to this. After 12 to 24 hours the cells move in such a way as to cover the damaged area (Odland and Ross 1968). Meanwhile the cells of the underlying dermis also proliferate, and polymorphs, macrophages, and fibroblasts proceed to operate as in the nerve regeneration already mentioned.

The response of the epidermis is especially interesting. In a small wound its cells continue to proliferate and to migrate until the form a complete covering. The migration is apparently triggered by the absence of contact between the cells and stops when the space is filled by contact inhibition which stops the fibroblasts. The cells then proceed to contract, drawing the separated edges of the wound together (James 1964). This process of wound contraction is distinguished from the 'wound contracture' by which collagen fibres may draw together the edges, at a much later stage, in a large, poorly healing wound leaving a bad scar.

The mechanism of skin-healing thus shows all the factors we have seen before: filling the space, adding new cells, and restoration of function. Here, however, there is some information about the stimulating factors. The stimulus to cell division is specific to the cells that are damaged. If a flap of mouse skin is cut beneath the dermis, then the overlying epidermis shows no increase in mitosis (Bullough and Laurence 1960). Conversely, damage to the epidermis causes mitosis only there and not in the dermis.

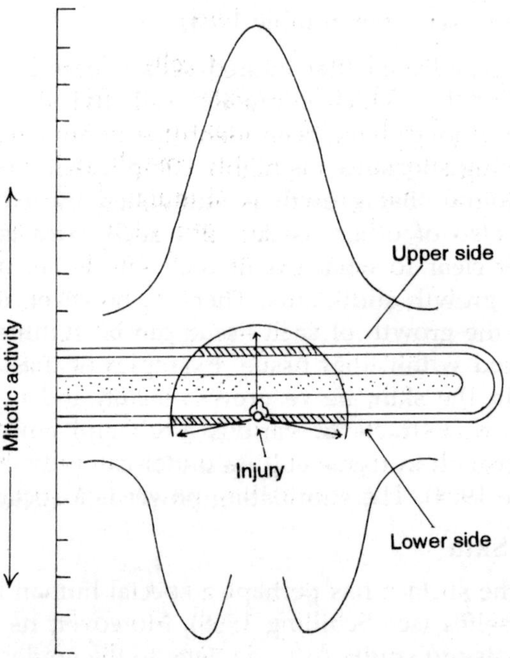

Fig. 7.6: The relation of epidermal hyperplasia to a wound. An injury inflicted on one side of a mouse ear causes proliferation in adjacent epidermis as well as in that on the opposite side of the ear, within a radius of 1 mm. The patterns of mitotic activity are therefore bimodal and unimodal respectively. (After Bullough and Laurence 1960.)

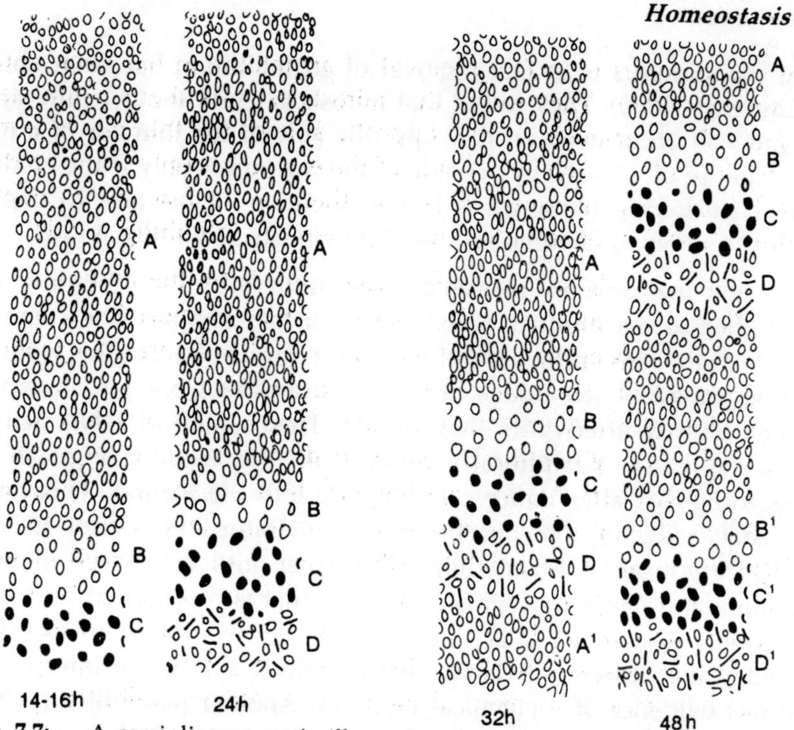

Fig. 7.7: A semi-diagrammatic illustration, in surface view, to show the spatial sequence of events in rabbit lens epithelium after infliction of a wound at the region shown at the lower side of eh figure. Zone A, unaffected epithelium. Zone B, area of diffuse cell population density. Zone C, region of DNA synthesis; cells labelled with tritiated thymidine are black. Zone D, mitotic proliferation. Zones A^1, B^1, C^1, D^1 represent a second wave of reaction, 2 days after injury. (From Goss 1964, after Harding and Srinivasan 1961.)

There is some evidence that the stimulation is produced by an extractable *epidermal growth factor*. Extrats of epidermis have been reported to increase the rate of epidermal wound healing when applied either locally or by in traperitoneal injection. Particularly convincing evidence has been obtained in the bug *Rhodnius*. A small skin wound is surrounded by a zone of mitotic cells and this area is increased by the application of an autolysed epidermal cell extract (Wiggles worth 1937). However, the effect is not specific and is produced by a variety of degarded protein extracts. It may be due simply to damage of the cells of the recipient by the extract. So we have again the same problem that the healing response appears only when tissues have been removed.

A somewhat similar phenomenon is found in mouse skin. After an epidermal wound the hair follicles at some distance away show mitotic stimulation (Argyris and Argyris 1959). This again suggests a chemical effect, but brings up the question of whether this is the production of a stimulant ('promotor', Abercrombie (1964) or the depletion of an inhibitor ('depressor') due to loss of tissue. The theory that removal of depressors is involved has many attractions, since it explains the otherwise puzzling phenomenon that the tissues respond to a *loss* in their neighbourhood, and also suggests the negative feedback by which growth is stopped as the defect is repaired.

Evidence that the stimulus is in fact removal of an inhibition has been obtained by Bullugh and Laurence (1960). They found that mitosis extends about 1 mm around a small wound and will occur even on the opposite side of the thin ear of a mouse (Fig.7.5). But the response of, say, the upper side of the ear occurs only if the epidermis of the under side is *completely* removed. It is not, therefore, a reaction to damaged epidermal cells but to removal of some influence previously inhibiting mitois.

The concept is that the; cells are prevented from mitosis by the inhibitory action of substances they themselves produce ('chalones'). Further support has come from the study of wounds of the lens epithelium of the rabbit. Twelve hours after wounding, cells in the neighbourhood of the injury begin to synthesize DNA (as shown with tritiated thymidine). Shortly afterwards they divide. The region of DNA activation and division then spreads peripherally away from the wound (Fig. 7.6). The explanation seems to be that after injury cells migrate into the wound. This depletes the neighbouring zone, reducing the concentration of inhibitor and setting in motion the procedure of DNA replication. Meanwhile cells further out have moved in, leading to a wave of depletion and hence to a wave of mitosis that is propagated outwards at about 17 μm/h. Twenty-four hours later the cells near the wound again incorporate DNA and a second wave passes outwards. This is indeed a possible interpretation, but there is no direct evidence of a chemical inhibitor. Another possibility is that the cells begin to synthesize DNA only when they have lost contact with their neighbours, as happens in some tissue cultures (R. Weiss, personal communication).

We thus have the outlines of a hypothesis about the control of repair in epidermal structures. Once again, the replication of DNA is seen to be the central phenomenon. The process of repair appears to be triggered by the most obvious stimulus, namely the absence of a component previously present in the neighbourhood. If this is due to the absence of a substance normally produced by the cells themselves, then we have a mechanism by which the repair is terminated when the appropriate concentration has been restored. It is unfortunate, however, that the specific substances concerned have not been isolated or identified. But there is enough evidence to make it well worth while to continue the search for suitable sources of them. We now have to inquire whether similar principles apply to replacement of larger quantities of material and to regeneration of whole organs.

REPLACEMENT AND REGENERATION OF PARTS AFTER LOSS

1. Regeneration of the Liver and Other Glands

The replacement of a single mass of a tissue such as liver is easy to study experimentally, although perhaps a not very common event in nature, except through disease. In tissue that renew by cell multiplication (such as the epidermis) the increased production needed for regeneration is presumably started off by acceleration of whatever system controls daily replacement. In the so-called 'expanding' tissues, however, such as the liver, where mitotic activity in the adult is normally very low, special mechanisms must be brought into play to produce replacement after loss of a large mass. These mechanisms are indeed exceedingly powerful. After reduction of the liver of a mouse to as little as 30 per cent of its mass it will return to normal within 6-8 days. A kidney will enlarge if its fellow is removed, and a host of similar compensatory responses can occur throughout the body.

We may perhaps distinguish situations in which the loss affects a systemic function, and produces hypertrophy elsewhere inthe body, from purely local losses. Thus, after removal of the salivary glands of a rat on one side those of the other side hypertrophy.

Such examples are numerous and suggest that the response to removal is basically governed by functional consideratins (Goss 1964). This, however, does not give us an insight into the mechanisms concerned, except to suggest that they are specific to each tissue and have been highly selected. Indeed, action of these compensatory processes.

The liver of rats has been especially well investigated (Bucher 1963). After partial hepatectomy the cells and nuclei enlarge and at about 18 hours after removal mitosis begins. It quickly reaches a maximum of nearly 3 per cent of the cells (normal 0.01 per cent) and then after about 30 hours begins to decline towards normal. The original weight is recovered withina week. The functions of the liver are so varied that it has been impossible to decide which of them it is whose deficiency triggers the cell multiplication (see Teir and Rytomaa 1967). It is probably partly the lowering of the plasma protein and the induction of specific protein synthesis. Other factors may be the failure of the digestive functions, or simply a result of the distortion of the; pattern of vascular flow.

Similarly, after kidney removal there are many physiological repercussions, any of which may trigger the hypertrophy of the remaining kidney. This is not due simply to the need for the single kidney to produce more urine. One ureter can be diverted to enter the peritoneal cavity or the gut. Its secretion is absorbed and makes more work for the other kidney, but the latter does nto hypertrophy.

Simlar unexplained discrepancies are abundant inthe literature of replacement. Thus one effective way of producing hypertrophy of the salivary gland of the rat is repeated amputation of the lower incisor and the effect is unilateral on the salivary gland only of the same side. Here some very special mechanisms (presumably nervous) must be involved to relate the secretion to the wear on the tooth. The wear would usually be symmetrical and it is an acident that the mechanism has remained unilateral.

Compensatory growth of the endocrine organs is very marked and is normally produced by stimulation from other endocrines. Thus the pituitary stimulates the thyroid, adrenal cortex, sex glands, and others. In each case any deficiency produced by removal s timulates the stimulator, but the details of how it does so are obscure.

Increased activty alone does not necessarily produce increase, as is shown by stimulating an exocrine gland repeatedly, say wtih a drug. Neither the pancreas nor the salivary glands hypertrophy after repeated stimulation by pilocarpine. Curiously enough, another drug that stimulates salivary secretion, isoproterenol, does produce hypertrophy, so that the glands may become five times larger than normal (see Goss 1964). These examples really serve only to emphasize our ignorance of the mechanism that sets off the hypertrophy.

2. Functional Regeneration in the Nervous System

Special processes are available for repair of some parts of the nervous system in all animals and of major parts in simpler animals. This is the extreme example of repair of a tissue with fixed cell-number. In mammals, defects of the peripheral nervous system are made good by outgrowth from the existing neurons, without any production of new ones. It should be possible, therefore, to see the factors that are at work initiating and controlling the 'hypertrophy', uncomplicated by cell division. It is

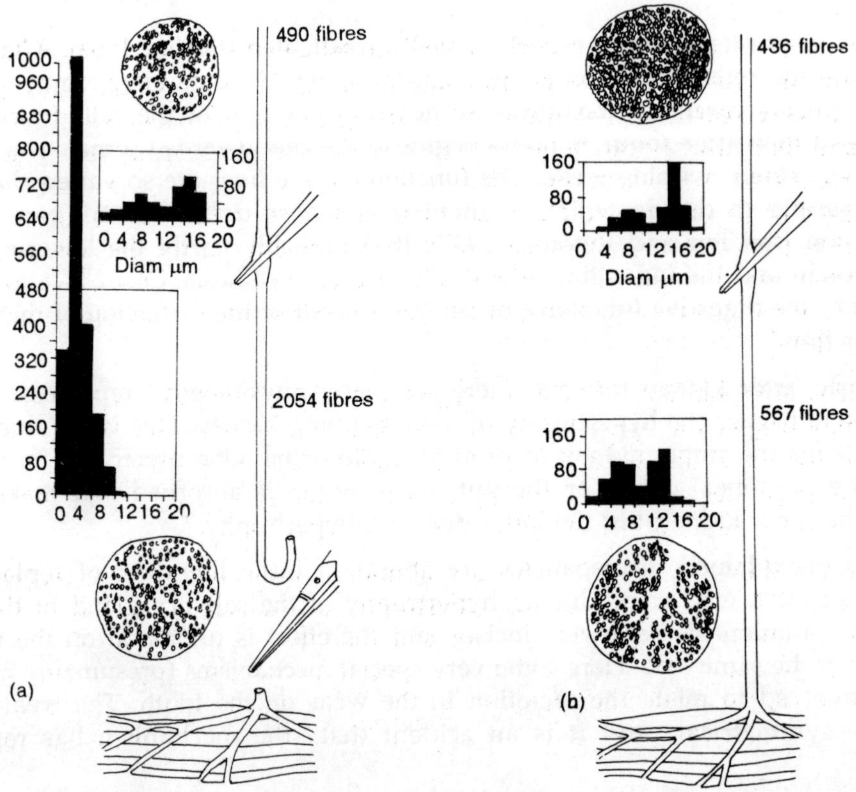

Fig. 7.8: The effect of peripheral connection on regeneration in the rabbit. Both nerves were crushed with forceps above but in (a) the nerve was also cut lower down and turned aside, whereas in (b) it was untouched. The period of recovery was 100 days in both cases .The photographs and histograms at the top show the shrnkage of the fibres of the central stump in (a). Below they show the conditions of the nerves 1 cm below the crush. There are many small fibres in the regenerated nerve of (a), while in (b) the new nerve is more nearly normal. (After Young 1949.)

indeed true that we can discern the principles of control but there is no detailed knowledge of their chemical basis.

Although nervous tissue is post-mitotic it probably undergoes continual replacement even in the absence of injury. As in other tissues, therefore, the regenerative processes are probably an extension of those normally at work. The axons of the central stump swell to make irregular masses of material, probably of amoeboid nature, from which filaments are spun along the; bands of Schwann cells to reach the peripheral stump. Once within the tubes of the stump they advance at a rate of 5 mm/day (in mammals) and are robably led by the tube all the way to a peripheral organ. However, the question of how correct connections are made is still highly controversial. The nervous system differs from all others in the extent of its division into specifically different units, which must be correctly connected to function properly. Effective regeneration therefore depends upon the outgrowing fibres making connection with the right peripheral tubes. Ths is assuming that there is no other means of finding the right peripheral organ or alternatively of remodelling the central connections. The latter is improbable, at least in adults of higher organisms. On the other hand, there is strong evidence of special mechanisms for ensuring correct connections at the end of the axon. The optic nerve can regenerate inmany vertebrates, from fishes to birds. The new nerve fibres may become much interwoven in the nerve trunk, but each of them ultimately comes to re-make connections with its original partners inthe optic lobe of the mid-brain (see Jacobson and Gaze 1965). Unfortunately there is really no satisfactory view as to how this astonishing and very important result is produced, and it would nto be profitable to discuss the interesting evidence here. It emphasizes that very precise mechanisms must be at work.

A little is known about the types of factor that operate for the control of the development of new fibres in a regenerating peripheral nerve stump. In a normal nerve the fibres with different functions differ in diameter. Thus the fibres that carry signals to the muscles to make them contract are large, and conduct fast. Afferent fibres that carry signals about the tension on the muscle are also large. Fibres signalling touch are rather smaller. Fibres controlling blood flow (sympathetic) are very small and conduct slowly, presumably because they change the state of the blood-vesels only slowly and maintain them in a given s tate for relatively long periods.

When new nerve fibres are first laid downtheyare all very thin, whether in the embryo or in a regenerating nerve. Some remain small, others grow larage and acquire sheaths, the process of 'maturation'. There must be specific factors at work to control the growth of each to the correct size. In fact each type of nerve fibre is highly specified from some early embryonic stage, probably by virtue of the connections made at the periphery, though we know little indetail abotu this. During regeneration the process is at least to some extent repeated. For example, motor firbes that do not reach to their normal periphery do not return to their original size (Fig. 7.8). Some influence must pass *up* the nerve from the muscle or other end organ to control the synthesis of new axoplasm. Unfortunately, once again we must admit to complete ignorance of the nature of this influence. New axoplasm is probably produced mainly in the nerve-cell body. After the axon has been interrupted there are quite explosive changes in the

RNA and synthetic mechanism in the cell, as indicated by the process known as retrograde degeneration or chromatolysis. This again shows that some reverse signalling takes place, by which the damaged lower end of the neuron evokes the production of extra supplies of axoplasm. Presumably a furthe set of signals switches off the synthetic proceses when sufficient axoplasm has been produced to return the fibre to its appropriate diameter, large or small (see Watson 1968, Leiberman 1971).

The end result of the nerve regeneration is thus to fill the space left as a result of the injury. However, the effectivensss of the repair lies in the capacity for useful signalling that it confers. Whether as pecific mechanism will be present to repair any part of the nervous system therefore depends on whether repair is likely to increase the survival value of the individual and of the species. Regeneration of the peripheral nervous system is no doubt party associated with a normal wear and tear process, for example at the tips of the nerves in the skin of the fingers. Interruption of a few small peripheral nerve fibres will not be sufficiently in capacitating to make life impossible, and repair processes to make the part useful again are well worth while. On the other hand, damage to the central nervous system is probably such a serious handicap that it is not worth while to attempt to make it good. A mammal or bird with the spinal cord severed is so severely handicapped that in nature it could never stay alive long enough for regeneration to take place. Indeed, it is surprising that it is worthwhile for many fishes to have the capacity to regenerate the CNS when the spinal cord is severed. Perhaps it is because the swim-bladder and the waterprovide support that is lackig on land. It is interesting that frogs also lack power to regenerate the spinal cord whereas this is present in newts and tadpoles.

Similarly, the higher nervous centres are so important to the life of a bird or mammal that survival is impossible without them and there is no mechanism for their repair. That the fore-brain and rectum can be regenerated in some fishes and newts thus really shows how little use the animal makes of them, since it can survive for a while without them! It may be asked what form of damage to the brain is likely to leave a fish in a condition in which repair is worth while. Presumably not after mechanical injury, for if the jaws of a redator damaged the brain, they would hopelessly disorganize the skull. It may be that the power of regeneration in these animals has developed to repair the damage of parasites rather than predators. The brains of fishes are often found to be packed with helminth (worm) parasites. The brains of higher vertebrates apparently have methods that prevent the entry of such invaders and it may be for this reason that they have lost the power of regeneration. But of course with increasing complexity and detail of connections the morphogenetic processes needed for regeneration must become exceedingly complex.

The capacity to regenerate the optic nerve in many vertebrates, up to and including birds, shows presumably that one eye can compensate for the missing one for a time, but also that to have both significantly increases the probability of survival. To understand this situation properly we should need to know more of how the animals use their eyes in the normal environment.

3. Regeneration of Limbs

The same principle applies to regeneration of limbs and of internal organs. Mechanisms able to produce repair have been evolved for organs that are useful but not indispensable for the life of the individuals. This raises interesting questions about the selective value of carrying whatever load is imposed by processes that are used only intermittently. Of course the rate of loss varies greatly. Ninety-three per cent of individuals of some species of crustaceans have been found to show loss of major parts. Siince crabs and lobsters have many liimbs they can afford to be without one while it is growing (Needham 1964). There is little dubt, from the fact that the power of regeneration is used so often, that is has survival value. It is sometimes suggested that this capacity is merely a 'casually persisting pristine power of morphogenesis.' It is much mor elikely that the basic powers of metabolic turnover have been specially selected and improved upon to produce regeneration of the more vulnerabel members.

Proof that these are specially selected powers comes from the animals such as lizards that have evolved autotomy mechanisms, regions at which a limb or tail is cast off when it is seized and remains wriggling to distract the attacker. There are special mechanisms to close the wound, and regeneration takes place faster following autotomy than after injury elsewhere. Autotomy may be useful even when not followed by regeneration. The crane-flies ('daddy-longlegs', tipulids) often cast off a leg when captured, but cannot replace it.

4. Regeneration and Evolutionary Advance

It is widely held that lower organisms have greater regenerative powers than higher ones. The attempt to produce quantitative scales either for evolutionary status or regenerative power is hazardous but has been bravely made by Needham (1964). He concludes that regenerative power is inversely related to evolutionary grade and that the reason is probably that morphological, and particularly histological, differentiation are directly related to the grade, so that if de-differentiation is necessary as a prelude to regeneration it becomes increasingly difficult or uneconomic. He recognizes, however, that the matter is very complicated and many factors may influence the presence or absence of regenerative power. Thus the alertness and nimbleness of the higher animals as potential prey saves them from serious wounds short of capture and total destruction, for which the appropriate biological defence is not regeneration but reproduction.

On the other side there are indications that the regenerative power is not *necessarily* lost in all higher animals. The capacity of stags to regenerate huge antlers each year is a conspicuous example. It seems unreasonable that the skin of a small area should have the power to produce the luxuriant velvet of new hair follicles, much more perfect tan the skin that is regenerated over wounds. Yet this beautiful new covering is then 'rubbed off in tatters soon after antler growth is complete' (Needham 1964). It is clear that we are still a long way from understanding the biological significance of many phenomena of growth and regeneration. We think that we can appreciate repair processes that are of value to ourselves, such as healing of wounds or of bones. But we know so little of the significant features of the life system of other species that some of their major repairs seem to us to be simply a waste. It is doubtful if this is a sound judgement.

Chapter—8

Ageing and Senescence

Cryptobiosis

One of the principal generalizations of biology is that living organisms depend on active expenditure of energy to maintain a steady state. In certain exceptional conditions, however, living organization can continue for a long time in a resting state of equilibrium. The exceptions prove the rule, because such arrested life has no repair mechanisms and its order is progressively eroded by random changes.

During active life the turnover of materials involves incorporation or anabolism, which is a form of growth, and catabolism and elimination, which is a sort of death, in the matter returns to the inorganic world whence it came. In the exceptions to this state of change all metabolic processes come reversibly to a standstill (Hinton 1968). Dehydrated bacterial spores, algae, moulds, lichens, and even some multicellular animals, such as rotifiers and nematode worms, can revive and grow normally after being reduced to temperatures as low as 0.008 °K. Moreover, seeds of the sacred lotus plant (*Nelumbium nuciferum*), shown by ^{14}C-dating to be over 2000 years old, can germinate well. Seeds frozen in permafrost for 10 000 years have also grown (Porsild, Harington, and Mulligan 1967).

There is therefore no doubt that the organization responsible for the processes of life can be maintained when reduced to a condition requiring no interchanges, but only certain spatial molecular dispositions.

In this cryptobiotic state most of the changes that make death more likely with passage of time do not occur, but begin again when it is revived. We can therefore enormously increase the calendar life-span by a period of cryptobiosis, but not the total duration of active life. There is no reason why this should not be possible for mammals, although, so far, the maximum period of freezing seems to be 45 min at –5°C for a hamster (Smith 1961).

Physiological ageing is thus evidently due to the actual operations of the cells and can be avoided in cryptobiosis. Nevertheless, the viability of the organism decreases even in that state. This is due to adventitious oxidations, reductions, and other changes, as can be shown by their dependence on temperature. Thus dehydrated eggs of the brine shrimp, *Artemia,* showed a steep fall in viability according to the time for which they were kept at a high temperature (103 °C). The injuries sustained in this inactive state are thus strictly cumulative. A normal organism in the active state effects repair, so that it can sustain a series of injuries over a span of time that would have caused death if simultaneous. In 'cryptobiosis only the total amount of damage is of concern and not the rate at which it is inflicted' (Hinton 1968).

2. Changes in Homeostatic Capacity with Time

These facts serve as an introduction to the study of senescence. They show that the proteins of the organism and the instructions incorporated in the DNA can remain intact (apart from adventitious changes), *provided that they are not used.* In the active state of organisms the instructions are used to produce controlled metabolism, and for this or other reasons the instructional molecules become damaged and the information is lost. Some of the damage may be repaired (see later), but any repair mechanisms are themselves subject to error, and the general conclusion is that any operating organism must inevitably gradually become impaired by use and by accident. The gradual accumulation of defects constitutes senescence. As mistakes become more numerous, homeostatic mechanisms are progressively impaired and the probability of death increases. The defects may occur in any part of the organism. Many workers have ascribed senescence to changes in the synthetic mechanisms of the cell, or alternatively to the accumulation of mutations in the DNA of somatic cells. These may take many forms, perhaps by depression of operons (Medvedev 1967). They may lead to a decay of various parts of the mechanism of protein synthesis (Maynard Smith 1962) and this may also occur owing to defects other than those of the DNA. In particular, the accumulation of errors in the systems of enzymes ensuring transcription and translation is more likely than errors in the DNA itself (Speyer 1965). Indeed, the DNA is presumably selected to operate without damage to itself. Perhaps there are also mechanisms for repair if damage occurs (Orgel 1963). Errors in the hereditary instructions and their transcription may be of especial importance in post-mitotic cells, such as those of muscle and nerve, where they cannot be corrected or eliminated by cell division. Degeneration of post-mitotic cells has also been widely considered as characteristic of senescence (see Alexander 1967), leading especially to the increasing probability of degenerative and autoimmune disease. These diseases may, of course, be produced by factors other than errors in protein synthesis. Defects due to cross-linkage of the chains of collagen molecule are particularly important in producing failure of skeletal and other structural organs of mammals.

There have been many attempts to define senescence, and to relate it to particular molecular species or organs. Perhaps the only general statement that can be made is that senescence consists of time-dependent changes tending to produce failure of some part of the homeostatic system so that death becomes more probable. As Strehler (1967) puts it, the changes of senescence are:

1. gradual,
2. harmful, and
3. universal (at least in all Metazoans except those that are clones, e.g. *Hydra*, sea anemones).

We must immediately emphasize that all these statements apply to the 'individual' as a unit of survival. There is no evidence that population as a whole undergo senescence. Indeed, life goes on indefinitely in spite of the ageing of the individuals, and it may be even that life goes on *because* individuals age and die, leaving place for others. In fact, we have to inquire whether some manifestations of senescence, far from being harmful, are actually beneficial to the race. Some phenomena of death of whole organisms or parts of them are undoubtedly of value to the species. In all animals and plants there are specific mechanisms that terminate the life of the individual or parts of it at some stage in the life-cycle. It does not follow that all ageing changes are of this specific sort. In particular, it may well be that the changes we call senescence in man are not part of a programme, and, indeed, occur because there is no selective force to prevent them. All this we shall discuss.

Estimates of the force of mortality have often been accepted as measures of 'ageing'. But liability to die is not necessarily a measure of senescence. Fig. 8.1 shows that dogs kept by private owners are more liable to die than those in laboratories, but this is due to accidents ,not to premature senescence. Again, heavy smokes die on average 7½ years younger than non-smokers, but this is mainly because they contract lung cancer and bronchitis, not because they age faster (Alexander 1967). However, smoking may also accelerate senescence by reducing cellular homeostasis in the lung.

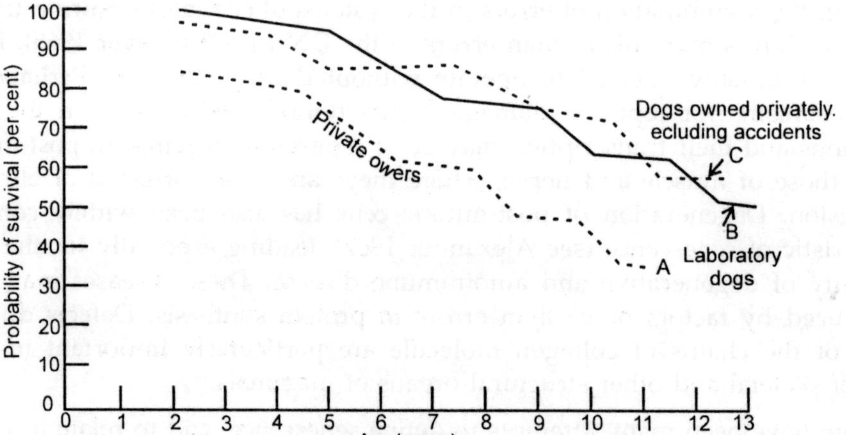

Fig. 8.1: Comparison of survival of two matched groups of dogs: A, kept by private owners; B, kept in laboratories; C survival in the privately kept animals *excluding accidental death.* (After Anderson and Rosenblatt 1965.)

To state that senescence is the accumulation of defects is no very original doctrine. What we need is understanding of the details of those defects that are the more common in man and how their effects may be ameliorated, if not avoided. From the general point of view, many questions can be asked. They mostly centre round problems about the selective effects of different rates of accumulation of defects leading to the characteristic life-spans for each species. It is not merely a question of what deficiencies each homeostatic system can stand, but of whether there is a genetically programmed rate of accumulation of them and hence a specific life-span, correlated with other characteristics of the species. In particular, we are interested in whether the human life-span is in some way related to other features of human life and what might be the effects of lengthening or shortening it.

3. Senescence and Differentiation

There is obviously some connection between the phenomena of development and those of senescence. Some people have maintained that these constitute one continuous set of events, the later changes being the continuation and completion of the earlier, all selected to occur in a manner advantageous to the species. This may be so in some species of animals and plants, but the relationship is usually more complicated, and is so in man. Throughout life every organism operates mechanisms for homeostasis on many time-scales simultaneously. It makes forecasts continuously for both short and long periods, it manifests not only homeostasis but also homeorhesis (Waddington 1968). For example, a growing foetus is mainly constructing equipment for a distant future, but its tissues must also make the short-term forecasts to produce correct levels of cellular respiration, metabolism, and repair processes. Converselly, a grown man may be mainly engaged in nervous and muscular work to find food for the immediate needs of himself and his family, but his tissues are also constructing new materials as medium-term forecasts (for instance in the bone marrow) or long-term ones in the testes.

Thus the processes of differentiation are directed towards producing a homeostatic individual, capable of maintaining the life of the species in its appropriate environment. Any such system is subject to errors, for which there cannot be an indefinite set of repair mechanisms. The most powerful system for securing correction is cell division, which involves abandoning the working parts and starting again from the instructions, eliminating cells that are incorrect. Cell division does, indeed, continue throughout life in many tissues, but it is not possible in the most highly differentiated cells, such as those of the nervous and muscular systems. Failure of the operations of these post-mitotic cells is one of the most characteristic features of senescence. They have their own repair systems, but these depend ultimately upon use of the DNA, and there are liable to be errors during transcription, and especially in RNA in translation, as well as by mutation induced by radiation and other agents. Indeed in a sense the DNA must be especially vulnerable to damage, since each part may be exist only as two molecules in each cell. Presumably it is in order to minimize this risk that nuclei undergo various forms of multiplication such as polyploidy and polyteny, by which extra 'working DNA' is provided as a template when a great deal of RNA has to be produced for cells that are engaged inactive synthesis. Indeed some of the most active parts of cells have their own DNA (e.g. mitochondria, chloroplasts). This avoids coding from the nucleus when there is rapid enzyme turnover.

There are probably other such devices for avoiding the possible effects of damage to the instruction system. It may be that in differentiated tissues there is a reduction of the information available, to avoid side-effects of damage to the unused parts of it. This may be by repression (though this leaves the danger of de-repression by mutation or in other ways) or perhaps by limitation of the transcription so that it will operate only for particular combinations of the various codons for each amino-acid. Each tissue, if it thus 'interprets' the DNA code only in its own way, will reduce its risk of confusion by changes in other parts of the script (see Gurdon and Woodland 1968).

Unfortunately, at present we understand rather little about these fundamental changes during differentiation. Hence we cannot say whether differentiation inevitably involves the ultimate failure of the cell, nor whether there are mechanisms by which such failure is delayed. It seems probable that many of the phenomena of senescence are attributable to the imperfections and errors inevitable in the functioning of any system. Ageing is thus the ultimate fate of all organized individuals. Selection will ensure the provision of such repair devices as prevent the imperfections from interrupting life before the programme appropriate to a given environment has been run. Selection may also in some cases arrange for 'imperfections' to kill the whole individual at a certain time, as it ensures the fall of the leaves from a tree in autumn.

4. Ageing in Plants

Plants show especially clearly that some phenomena akin to senescence are adaptive and are arranged for in the genetic programme of the species. We regard the death of annual plants in the autumn as so natural a phenomenon that we fail to recognize the implication that death may be an essential part of the life system of a species. Further, the fact that in many perennials of temperate regions a large part of the body, the foliage, dies every year, should remind us of the danger of equating death with annihilation of the whole individual as we are apt to do for mammals, though unreasonably. We should be in no danger of making this error if we think of the continual changes of metabolism, or of the death of many cells during development of the embryo, or in the continually replacing tissues. But the fall of the leaves in autumn is visible, while the death of thousands of red cells is not, although the latter is occurring in every one of us every minute.

In any case neither leaf-fall nor the death of red cells indicates the phenomenon we ordinarily call senescence. They show us that death, in the sense of degradation of differentiated material and its return to the non-living , is not by itself a sign of senescence. The mark of the latter is the increased probability of degradation of the whole homeostatic unit, where such can be recognized.

In plants, therefore, we have to consider two kinds of longevity and two kinds of senescence (see Woolhouse 1967), namely those of the whole plant and of its parts.

Thus the leaves of woody deciduous perennials show senescence each year. The whole tree gradually becomes old and decrepit from wear of its unreplaced or infected parts, and shows a reduced growth rate. Its chances of being destroyed are thereby increased. But conversely there is probably no long-term senescence at the cellular level, and cuttings from old trees will resume the growth rate of the parent tree when it was young. However, the question of whether there is a progressive decline in homeostatic capacities in long-lived plants seems to be still undecided. Many factors complicate the assessment of experiments in which there seems to be renewed vitality on transfer to new conditions. For one thing, new opportunities may be provided for growth and cell division and hence elimination of aberrant cells.

5. Senescence in Protozoa

The fact is that a great many different patterns of development and senescence are present and we cannot possibly survey them all here. Not all metabolizing organisms show the signs of deterioration that we call senescence. There are some strains of *Paramecium* that continue indefinitely, dividing without conjugation, and without the nuclear reorganization, called autogamy or endomixis, that substitutes for it. In other protozoa indefinite survival requires periodic meiosis, cross-fertilization, or endomixis. Moreover, many bacteria, yeast cells, and clones of plants propagated vegetatively do not show senescence. Furthermore, there are great variations in the effects of ageing among higher animals.

The case of *Amoeba* has been investigated in detail (Danielli and Muggleton 1959, Muggleton and Danielli 1968). Under normal culture conditions *A. proteus* forms a clone that continues to divide indefinitely However, by protracted treatment (with reduced nutrition, 3.5 weeks at 17 °C) growth can be prevented and the clones ultimately die out. Injections were made of nuclear or cytoplasmic material from such 'spanned' amoebae into normal cytoplasm, and vice versa. Transfer of a normal nucleus into spanned cytoplasm produced resumed logarithmic growth, and a spanned nucleus in normal cytoplasm produced the spanned condition. It was also shown, however, that small amounts of spanned cytoplasm (or a homogenate of it) would produce the spanned condition in normal amoebae.

The transplanted nuclei would of course carry some cytoplasm and it is impossible to say that either part of the cell is exclusively involved. It is clear, however, that the conversion of one type to the other involves the intracellular control mechanisms. Since changes in the DNA are not readily produced, it seems likely that the main factor involved in the 'spanned' condition is either them RNA or the mechanisms for its translation into proteins in the cytoplasm.

6. Is senescence a product of Selection or of its absence?

Evidently senescence is not the universal fate of living matter, as it appears to be for us humans. However, all these creatures show cell division, even if they do not need periodic sexual reproduction. We shall have to inquire whether there is a connection between methods of cell replacement and senescence, the failure of homeostsis, and the mechanisms by which the integrity of the instructions is preserved. There would be perhaps some satisfaction in thus finding a 'cause' for senescence. A more intellectually complete reason for believing it to be true would be found if the different patterns of ageing of different species could be explained as adaptations to particular ways of life. Unfortunately, further analysis shows reason for thinking that a main reason for senescence in higher animals and man is both simpler and less interesting. The essential point is as follows. Obviously genes that manifest their effects late in life escape partly or wholly from the force of natural selection (Medawar 1952). There is every reason to think that such gene sexist, for example genes producing cancer, and that they have not been eliminated because the individuals who bear them have already reproduced before the bad effects occur.

More generally, it is demonstrable that any genes that confer reproductive advantage relatively early in life will be selected, even if they are pleiotropic and have ill-effect later. Senescence may thus even be an inevitable consequence of natural selection (Hamilton 1966).

It is important to realize that this analysis involves many assumptions about the incidence of the force of mortality. This must be so since we have to reconcile the facts that

(a) organisms have very different life-spans, and

(b) 'in wild populations generally, it probably seldom happens that senescence has more than a slight modifying influence on mortality' (Williams 1957).

Animals fall a prey to some natural accident before they become senescent. But this is only to say that the selective forces operate so as to produce a developmental programme of a length appropriate to the circumstances of the species. There are, indeed, indications that low adult death-rates are associated with low rate of senescence (and Williams 1957).

It is very difficult to make reliable inferences in such matters. It is certainly demonstrated that genes giving an early reproductive advantage are likely to be selected. But there remain all the influences that must counteract this effect, since lifetimes vary. In man, in particular, selection has obviously *not* chosen those genes that ensure maximum fertility at the earliest possible age. It may be that a process of natural selection inevitably implies senescence, yet the rate of senescence of each species must be individually determined by particular selective factors. It seems likely that these differences are somehow connected with the means adopted by the species to maintain its integrity.

7. Programmed Cell Death

Just as there is constructive synthesis on all time-scales, so there is programmed death at all times throughout the life-cycle. Statisticians like to study only the death of the whole organism, and are happy that each dies only once. But there is death of cells at all ages. And certainly often this is not simply a degenerative phenomenon of senescent failure but a part of the constructive actions of morphogenesis (Glucksmann 1964). This, waves of cell death occur as the brain is formed (Kallen 1955). Certainly, therefore, there are systems in the developing organisms that ensure the death of a proportion of the cells, as part of the programme that is to lead to the continued life of the whole.

Thus the periodic rejection of material in catabolism and of whole cells and organisms by death is a universal feature of living things. Death is as universal as life, and in this sense is essential to life. Like other phenomena manifested by organisms, it somehow serves to assist in the homeostasis of the species. Of course we must recognize the possibility that death, being the negation of life, is not 'adapted' as are all other aspects of living systems. It is sometimes said to be 'outside the programme' of the genetic instructions. It is even denied that phenomena of wear or failure of repair are involved in senescence. William (1957) considers that the concept of wear is inappropriately borrowed from static man-made machines and should not be applied to self-replacing living systems. But the difficulty of avoiding the effects of random influences on an organized system does not depend on calling it a 'machine'. Nor are self-replacing systems made invulnerable to the inevitability of random errors by the fact that in our language they are not 'machines' in quite the usual sense. Our task is to define the changes that take place with time in living systems and to show how they are connected with the outstanding fact about life—that it continues.

Living organisms maintain their lives in many diverse ways, and the usual biological principal is that their differences have been tailored by natural selection to allow continuance in varied environments. We shall ask whether this is so also of the pattern of ageing. Curiously enough, this is not a popular question for gerontologists, who perhaps feel that it over-simplifies the problem. Certainly a first task must be to define more clearly what changes with age we are considering. No one would deny that the age changes by which an embryo builds its enzymes are 'adapted', nor would this be denied of the weaves of cell death in neurogenesis, discussed above (Kallen 1955). But are the deaths of the nerve cells during senescence properly to be compared with those in the early embryo? Evidently we are in the danger that words like *ageing, senescence,* and even *death* are inexact and carried over from everyday speech. We must define more precisely the phenomena that are to be included.

8. General Definition of Senescent Changes

The changes of old age affect so many parts of the body and are superficially so familiar that it is not easy to classify them in a biologically meaningful manner. What is the connection between the facts that old men and women have wrinkled skin, are foolish, pant when they exert themselves, are no longer able to reproduce, and are increasingly prone to die? It is not our purpose to present a detailed catalogue of all the aspects of the misfortune that we call senescence. In order to try to bring out the essential features of the process we have to see whether it is possible to go behind all these particular manifestations and to classify them all into a few categories. It would presumably be the view of Williams, and perhaps Medawar, that no such classification is possible. According to them, the changes of senescence are a mixed bag of side-effects of genes that happen to have favourable effects in younger stages. On the contrary, Comfort (1966) provides a very concise classification of senescent changes divided into three categories:

1. changes in dividing cell quality ("faulty copying");
2. loss or damage among fixed post-mitotic cells; and
3. colloidal change in molecules, including extra-cellular structural molecules (such as collagen) and the intra-cellular information store.

Comfort goes so far as to say that all the changes that we associate with senescence can be placed under one or several of these headings. Thus, collagen and elastin, which are molecules with a small rate of turnover in the adult, become increasingly changed with old age. New combinations as appear in the molecules, serving to cross-link the polypeptide chains and alter the strength and elasticity that are their valuable properties for the body. Many of the more obvious changes in old age are due to alterations in the connective tissue, for example some of those in the arteries and the skin.

The fixed (post-mitotic) cells, such as those of muscle and nerve, seem to show especially severe effects of senescence. Rate of change of psychological capacities is said to provide the best index of expectation of life (see Jarvik and Falek 1963). Neurons are probably lost steadily in the later decades of life. A figure of 100,000 dying per day is often mentioned, but there are no good data, or information as to whether the rate of loss increases. The nervous system depends upon the redundancy of its numerous channels, with which it can make adequate predictions and communicate them to the effectors, in spite of the high noise-level and relatively slow speed of operation and conduction. Reduction in the number of channels would have the effect of reducing the reliability of prediction. This may be especially important in such key regions as t' hypothalamus, whose neurons influence so many homeostatic processes, including several systems that operate as clocks. The hypothalamus and other lower centres contain relatively few neurons (but there are no exact data about their loss in senescence).

The post-mitotic cells are presumably especially vulnerable to the probability of damage by use and by mutation. Indeed many biologists believe that somatic mutations (in the widest sense of any genetic injury) alone provide a sufficient explanation for senescent change (see Curtis 1966). For example, in old mice 70 per cent of liver cells have gross chromosome abnormalities. Evidence has recently appeared that chromosome abnormalities, especially in the *X* and *Y*, are increasingly evident in man after age 45 (Hamerton *et al.* 1965). There is evidence, therefore, of declining mitotic efficiency. Further, radiation, which increases the mutation rate, also shortens life, in a manner that some hold to be similar to senescence (radiation ageing). However, a careful study of the effects of radiation on the population of Hiroshima showed no correlation between senescent changes, measured in various ways, and closeness to the centre of the atomic explosion. Moreover, there are other reasons for supposing that somatic mutation is not by itself the cause of all senescent changes.

On the other hand, as we have seen, degenerative diseases play a prominent part in the increasing probability of death with age. It is suggested that they may be due to stepwise changes, each produced by a mutation with its own ageing probability (Fig. 8.2) (Curtis 1966, Burnet 1965). If this were so the relation between death-rate for any disease and time should be a power function, the power depending upon the number of steps.

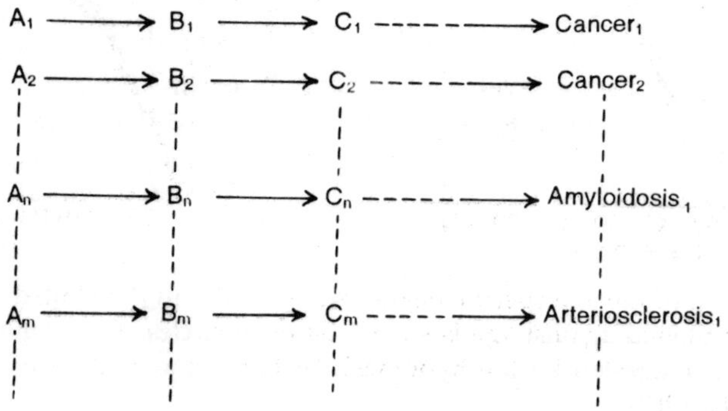

Fig. 8.2: Representation of the ageing process as the result of a series of step changes. Each of the degenerative diseases is the result of a series of processes each with its own rate constant. One or more of the steps from one process to the next is almost certainly a mutation, but equally not all are mutations. (After Curtis 1967.)

$$\frac{dN}{dt} = Kt^{(10\rho-1)},$$

where *N* is the percentage of persons contracting the disease, *t* the time, *n* the number of steps in the process, and *K* a constant depending upon the mutation rate, the environment, and other factors. This equation fits the facts very well (Fig. 8.3). The chief exceptions are diseases strongly influenced by environment, such as cancer of the lung, or by hormones, as cancer of the breast. Somatic mutation, then, probably plays a large part in the causation of diseases of the old, but these do not in themselves constitute senescence, or at least certainly not the whole of it.

But of course radiation mutation, presumably damage to the DNA, is only a conspicuous example of the hazards that affect all parts of the system from the operation of random influences. We are assuming that any system in continued operation will be subject to such hazards and therefore be in need of repair or correction. This can be achieved only by one of the two methods we have discussed, namely comparison with some standard to effect a repair, or rejection of the damaged part. There are specific repair mechanisms in the post-mitotic cells, including synthesis of new DNA, but these mechanisms must themselves ultimately be subject to wear, and so on. The regress can only be evaded by some special procedure, which seems

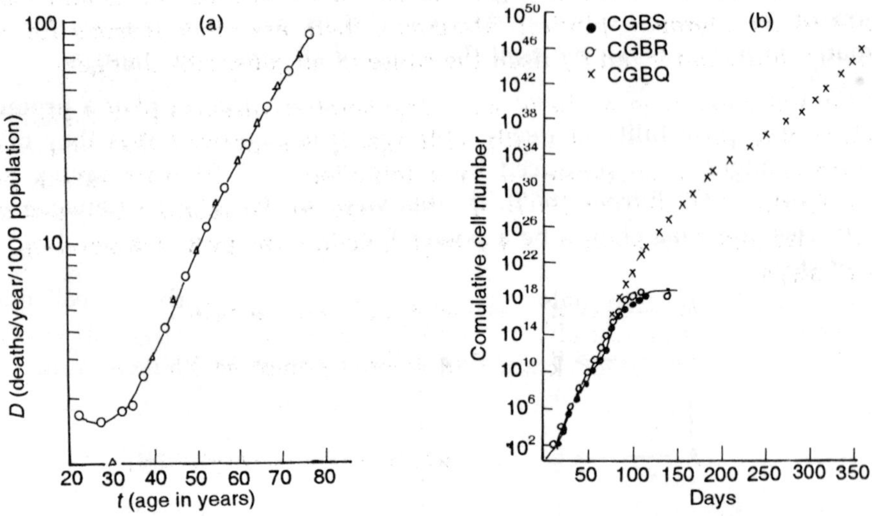

Fig. 8.3(a): (a) Log of the age-specific death rate of males in the United States from all causes in 1963 plotted against age is shown by open circles. The triangles show the theoretical curve suggested by the hypothesis of Fig. 8.2 and derived from the equation (after Curtis 1967)

$$\frac{dD}{dt} = 3.0 \times 10^{-8} t^5$$

Fig. 8.3(b): (b) An established cell line of goose fibroblasts. Cumulative growth curves of cells of three geese grown in tissue culture. Two of the cultures, CGBS and CGBR, were typical cell strains and became senescent after 65 generations, but the third has continued to grow as a line for more than 150 cell generations. The chromosomes of these birds are difficult to study, but so far as can be seen those of the line CGBQ are normal, although they may perhaps be polyploid. (From Coon, H. G., and Williams, I., *Carnegie Institution Year Book 67*. 1967-8.)

to be connected with cell division, and perhaps involves simply elimination of errors by rejection of the unfit.

9. Ageing in Dividing Cell Populations

It is therefore especially important to find that the dividing cell populations also show signs of ageing. It is probable that cells growing in tissue culture survive for only a limited number of generations, *unless they undergo chromosome changes and become cancer cells* (Hayflick 1965). It has been very widely held in the past that cells in tissue culture multiply indefinitely, a view first proposed by Carrel (1924), one of the pioneers of this technique. The secret of such apparently immortal cultures may be either that new cells were unwittingly introduced with the embryo extract that was used in the medium for the repeated subcultures, or it may be simply that the cells had mutated and become cancerous. It has become the practice to discriminate between *cell strains*, which are of limited life, and *cell lines*, which are unlimited but abnormal.

The cells in cell strains keep their normal morphological and immunological properties and normal diploid chromosome number. They can divide only a finite number of times. Cell lines will multiply and produce tumours when inoculated into members of their parent species, which cell strains will not. The cells of cell lines have abnormal chromosome numbers. They will continue to multiply indefinitely —some have lasted more than 20 years, dividing every day. There are, however, a few cell lines known whose cells, though abnormal in karyotype and in other ways, are not cancerous (e.g. the $_3T_3$ strain of fibroblasts of mice, Todaro and Green 1963, and the line shown in Fig. 8.3(b)).

The cells of different animals vary in their capacity to produce cell lines. Thus, cultures of mouse cells almost always become changed and malignant, whereas those of chicks do not do so. It is not known why some clones are stable and others not.

Human embryonic somatic cellls (fibroblasts) cultured *in vitro* survive for about fifty cell generations; those from adults for less, usually about twenty generations. It has not been possible to establish an exact correlation between the age of the donor and length of life of the strain. The strain can be preserved in an arrested state at sub-zero temperatures indefinitely. When warmed it will recommence division, but finally this ceases after a total of about fifty divisions before and after cooling. Cultures of human cells are longer-lived than those of most mammals, and this is presumably correlated with man's longer life-span.

These facts suggest that mammalian somatic cells have a fixed life-span, in terms of cell divisions. This is confirmed by the facts of heterochronic grafts between inbred hosts (Korhn 1962). The grafted tissue survives for a time that is set by its own life-span, not that of its surroundings.

Other evidence of a finite life-span, for cultured tissues is that the latent period before migration of cells from chick-heart cultures increases with age (Lefford 1964). The decine in 'growth energy' of cells of the chick has already been mentioned and a similar phenomenon has been shown for explanted liver of the rat (Glinos and Bartlett 1951).

However, some plants show indefinite asexual reproduction, and we have used these examples to suggest that there is not inherent principle of senescence. Presumably cell division and cell death may be able to supply the corrective measures that repair inevitable damage, particularly to the DNA (see below). The existence of such instances supports the view suggesting that senescence is an unfavourable feature (in relation to homeostasis) *inevitably incorporated into organisms that develop a soma that is not passed on in reproduction*.

The fact that there is a finite life-span even among dividing cells of a metazoan agrees with the view that there is a species-specific age determined by the genotype. It is not difficult to understand that in post-mitotic cells the metabolic processes must ultimately go out of control. Individual enzyme systems may be repaired or renewed, but the mechanisms for reading out the instructions must also be liable to damage by use. Even the DNA instructions themselves cannot be immune from the liability to wear as they are used to produce messenger RNA. These molecules are very stable but they have no magical inherent property that insures them against all possible damage. However, they must be capable of repair when damaged, for we know that the DNA is passed on from generation to generation by the germ cells, unchanged except by mutation.

10. Damage and Repair of Instructional Molecules

There is still no clear agreement on changes in DNA itself with age. DNA prepared from the thymus of old cows is more resistant to heat than that from calves (Hahn and Verzar 1963). This has been compared to the structural ageing of collagen by cross-linking. This protection of the DNA from heat may be due to increased binding of the DNA to protein (probably histone, Hahn 1966), which interferes with its power to act as a template for RNA-polymerase action (Devi *et al.* 1966). It remains to be discovered whether this cross-linking is a result of the use of DNA for manufacture of messenger RNA.

The interpretation of all such observations is still uncertain since the changes may be in any of various cell components. Thus the difference in thermal stability of cow and calf nucleo-protein disappears if the preparation is made in such a way as to reduce the amount of non-histone protein and RNA in the preparation (Pyhtila and

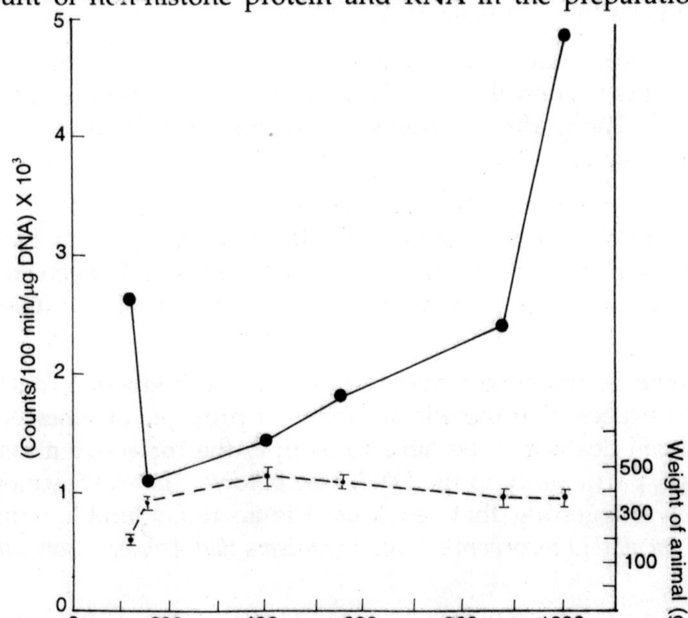

Fig. 8.4: Specific DNA activity (continuous line) and weights of animals (broken line) after ^{3}H-thymidine uptake by rats, as a function of age. Each point is from one animal only. The youngest shows a high specific activity, presumably associated with growth. There is a striking increase at the end of life. (After Samis *et al.* 1966.)

Sherman 1968). The age-change in template action also disappears, and in fact no change in the DNA itself could be detected.

On the other hand, in rats, the amount of ^3H thymidine incorporated into liver is said to increase linearly from 154 to 881 days and thereafter still faster to 1007 days *without increase in net DNA per nucleus or in mitotic index* (Fig. 8.4, Samis *et al.* 1966). The suggestion is that the RNA polymerase system is able to detect a nonsense sequence and that degradation of the damaged segment by DNAase and repair by DNA polymerase follows, using the information in the unaltered homologue. This is as yet a rather vague and untested hypothei, but DNAase and DNA polymerase certainly exist in tissues that have a low mitotic activity. In any case increased DNA synthesis might well result from causes other than degradation of DNA itself. Thus it may well be that the cytoplasmic systems for RNA translation and protein synthesis function less well with age and this produces a demand for greater DNA template activity.

A further suggestion is that there is in the chromosomes working DNA, distinct from the master copy, which is not used to produce RNA but only to make the working DNA (Callan 1967).

11. Cell Division and Cell Death as Guarantors of the Instructions

If our general view is correct there is a tendency for faults to occur in all parts of a living system. They occur as the result of random operating errors, but their frequency may well be greater in certain parts that are especially liable to wear. Many of these effects will be small and they will effect homeostasis only if they accumulate.

In order to remain alive the system must somehow compensate such errors. In order to do this it must have a mechanism for detecting them, and we have already considered possible means by which defects in the DNA are detected and repaired.

It is suggested that further important mechanisms for eliminating errors are provided by mitosis and perhaps by meiosis. During the period when a cell is dividing its DNA is unable to produce RNA and its enzyme systems must function without renewal. This is therefore a time of stress, which may serve to test the adequacy of the cell machinery. Cells are known to be especially susceptible to radiation while they are dividing (though this may be due to the state of the nucleic acid molecules at the time). The point is that mitosis provides a filter by which altered cells are detected and destroyed.

Filters of some sort are evidently important mechanisms that protect the integrity of the information store. An essential part of the process is reduction of the organisms or part of it to a minimal state in which selection can operate. The necessity for this periodic reduction and selection might be said to be the reason for senescence and death, but does not explain to us the; factors that control the pattern of senescence as it operates indifferent species.

Although this selective action during cell division is important there may also be other factors at work. With the breakdown of the nuclear membrane the chromosomes are placed in contact with the enzymes and other contents of the cytoplasm. These may serve to alter the pattern of repression of the genes, suppressing some that have been operating and releasing others. This reprogramming of the read-out of the DNA may serve to bringing to play a new complex of enzyme systems, appropriate to changed circumstances (Gurdon and Woodland 1968).

These suggestions are all relevant to the basic problem of how the information store is preserved and whether there is a connection between the failure of the mechanisms that ensure this and the ultimate death of the individual. It seem certain that repair mechanisms such as the above exist, and that they are developed to different extents in organisms with different life-spans. It is equally certain that the cannot be ultimately successful, for this would involve the regress of the repairers that we have considered so often. The only ultimate guarantee of homeostasis is to allow for the failure and elimination of the incorrect. This is the function of death, whether of cells or organisms.

LIFE TABLES AND THE PATTERN OF SENESCENCE

1. Changes with Age in the Expectation of Life

In order to study patterns of senescence we must have means of measuring it. Obviously many aspects of old age could be measured, but the most useful one has been found to be the liability to die. When we say 'he looks older' we do not mean only 'we think his chances of dying have increased,' but the remark carries this implication. Biological statisticians find it easier to measure the expectation of life than the wrinkles, inanity, and other feature by which we judge old age. Their studies of death-rate shave provided a means of measuring the rate of senescence and they thus provide some basis for a scientific approach.

The rate of mortality is expressed as the number per 1000 of a given population living at time x, who will have died by time $x + 1$, from which we can derive an expression for the force of mortality (Fig. 8.5). Senescence is measured as the progressive increase in the probability that the individual will die during the next interval of time. Each organisms dies only once, therefore a quantitative estimate can only be determined statistically in a population.

Living organisms are highly complex systems, which are so controlled as to maintain their steady state by use of information received from the past. If the efficiency of the control and the amount of information available to an individual remained constant throughout its life, the chances of its dying would be the same at all ages. If it acquires more information (or better control) its chance of life improves, and if it loses information or control they decrease. All three conditions can be found. We ordinarily call the increase of information 'learning,', or becoming wiser, but it also

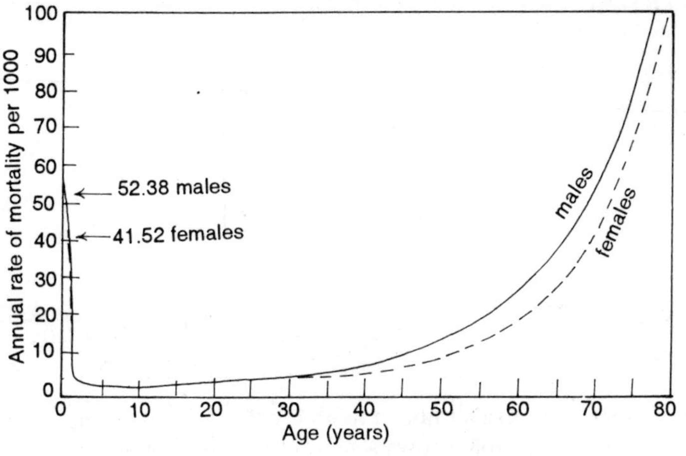

Fig. 8.5: Annual rates of mortality per thousand, by sex (United States 1939-41). (After Comfort 1964.)

includes acquiring immunity, growing suitable muscles and bones, and the many other sorts of information storage we have discussed.

2. Life Tables

If the chance of death is independent of age the proportion of survivors will decrease steadily, as shown in the logarithmic curve of Fig. 8.6(a).

If they die only because of senescence the curve will approach the 'rectangular' form of Fig. 8.6 (b). In this case as life proceeds individuals die from causes that would not have killed them before, the exact shape of the curve depending upon the rate of senescence. Fortunately, it is relatively easy in man to record life tables. It is also possible for animals kept in the laboratory, especially if their lives are short. But it is very hard to get life tables for animals in the wild state, which are the ones we most need to study since they are exposed to natural risks. The life table available show the above possibilities and various mixtures of them. Thus the slug and *Hydra* of Fig. 8.7(b) showed death-rates almost independent of age. In other species, such as the minute rotifers that live in ponds, the onset of senescence is sudden and death comes to the majority of the population at the same age (Fig. 8.8). When it is more gradual we get the curves of Fig. 8.7(a) and this is the general form of human life tables (Fig. 8.9).

When death occurs in most individuals at about the same age this is called the specific age, and it is presumably determined by selection. For the rotifers it is probably an advantage to kill off all the individuals when they have reproduced, in order to allow for several generations during the short persistence of the pond. In other animals there may be different reasons for terminating life abruptly. The life-span, like so many other features, is determined for each species, not by any general factor. In man the shape of the life table is obviously the result of the operation of several influences.

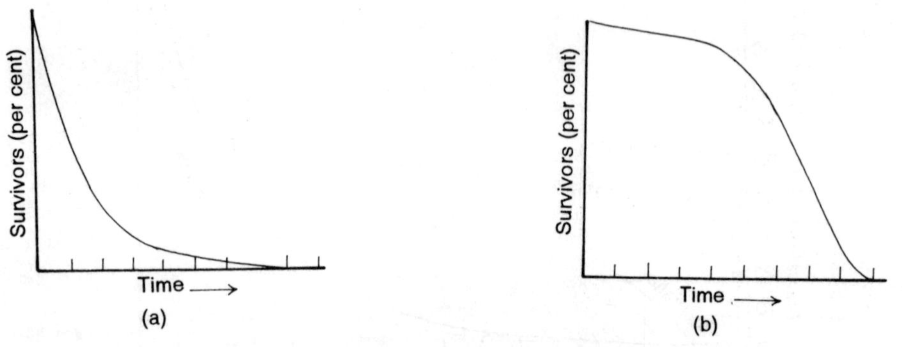

Fig. 8.6 (a): Survival curve at a constant rate of mortality (50 per cent per unit of time). (b) Survival curve of a population that shows senescence. (After Comfort 1964.)

The chance of survival with age neither decreases nor increases steadily and we have to try to identify the overlapping influences that affect the death-rate at each part of the curve.

1. Young children are very liable to die, largely from infectious diseases, for which they have not yet acquired immunity.

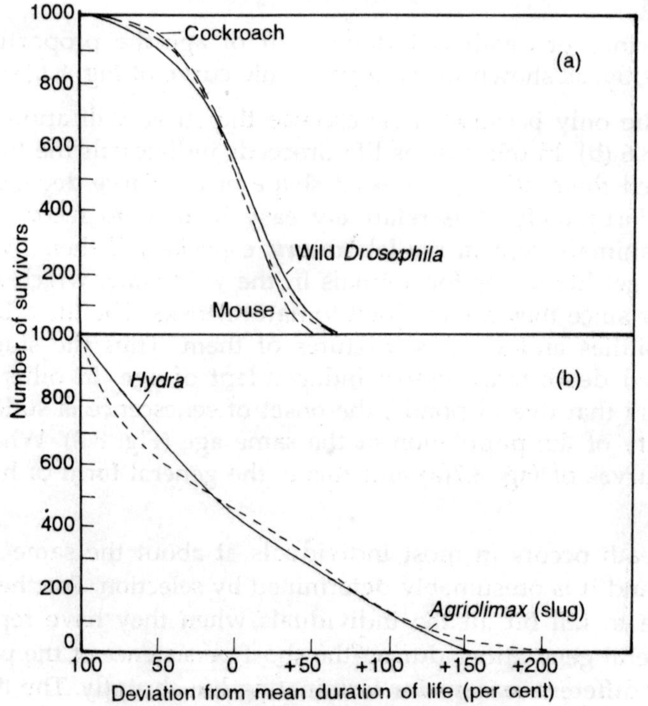

Fig. 8.7: Types of survival curves illustrated by different organisms. (After Pearl and Miner 1935.)

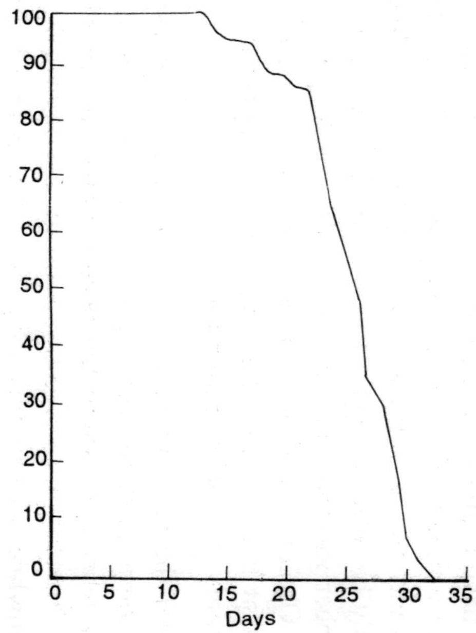

Fig. 8.8: The life-span of a rotifer, *Philodina citrina*, in a culture. (After Lansing 1947.)

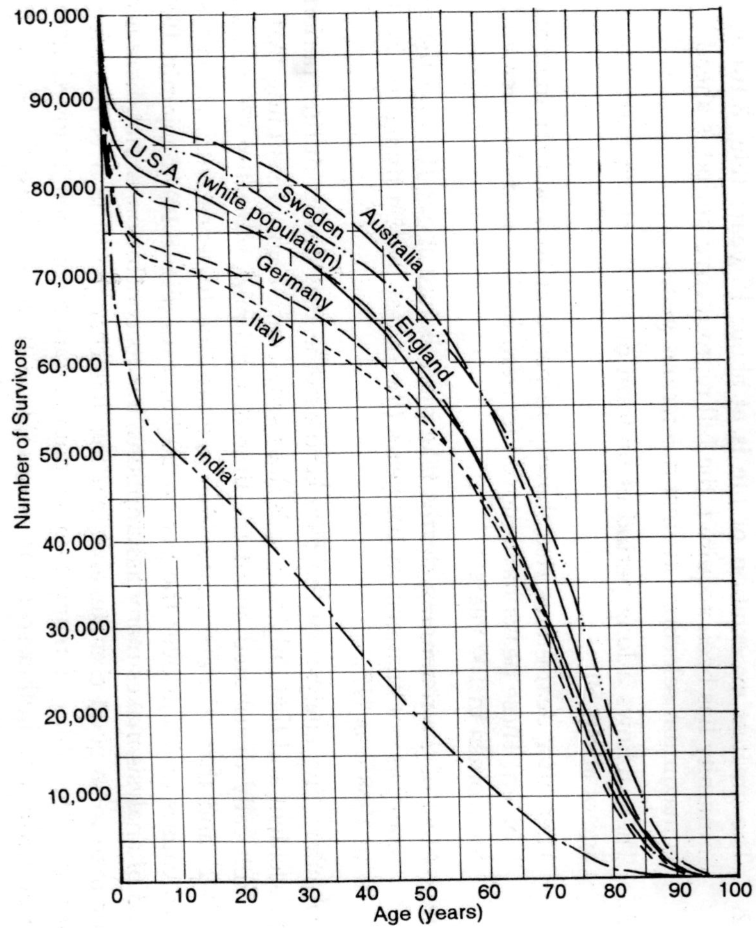

Fig. 8.9: Survivors of 100 000 males born alive in various countries, 1901-10. (After Pearl 1940.)

2. The greatest expectation of life is at about 10 years (Fig. 8.10) after much immunity has been acquired, the full reproductive period is ahead, and there is maximum selection against deleterious genes.
3. There follows a long period of uniform or slowly rising mortality up to the age of about 50.
4. Thereafter death-rates rise sharply to peaks at 76 years for men and 80 for women. These peaks have moved forward by 3-4 years for men and 7 years for women in 100 years.
5. Most of the remaining deaths take place before about 100 and the total life-span has not changed in 100 years (the oldest person living in Great Britain in 1968 was 112).

An interesting analysis of the data is to assume that deaths during period (3) are due to accident and that only after this does senescence begin. Assuming that its onset is symmetrically distributed about the time of maximum mortality (76 years for men in 1950–2) and that all deaths after that age are due to senescence, we can separate the two causes of death over the period 50–76 years (Fig. 8.10) (Benjamin 1959). This method of analysis makes many assumptions, but its conclusion is that there is a *specific age* for man in a good climate, and that this has changed little over recorded human history. The life tables of many human populations have been changed greatly as a result of medical and other social advances (Fig. 8.9). Before these changes populations approach the logarithmic curves of constant death-rate. With better conditions the rectangular shape becomes increasingly apparent, emphasizing that there is a specific age for man.

There are no accepted hypotheses as to how this situation has come about. One explanation may be that the limit is related to the time of menopause in women. It is suggested that this may have been determined at about 45 as the time at which she and her mate would have produced the maximum of offspring that could be properly cared for. If the maximum age of reproduction is thus fixed, deleterious genes manifesting later will only be selected against as long as they seriously prejudice the upbringing of the younger members of the family. This would put the time of the specific age at about 60. Obviously no such simple single influence has been at work. It is even just possible that the brains of the longer-lived people have helped the survival of their descendants in social communities!

It emerges clearly that the phenomena of senescence are not simply a matter of wearing out, but bear a complicated relationship to the pattern of life of each species. We can see only vaguely what this pattern has been for man and are not yet able to forecast what it will be. Probably our successors with greater knowledge will see the future more clearly. Meanwhile, we can only look for situations in which it is possible to study the expectatiosn of life under various conditions.

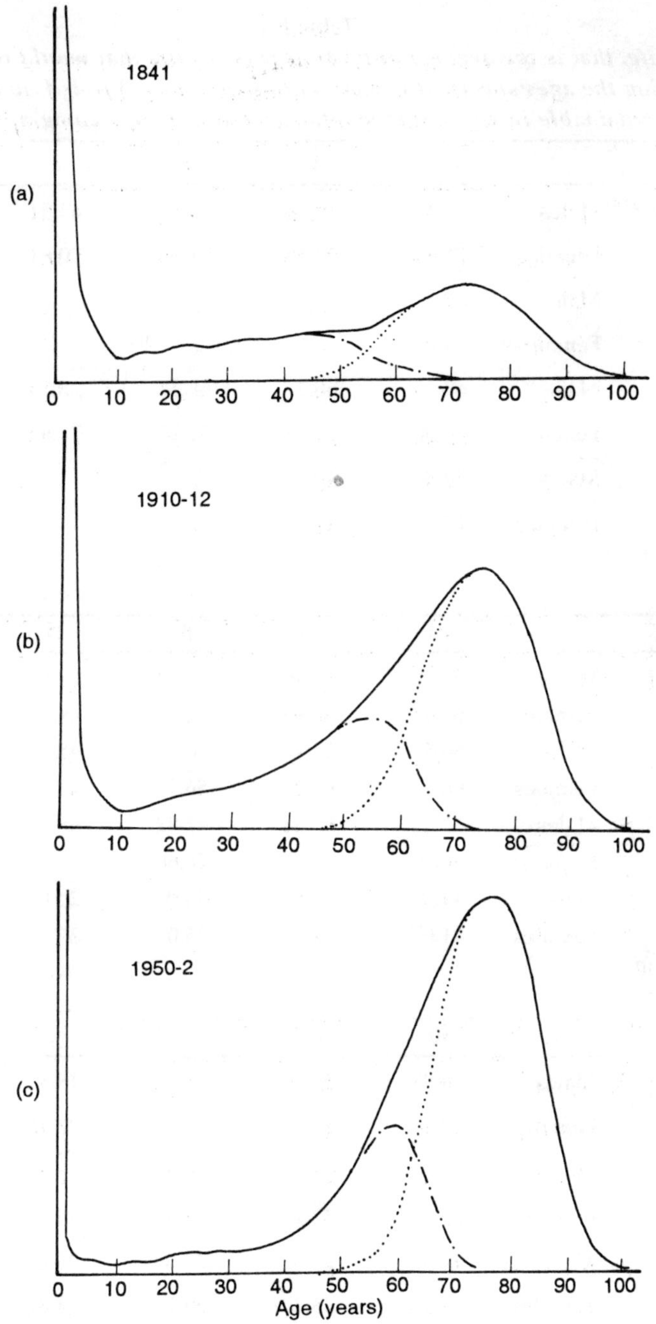

Fig. 8.10: English life tables, males (curves of deaths). The ordinate for (a) based on 51 000 births, those for (b) and (c) on 100 000. The senescent deaths and anticipated deaths are calculated as described in the text. (—total deaths;...senescent deaths;—.—.—.— anticipated deaths). (After Benjamin 1959.)

Table 8.1:

Expectation of life, that is the average number of years of life that would remain for males and females reaching the ages specified if they continued to be subjected to the same mortality conditions as obtainable in the period mentioned (** from **U.N. Demographic Yearbook 1966)

Age		0	1	2	3	4	5
United Kingdom†	Males	66.42	67.66	66.82	65.91	64.98	64.04
1950–2	Females	71.54	72.38	71.53	70.61	69.67	68.72
1963–5	Males	68.3	. .	. .	. .	. .	65.1
	Females	74.4	. .	. .	. .	. .	71.0
Albania	Males	63.69	68.2	68.58	68.24	67.56	66.74
1960–1	Females	66.00	71.13	71.99	71.93	71.34	70.54
Africa	Males	34.3	40.0	. .	. .	. .	42.0
West Cameroon African population 1964–5	Females	37.2	41.0	. .	. .	. .	44.0

Age		10	15	20	25	30	35
United Kingdom†	Males	59.24	54.40	49.64	44.96	40.27	35.60
1950–2	Females	63.87	58.99	54.17	49.41	44.68	39.97
1963–5	Males	60.3	55.4	50.6	45.9	41.1	36.3
	Females	66.1	61.2	56.3	51.4	46.5	41.7
Albania	Males	62.27	57.54	52.82	48.15	43.51	38.90
1960–1	Females	66.19	61.43	56.66	51.99	47.40	42.83
Africa	Males	41.0	38.0	35.0	31.0	28.0	24.0
West Cameroon African population 1964–5	Females	41.0	39.0	35.0	32.0	28.0	24.0

Age		40	45	50	55	60	65
United Kingdom†	Males	30.98	26.49	22.23	18.31	14.79	11.69
1950–2	Females	35.32	30.76	26.34	22.10	18.07	14.43
1963–5	Males	31.6	27.1	22.8	18.7	15.1	12.0
	Females	37.0	32.4	27.9	23.6	19.4	15.6
Albania	Males	34.33	29.83	25.56	21.64	18.09	14.74
1960–1	Females	38.26	33.3	29.21	24.81	20.67	16.82
Africa	Males	21.0	18.0	15.0	13.0	12.0	9.0
West Cameroon African population 1964–5	Females	20.0	17.0	15.0	13.0	12.0	10.0

Age		70	75	80	85
United Kingdom†	Males	9.00	6.70	4.86	3.48
1950–2	Females	10.97	8.10	5.83	4.20
1963–5	Males	9.4	7.1	5.3	4.0
	Females	12.1	9.1	6.7	4.9
Albania	Males	11.65	8.94	6.63	4.73
1960–1	Females	13.34	10.19	7.56	5.43
Africa	Males	8.0	..	..	..
West Cameroon African population 1964–5	Females	8.0	..	..	..

† England and Wales only.

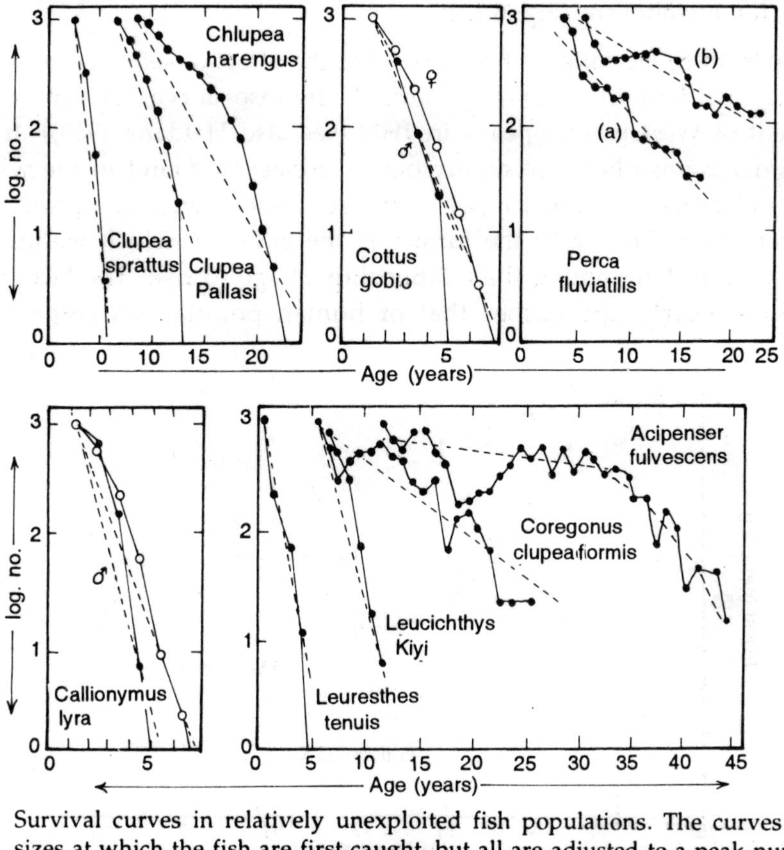

Fig. 8.11: Survival curves in relatively unexploited fish populations. The curves begin at the sizes at which the fish are first caught, but all are adjusted to a peak number of 1000. The scale is logarithmic: a straight line shows constant mortality; a downward curve shows increasing mortality with age. The dotted lines are to help the eye. (After Beverton and Holt 1959.)

As life tables become available, using new methods for determining the age of animals (for example by weighing the lens of the eye or studying the tooth structure, see Linhart and Knowlton 1967, Sikes 1966), it is possible that wide differences may be found between longevity in laboratory and wild populations.

Survival curves are available for various species of fish, whose age can be determined from their scales (Fig. 8.11) (Beverton and Holt 1959). The shorter-lived species mostly show clear signs of departure from constant mortality, that is, they have specific ages. The longer-lived forms (sturgeon, whitefish, and perch) seem to show constant mortality rates at least over long periods, though in the oldest of all (sturgeons) the death-rate increases in the end. Many other factors enter into the question (i.e. growth and size, and reproduction) but it seems that in some fishes life is not terminated at any specific age. On the other hand, some have a definite maximum age, though even this may be influenced by external factors, such as temperature, which affect the metabolism (Fig. 8.12).

Again wild mice are killed off so early that the effect of senescence never appears, but it is very evident in laboratory animals (whose curve, as Comfort observes, resembles that of Western Europeans in 1900 (see also Haldane 1953)). It thus seems that we can distinguish between an ecological longevity found in the wild state and a physiological longevity that can be achieved under better (? optimal) conditions (Bodenheimer 1938). Obviously the former is the entity in which we are, in general, more interested, in determining the relationship of age to risk. The fact that the latter condition more nearly approaches that of human populations emphasizes that in

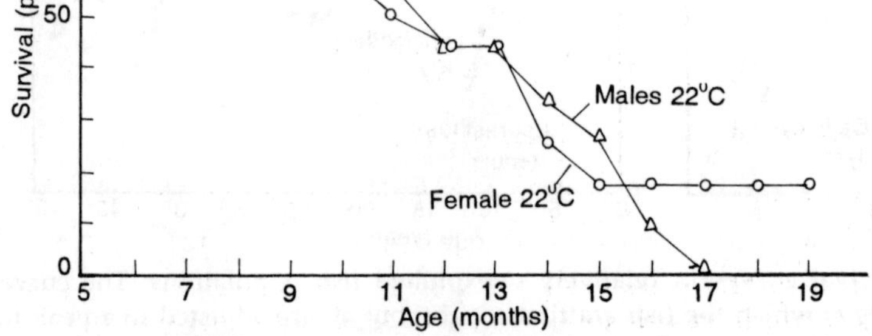

Fig. 8.12: Percentage of survivors of the fish *Cynolebias adloffi* kept at different temperatures. (After Liu and Wallford 1966.)

matters of ageing man is unusual—as a result of his exceptional capacities for obtaining and storing information and exerting control.

3. The Pattern of Senescence

If the pattern of senescence is determined by the adverse secondary late manifestation of genes there should be some correlation between senescence and the age distribution of reproductive potentiality in relation to the period of differentiation of the body. The following indications of this are given of Williams (1957):

1. Unicellular organisms, having no soma, may show no senescence. This is not of course to say that individuals are immortal. The fact that some clones die out while related ones survive shows that selection is at work.
2. Animals with a relatively low adult mortality show a slow rate of senescence. The tortoises and turtles show further examples. Their shells give them a special protection and they have a longer life-span than any other vertebrate. Converely, animals that have a very high mortality before maturity also do not show especial manifestations of senility (e.g. marine lamellibranchs or crustaceans).
3. Organisms that increase in fecundity after maturity should not show rapid senescence. This is markedly true for those animals that continue to grow after maturity and hence produce more and more eggs, notably many molluscs, crustacea, and aquatic vertebrates. An old carp produces ten times as many eggs as one just mature. This increased fecundity *increases* the reproductive probability for later ages and hence produces selection against senescent changes. Perhaps when we have means for measuring the various senescent changes in these animals it will be possible to make quantitative correlations.

   Such continual size increase is much more easily achieved by aquatic and especially by marine animals because their growth is not limited by gravity considerations. Indeed, since the resistance to movement in the water is mainly the skin friction, a large animal (or ship) is more efficient than a small one, because area increases only as the square of the linear dintensions. Conversely, land animals must be provided with mechanisms the limit their growth if their limbs are to carry their weight. It was noticed by Bidder (1932) that this size limitation is accompanied by senescence. The association may be dependent on the different patterns of fecundity that are correlated with the dimensional factors.
4. If senescence results from a failure of selection then it should supervene very rapidly in animal species that breed only once. Certainly this is true of some of them. Conspicuous examples are the octopus and some other

cephalopodds. The female, after laying her eggs, guards and aerates them, but when they have hatched she stops eating, and dies soon after (see Vevers 1961, Wood 1965). A classic instance among insects is the imago of the mayfly. Among vertebrates we could name the lamprey and other anadromous migrants. They return to the rivers to breed and die immediately thereafter. Such a pattern would tend to evolve from one with several breding seasons (Williams 1957). Individuals that returned to the sea would suffer further risks and their reduced fecundity would accelerate the development of faster senescent changes in the population.

In all these species we recognize immediately that the timing of the sequence of the life-cycle is specifically determined by the genotype. Indeed, in some cases we know the mechanism by which this is achieved. In *Octopus* it is hormonal; there is an optic gland that stimulates the ovary (or perhaps releases it from inhibition—see Wells and Wells 1959). The optic gland is itself controlled by changes in illumination (Richard 1967), which, as in so many animals, play a central part indetermining the onset of reproduction, presumably because appropriate timing in relation to the seasons is essential in many environments. If the optic gland of the octopus is denervated the female at once begins to mature rapidly and she produces fertile eggs when she is one-tenth of the normal weight. She will then lay her eggs and usually dies, though occasionally such artificially stimulated females survive for a while.

Many similar mechanisms are known in arthropods and vertebrates. They tend to be controlled from a centre that lies relatively high up in the nervous system (e.g. supraoesophageal ganglion and hypothalamus respectively). This is presumably an indication of some connection both with the other control centres in these regions and with the mechanisms for storing information in the neural memory. There are evidently complicated inter-relations between the length of life and the means of collecting information by the individual and the species. Where the genetic mechanisms is particularly important, lifetimes tend to be short. On the other hand, it is no use having a good brain and filling it with a lot of information if this is then lost by an early death. In *Octopus* the memory-storing centre develops relatively early and with the information acquired the individual then proceeds to grow rapidly and at the right time produces its numerous offspring.

5. The view that senescence is the result of failure of selection seems at first incompatible with the continued existence of individuals after reproduction. Selection should, one might think, always produce a programme that ends with reproduction, as it does in the octopus or mayfly. Obviously this is not so in man or in domestic animals. This may be partly because 'reproduction'

in man includes the period of bringing up the young. As Williams (1957) puts it, 'No one is post-reproductive until his youngest child is self-sufficient'. This prolonging of the usefulness of the individual within the family may have led to continued selection for longer life because of the value of the information store to a larger group. Probably such survival is a recent development.

The sudden cessation of human reproduction at the menopause at about 45 years of age is one of the striking features of human biology. As noted above it must surely have arisen under some special selective influence. Williams (1957) suggests that this was the advantage produced by devoting the woman's care to her (already numerous) children, rather than to be interrupted by further, increasingly hazardous, pregnancies. This concept is in agreement with the fact that a woman's capacity for orgasms does not stop at menopause, but continues into old age.

The theory we have been considering predicts that there must be a declining effectiveness of homeostasis because of the selective advantage of genes that manifest their effects in the young. Any change causing prolongation of the life-span must be produced by some selective advantage that accrues inspite of this effect.

6. Many other aspects of the human and other patterns of senescence can be understood on this theory. Males have shorter lives than females in many animals, as in man. This would result from their increased mortality owing to such factors as more conspicuous colour, more or less ritual social fighting, defence of territory, and repelling of enemies. These factors, by decreasing the fecundity of males at later ages, emphasize the failure to select against any deleterious genes that then appear.

4. External Influences on Ageing

Environmental conditions certainly affect the rate of senescence. Thus lizards in Florida live for only 1 year whereas in Maryland, where they are less active, they live for 4 years. Of course, without experiment one cannot be sure that such differences are not due to different genotypes (see Bourliere 1957). Possibly similar effect are responsible for the facts that bats live for up to 15 years (they are asleep and cold for much of the time) whereas shrews are very active and short-lived. Such effects are related by some people to the concept that 'Most organism seem unable to transform more than a certain fixed quantity of energy during their specific lifetime and the rate of energy turnover appears to determine the rate of ageing' (Bourliere 1957). This, of course, assumes that there is a specific lifetime. The life-span of fruit flies (*Drosophila*) is ten times greater at 10°C than at 30°C. On the other hand in mammals the increased metabolism produces shorter lives at lower temperatures (in rats). This agrees with the suggestion that the life-span is proportional to the total metabolism per gramme. Thus in man the surface-volume ratio is lower than the mouse and he lives some thirty times longer. But both expend about 700 calories/gramme/lifetime.

Nutrition is the external factor that is most obviously related to senescence. Well-authenticated experiments show that longevity can be increased by reduction of diet (McCay *et al.* 1939). Young rats were given a diet containing all essential nutritional elements but lacking in calories. They remained immature for up to 900 days. Growth could then be restarted by increased feeding. The animals matured and had a life-span up to 200 days longer than controls. There was retardation of such specific senescent changes as collagen-ageing.

These experiments are often interpreted as supporting the view given above that the life-span allows for a fixed energy expenditure, but the lowering in McCay's experiments was much less than the nearly doubled lifetime. There are, however, other interpretations. Domesticated rats probably store more fat than is appropriate, and the starved rats might have a avoided this.

The effects of qualitative differences in diet are even less well understood. It is generally assumed that old people require a lower calorie intake, but it has not been shown in general that they require combinations of amino-acids different from those needed by young people. Nor are patterns of excretion different enough to provide clues as to metabolic differences. Although information is very scarce we can certainly assume that many human diets are far from optimal. Thus if they contain adequate amounts of the rarer amino-acids diets must often contain toxic amounts of the commoner ones (e.g. tyrosine, phenylalanine, or tryptophan). These are , however, general questions of diet; such imbalances can hardly be the reason for senescence.

There have been reports of a few particular differences in amino-acid requirements.Thus old people are said to require more methionine than young (see Schlettwein-Gsell 1966). All sorts of possible influences of diet on ageing have been reported. Many of them are obvious rationalizations of local customs (but not therefore by any means to be neglected). The most famous was perhaps the suggestion of Metchnikoff (1907) that drinking yoghurt was the secret of long life. He associated this with a theory of auto-intoxication by products of the intestinal bacterial flora. More 'scientific' is the suggestion that senescence may be accelerated by tanning agents such as tea, which increase cross-linkage between molecules. But this was proved to be false by McCay in rat experiments where only coffee was given as drink.

Other possible dietary influences on ageing are those that influence calcification, especially in the vascular system, thus the physician Osler's famous dictum was 'a man is as old as his arteries'. Selye (1962) recognizes a condition of 'calciphylaxis' in which tissues respond to certain 'sensitizers' in the diet by clacification. Some of the

D vitamins act as sensitizers and they operate especially strongly in the presence of 'challengers' in the diet, such as those found, for example, in eggs.

Study of particular factors may be of great value in understanding particular phenomena of senescence, among which calcification is certainly conspicuous. It does not seem, however, that we can find the 'causes' of ageing in the environment. Our view is that there is no one cause, but environmental stresses are certainly the source of both the wear and of some of the random errors that characterize senescence.

5. Genetic Basis of Ageing

Whatever the basis of senescence may be, it certainly includes a large heritable factor. The best way to attain old age is to have old parents. Conversely, strains of mice have been bred in which all individuals die of leukaemia at the age of 9 months, as against the normal 36-month lifetime of mice. Here of course artificial selection has operated to produce a particular pattern of death. It may be that natural selection has operated in a similar way to determine the life-span of each species. However, as we have seen, this is only part of the 'explanation' of senescence—much of it is due to random influences, but even the liability to somatic mutation and capacity to tolerate and repair it have themselves genetic backgrounds.

HUMAN FERTILITY AND MORTALITY

1. The Importance of Numbers

One of man's many paradoxes is that although with him each individual organism is more important than in other animals as an information-gatherer for the species, yet his manner of life is largely controlled by his fellows, and particularly by how numerous the are. It can be argued that as men became more clever in avoiding death the crude pressure of population has been an increasing determinant dictating the changes in human technology. Today many people believe that the problem posed by our capacity for reproduction is even greater than that generated by our potential for mutual destruction. Certainly the pressure of population is great and is growing.

The study of human population change must therefore be one of the central themes of a study of human homeostasis. We all have the closest possible interest in understanding the changes that have occurred and especially forecasting numbers for the future. The provision of our daily bread depends on knowing for how many we have to provide. Everyone can gain from learning where the needs of the future will lie. Plans can be made by politicians, farmers, fishermen, business men, engineers, civil servants, soldiers, hospital authorities, medical men, and the rest of us only if we know the numbers and characteristics of the populations to be expected. Indeed plans are essential to meet the multitudes who are coming.

In this chapter we shall consider simply the question of number and we shall raise the suspicion that this alone is a feature of human life that is almost as important as the characteristics of the individuals. It is not that the contributions of individuals are irrelevant: on the contrary, single brains and the inventions that they make have been essential features of man's astonishing progress. But are individuals able to produce the changes as a result of the pressure of population growth? Or is it the other way round?

These are huge historical questions and we shall not presume to reach final answers. But we can look at least at some biological aspects of the problem and discuss some of the methods for studying population numbers and the facts that have been revealed about the past.

Chapter—9

Evolution of Animals

As pointed out in previous Chapter 9, the Protozoa, the most primitive phylum of animals, appears to have been derived from primitive flagellate algae, and within single genera among the euglenoids there may be some species which show predominantly plant characters and others which show predominantly animal characters. Thus there is a high probability that the flagellates are close to the point of separation of the two kingdoms, if, indeed, the separation is complete here, for there are flagellates like *Trypanosoma* which are undoubtedly animals; there are others like *Chlamy-domonas* which are undoubtedly plants; and there is the great intermediate group, typified by *Euglena,* which defies any indisputable assignment to the kingdoms.

Diversification of the Protozoa

Once the Protozoa were established, they became greatly diversified, and their relationships are by no means certain. The class Flagellata itself includes a wide range of structural and ecological types even when only indisputably animal types are considered. Feeding may be by engulfing food with pseudopodia, or by a simple mouth. The parasitic forms are generally saprozoic. Several orders of flagellates deserve especial mention. The order Protomonadina includes a wide variety of small, colourless flagellates, typically with two flagella, one of which trails along the side of the animal. Reproduction is always asexual. In the free-living members of the order, this is by simple fission, but among the parasitic species there may be multiple divisions which are difficult to distinguish from spore formation of the Sporozoa. It seems highly probable than the Sporozoa, all of which are parasitic, were evolved from this group. But sexual reproduction was also developed among the ancestors of the Sporozoa. The best known protomonads are the members of the genus *Trypanosoma,* all of which are blood parasites of vertebrates. Quite as important for evolution, however, are the choanoflagellates. These are protomonads which have a protoplasmic collar encircling the base of the flagellum. Its theoretical importance lies in that fact that sponges have similar cells, called choanocytes, and that colonial choanoflagellates, such as *Proterospongia,* resemble simple sponges.

The most highly specialized flagellates are probably the orders Polymastigina and Hypermastigina. The polymastigotes are generally parasitic on the digestive tract of arthropods or vertebrates. The simpler members generally have four flagella, but the more complex members may have large numbers. There may be only a single nucleus, or there may be many. Complicated cytoplasmic organnelles may be present. Reproduction is generally reported to be by simple fission, but Cleveland has reported sexual reproduction in some species, including the unique feature of reduction of the chromosome at a single division. The hypermastigotes are morphologically the most complex flagellates. There are numerous flagella arranged indefinite ways. Cytoplasmic organelles are very complex (Fig 9.1). All hypermastigotes are parasites or commensals in the digestive tract of termites or cockroaches. They are essential to the nutrition of the hosts for which they digest cellulose, for it has been shown that the host starves if the protozoans are removed experimentally.

The order Rhizomastigina is of especial interest because the members of the order appear to be intermediate between the Flagellata and the Sarcodina, amoeba and its allies. Many flagellates are capable of ameboid movement, but the members of this order, although they do possess a flagellum, are also permanently ameboid, and so they make a nice connecting link between the Flagellata and the Sarcodina. *Mastigamoeba* is a good example.

Ameba and Its Allies

The class Sarcodina is also highly diversified, and it is by no means certain that it is really a unified group. That is different orders of the Sarcodina may have arisen from different flagellate ancestors. By far the best known member of this class, or perhaps of any protozoan class, is *Amoeba proteus,* the familiar study material of every elementary biology laboratory, and the traditional example of a primitive animal. *A. proteus* is quite typical of its order, the Lobosa. But, in addition to such simple, free-living, ameboid organisms, the Lobosa includes parasites like *Entamoeba histolytica,* which parasitizes the digestive tract of man; and there are free-living but shelled organisms such as *Diffugia.* The nutrition of all of these, including the parasites, is holozoic. Reproduction by binary fission is the rule, but sexual reproduction has been reported for one species of *Amoeba,* and it is common among the shelled species. The order Lobosa is generally regarded as a terminal group in evolution, yet the idea that the Myxomycophyta or the Eumycophyta or both may have been derived from this group is gaining prestige among mycologists. And some zoologists treat the Myxomycophyta as protozoans under the name of Mycetozoa.

The remaining orders of the Sarcodina are all terminal groups which are characterized by elaborate and often beautiful calcareous or siliceous shells, and by many slender, semipermanent pseudopodia, the characteristics of which are peculiar to each order. These are the Foraminifera and the Radiolaria, both of which are abundantly represented in the fossil record even into pre-Cambrian times, and the Heliozoa, which are less adequately known as fossils.

Fig. 9.1: Representative Protozoa, *A, Trypanosoma; B, Proterospongia; C, Macrospironympha; D, Difflugia; E, Amoeba; F, Vorticella; G, Plasmodium* infested red blood cell; *H, Acineta, A—C* are flagellates; *D* and *E* are sarcodinans; *F* is a ciliate; *G* is a sporozoan; and *H* is a suctorian.

Ciliata and Suctoria

In contrast to the classes Flagellata, Sprozoa, and Sarcodina, all of which are fairly clearly interrelated, the two remaining classes of Protozoa show no evidence of relationship to the above named classes. These are the classes Ciliata and Suctoria, which some protozoologists would prefer to assign to a separate phylum. The ciliates have a definite shape, maintained by a pellicle, as do some flagellates. There is a definite anteroposterior axis, and the animal is symmetrical, sometimes radially, sometimes bilaterally, and sometimes an irregular deviant from one of these types of symmetry. Nutrition is holozoic, with other minute organisms being ingested by a mouth. Cilia are commonly arranged in specialized tracts which sweep a food-bearing current toward the mouth. It is often said that the Protozoa are the simplest of animals. While this may be debated, there is little doubt that ciliates are the most complicated of cells. The cilia themselves may be arranged as coordinated tracts, or they may be fused in sheets to form undulating membranes, or they may be fused in tufts to form appendage-like cirri. Whatever the arrangement of the cilia, their movements are coordinated by a complicated network of fibrils, the neuromotor system, which exceeds in complexity the simplest nervous systems of Metazoa. The trichocysts, small bodies which can be discharged to produce filaments, underlie the pellicle. Their function is uncertain, but defense and attachment while feeding have been the principal alternatives suggested. The food vacuoles, formed at the mouth, pass through the body by a regular route and leave at a definite point. In some, it would scarcely be an exaggeration to say that there is a digestive tube and an anus. There are two nuclei, a micronucleus which is concerned with heredity and a macronucleus which is concerned with metabolic functions. This led Kofoid to state that it would be as logical to call a whale unicellular as to call a *Paramecium* unicellular. Finally, the ciliates have developed a unique type of sexual reproduction called conjugation. The details are quite complicated, but in substance it amounts to this, that the maturation divisions result in a stationary nucleus and a wandering nucleus in each of a pair of conjugants, and the wandering nucleus of each fertilizes the stationary nucleus of the other.

The most primitive order of ciliates, the Holotricha, is completely clothed in cilia, and the animals are strong swimmers. But feeding by means of currents seems to lead to a sessile mode of life, for specialization in the ciliates has generally resulted in attachment to the substrate, restriction of the cilia to limited areas usually related to feeding, and specialization of the ciliary tracts to form membranelles, undulating membranes, or similar structures. The well known *Vorticella* is a good exemplar of such tendencies. In a single order of ciliates, the Hypotricha, specialization has been in the direction of increased efficiency of locomotion. In this order, a band of typical cilia still creates a current for feeding purposes, but locomotion is based upon the use of fused tufts of cilia, the cirri, which function in a leg-like fashion.

Finally the class Suctoria is a small group which is undoubtedly related to the ciliates. Adult suctorians have no cilia, nor have they any other locomotor organelles. Nor do they have a mouth, for they use protoplasmic tentacles for the capture and ingestion of food. They are generally permanently attached to their substrate. Up to this point, they show no affinity with the ciliates. However, they do have the two types of nuclei and sexual reproduction by means of conjugation, phenomena which are unknown elsewhere in the world of life except for these two classes. Further, the suctorian zygote becomes a free-swimming, ciliated organism which only later settles down on the substrate and adopts the typical suctorian mode of life. This is interpreted as embryological recapitulation of ancestral history.

ORIGIN OF THE METAZOA

Hyman has said that there is no direct proof of the origin of the Metazoa from the Protozoa, yet the discussion of the origin of the Metazoa (multi-cellular animals) usually revolves around the question of which protozoan stock would seem to be the most probable progenitor of the Metazoa. Two broad possibilities exist by which the Metazoa could have been formed from the Protozoa. The first is that repeated nuclear division without cytoplasmic division might have led to formation of a plasmodium, like some of the Heliozoa. Formation of cell membranes would then result in multicellularity, and differentiation might then lead to the true multicellular individual. The second method is the differentiation of cells within a colony or Protozoa, comparable to *Volvox*, leading to interdependence and individuality.

A very different, third possibility has been suggested. In protozoan colonies each cell ingests food, but even in simple metazoans, a new method of feeding is used, with a digestive tract feeding for the; whole organism. This transition might be difficult. Hardy suggests that simple plants, like *Volvox*, living in an environment deficient in nitrates and phosphates, may have satisfied the deficiency by capturing smaller organisms. Increasing utilization of this nutritive pathway, together with loss of photosynthesis, would then lead to a simple metazoan. Insectivorous plants show the feasibility of such a nutritive mechanism, and the fact that unicellular plants seem to have given rise to protozoans more than once lends plausibility to the suggestion that multicellular plants may have achieved animalization at least once.

Paleontology is of no help in this problem, for the Metazoa were already well established at the beginning of the Cambrian. It is therefore probable that the origin of the Metazoa will always be speculative. But most zoologists favour the flagellates as the most probable progenitors of the Metazoa. The reasons for this are many. The flagellates are a highly variable group which appear to have given rise to many other groups of plants and to several, perhaps all, other groups of Protozoa. Further, some groups of flagellates show a tendency to form colonies of ever-increasing size and complexity. The evolution of sex has occurred here, and the colonies are definitely divided into somatic and germinal "tissues." Oogamous reproduction is the rule for such colonial flagellates. The sperm are at once similar to some simpler, noncolonial flagellates and to the typical sperm of Metazoa. Such colonies may also show anteroposterior differentiation. These large, highly specialized colonies occur principally among the plant flagellates, yet colonies of a highly suggestive character also occur among the animal flagellates. None of these reasons is conclusive, yet collectively they carry considerable weight.

However much the origin of other Metazoa may be disputed, it seems almost certain that the Porifera (sponges) were derived from choanoflagellates. It is a short step from the structure of the colonial choanoflagellate *Proterospongia* to that of the simplest sponges. *Proterospongia* (Fig. 9.1) consists of a small mass of gelatinous material in the surface of which are imbedded choanocytes (collared, flagellate cells), and in the interior of which are ameboid cells. The choanocytes can withdraw the collar and flagellum and move into the interior to become ameboid cells. In order to change this to the structure of a simple sponge it would be necessary only to develop a system of channels through the gelatinous mass, to let these channels be lined with choanocytes, and to let the outer surface be covered by a simple epithelium. No organ systems are present, the functions of the sponge being carried on by the component cells individually.

Phylum Porifera

Cooperative activity is at a minimum, and it has often been debated whether a sponge is a true organism or a colony of unicellular organisms. There are enough evidences of cooperative activity to swing the balance of zoological opinion in favour of the former alternative. Thus, while some sponges have no definite shape, many have rather complex shapes. Most sponges produce skeletal elements in the form of spicules which may be quite complex, and so exactly formed in every case that they are among the best taxonomic characters. Further, the beat of the flagella in the canals is not obviously coordinated, yet the water current is unidirectional, and so there must be at least a limited control of the flagellar beat.

Sponges can reproduce asexually by formation of groups of cells, the gemmules, but they also reproduce sexually by means of typical eggs and sperm. The zygote develops into a free-swimming larva which is flagellated. After swimming for a brief time, it settles down to become a sponge. Because this larva has typical (not collared) flagella, some zoologists believe that the sponges were derived from some typical flagellates, with choanocytes being a secondary development. Yet this is not unexpected in view of the recapitulation principles, for the choanoflagellates themselves must have been derived from typical flagellates.

Evolution within the Porifera has taken the forms of elaboration of the canal system and of the supporting spicules or fibers. As relationships within the phylum are not at all clear, it may be more profitable to go directly to the problem of the relationships of the Porifera to other animals. The Porifera are a terminal group. Further, they are so different from all other Metazoa in the absence of unified tissues, in the physiological independence of the individual cells, and in their embryology and adult anatomy, that it is generally believed that their origin from the Protozoa must have been independent of that of the rest of the Metazoa. The Profiera are therefore assigned to a separate branch; of the subkingdom Metazoa, the branch Parazoa, in contrast to the Eumetazoa.

The Gastrea Theory

Speculation on the origin of the Metazoa has been dominated by the Gastrea theory of Haeckel, a theory based on the literal application of the Biogenetic Law. In its original form, Haeckel interpreted the egg as corresponding to an ameboid ancestor, possibly to *Amoeba* itself. As evidence in favour of this, he pointed to the ameboid eggs of sponges and of some coelenterates. Other types of eggs he assumed to be secondary specializations. The egg, of course, undergoes the cleavage divisions which result first in a solid morula and then in a hollow ball of cells, the blastula. The morula was interpreted as corresponding to a simple hypothetical ameboid colony, the *Synamoeba,* while the blastula was supposed to correspond to a colonial ancestor, the *Blastea,* more or less comparable to *Volvox,* but ameboid rather than flagellate. The modern exponents of this theory assume flagellate rather than ameboid ancestors for the reasons stated above. There being only a single layer of cells in the *Blastea,* all cell functions were at first shared by pall cells, but then a division of labour occurred, with the posterior cells assuming the nutritive functions. These cells then invaginated, with the result that the organism became a two-layered gastrula, having an outer layer of flagellated cells (ectoderm) and an inner layer of digestive cells (entoderm). Haeckel called this hypothetical organism the *Gastrea,* and he believed it to be ancestral to all Eumetazoa. Some of the coelenterates he regarded as living gastreads. Next, the *Gastrea* developed a third cell layer, the mesoderm, between the first two, and he regarded all of the structures of the higher phyla as derived from these three layers. The development of the bottom-feeding habit led to elongation of the body and the formation of primitive worms, similar to the living Turbellaria. From these the higher phyla were developed.

The Gastrea theory is a beautiful simplification and synthesis of a vast amount of embryological, morphological, and taxonomic data, and it is almost without a serious competitor. As a result, it was, until recently, presented in nearly every elementary textbook of zoology. Unfortunately, however, as Hyman* has pointed out, "it is probably one of those simplification that are too beautiful to be true." Embryology is not a safe basis for construction of pedigrees, especially if comparison is made between embryos of advanced species and adults of their supposed ancestors, as Haeckel did in this case. At most, embryology should be treated only as one of several corroborative lines of evidence. But there is the further difficulty that even the embryological evidence does not give unequivocal support to the Gastrea theory. For in the coelenterates, the group which is closest to the hypothetical *Gastrea,* gastrulation ordinarily occurs not by simple invagination of the posterior cells, but by the in wandering of many cells from all parts of the blastula. And this does not result in the immediate formation of a typical gastrula, but rather it consists of a ball of ectodermal cells. This type of larva is called a *planula.* Only later does this entodermal core hollow out and a mouth (blastopore) break through to form a typical gastrula. While the type of gastrulation with which Haeckel dealt is known, for example in the starfishes, it is not widespread

---

* By permission from Hyman, L. H., "The Invertebrates," Vol. I, McGraw-Hill Book Co., Inc., 1940.

in the Animal Kingdom, and it appears to be a secondary modification rather than a primitive character.

It is plausible, then, that the ancestor of the Eumetazoa may have been a blastula-like colonial flagellate, in which there occurred a differentiation between somatic and reproductive cells, as in living *Volvox* and then a differentiation between digestive cells and locomotor cells, with the former type moving into the interior of the organism to form either a gastrula or a planula. Yet it is probable that decisive evidence on this basic question will never be obtained. That this primitive metazoan was not identical with any living type is almost certain.

The three most primitive living metazoan phyla are the Mesozoa, the Coelenterata (= Cnidaria), and the Ctenopora. These phyla are generally radially symmetrical (having one differentiated axis), or else biradially symmetrical (having two differentiated axes). Their general grade of organization is more advanced than that of the Porifera, for, while there are no organ systems, there are two well-defined tissues, the ectoderm and entoderm (or epidermis and gastrodermis). In most coelenterates and in the ctenophores, there is between these layers a jelly-like mass, the mesoglea, which also includes some cells. Thus it is not strictly true, as is often stated, that these phyla have only two cellular layers.

Phylum Mesozoa

The correct phylogenetic position of the Mesozoa is a much vexed question. From a structural view point, they are the simplest of metazoans, consisting of an outer, generally ciliated, layer of cells enclosing a core of internal, reproductive cells (Fig 9.2). They thus resemble a planula, but the internal cells are not digestive cells. Van Beneden named the group in 1877 with the intention of indicating his judgment that the group was extremely primitive, intermediate between the Protozia and the rest of the Metazoa. On the other hand all Mesozoa are parasitic, and their life cycles are somewhat similar to those of the digenetic trematodes. Many zoologists treat the group as a degenerate offshoot from the flatworms. It is could be shown decisively that the characters of the Mesozoa are primitive rather than degenerate, the group would assume great phylogenetic importance, for it could then be reasonably argued that the group must be but little changed from the remote, pre-Cambrian ancestor of the Metazoa. It would prove that the Metazoa were derived from a planula-rather than from a gastrea-type ancestor, and it would leave the Gastrea theory very badly damaged. But the available evidence does not furnish a basis for a final decision on the taxonomic position of the Mesozoa. W. K. Brooks has said that "suspended judgment is the greatest triumph of intellectual discipline," and this appears to be an appropriate place for that achievement. Hyman sets them off as an independent branch of the Metazoa, the branch Mesozoa.

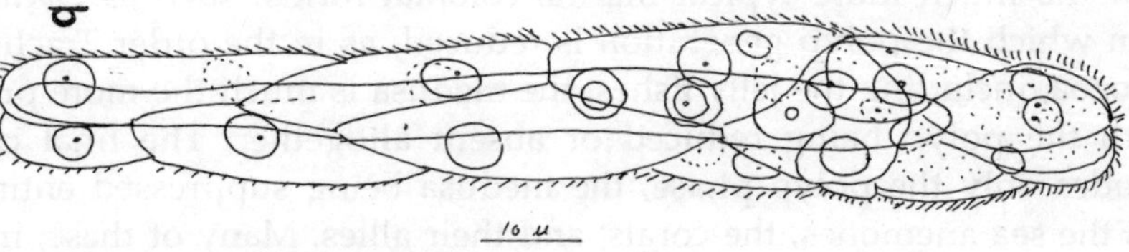

Fig. 9.2: *Conocyema deca,* A Typical Mesozoan. (From MacConnaughey, *J. Parasitol,* V. 43, 1957.)

Phylum Coelenterata

The Coelenterata have traditionally been treated as the most primitive of the Eumetazoa. Haeckel regarded them as the source of the flatworms, and hence of all higher phyla. The principal reason for his viewpoint is the obvious resemblance of a hydroid polyp to a gastrula. For the hydroid consists of two simple cell layers, with no organ systems and with only a trace of noncellular mesoglea. The anatomical structure of the polyp could be derived from that of the gastrula simply by the elongation of the body and the drawing out of a circlet of tentacles around the mouth. Food is taken in and waste residues expelled by the mouth, which is simply the blastopore of the balstula. Food is still digeted by the protozoan method, that is, the cells of the gastrodermis engulf food particles, and digestion is carried on intracellularly. But enzymes are also secreted into the gastrovascular cavity, and much digestion occurs there. Muscle tails may be formed in connection with either the epidermis or the gastrodermis. A nerve net is formed from epidermal elements. Thus there is a high degree of tissue differentiation, but no organ systems. Outstanding specializations of the coelenterates include the development of nematocysts, organelles for food procurement and defense; and the alternation of a free-swimming, sexually reproducing medusa generation with a sessile, asexually reproducing polyp generation. This is unrelated to the alternation of generation of plants, for both generations are diploid. Superficially so different, the polyp and the medusa are structurally very similar, for one can derive the medusa from the polyp by simply inverting the latter, greatly increasing the amount of mesoglea and its cellular contents, and drawing the circlet of tentacles away from the mouth, as illustrated in Fig. 9.3.

Haeckel assumed that the anestral coelenterate was a polyp, because of the ease with which a polyp can be derived (in theory) from the Gastrea. But study of what appears to be the most primitive order of hydroids, the Trachylina, has led to the conclusion that the medusa phase is primary and the polyp derived. Once formed, the Coelenterata diverged along three major lines, each comprising a class of the phylum. In the most primitive class, the Hydrozoa, both generations are generally well developed. The class includes hydroids such as the *Hydra* of elementary laboratories and the much more typical marine colonial forms, such as *Obelia*. There are also some in which the polyp generation is reduced, as in the order Trachylina. In the class Scyphozoa, including the jelly fishes, the medusa is much the more prominent generation, with the polyp being reduced or absent altogether. The final class, the Anthozoa, includes only the polyp phase, the medusa being suppressed entirely. The class comprises the sea anemones, the corals, and their allies. Many of these, including all of the corals, secrete a calcareous exoskeleton, because of which they have left an excellent fossil record going back to the Ordovician. All three classes are ancient, and it is probable that they diverged in pre-Cambrian times from a primitive hydrozoan type not dissimilar to the Trachylina.

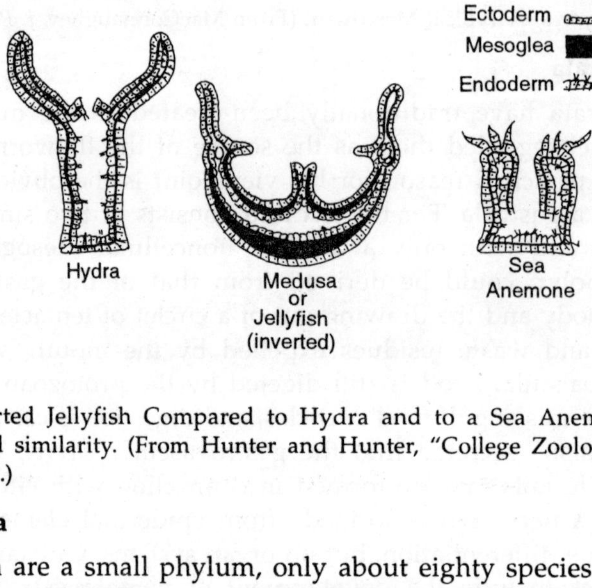

Fig. 9.3: An Inverted Jellyfish Compared to Hydra and to a Sea Anemone, showing their structural similarity. (From Hunter and Hunter, "College Zoology," W. B. Saunders Co., 1949.)

Phylum Ctenophora

The Ctenophora are a small phylum, only about eighty species being known. All are small, marine animals. They are commonly known as "sea walnuts" or "comb jellies" because of the presence of comb-like plates of cilia. They share some important characteristics with the coelenterates. Thus they are radially or biradially symmetrical in contrast to all higher phyla, including the Echinodermata, in which bilateral symmetry is primary, and radial symmetry derived. They are at the tissue grade of construction, with an abundant mesoglea separating the epidermis from the gastrodermis. In both phylla, the gastrovascular cavity is present, and its branches distribute food through the body. But the ctenophores lack nematocysts, and there is no alternation of generations. Unlike the coelenterats, they are hermaphroditic. There seems to be little doubt that the coelenterates gave rise to the Ctenophora, yet all attempts to relate them to any of the classes of living coelenterates have failed. It seems most probable that the Ctenophora were derived from the same ancient, pre-

Cambrian, trachyline stock that gave rise to the three classes of the Coelenterata, and at about the same time.

PRIMITIVE BILATERAL PHYLA

Some of the ctenophores have become elongate and flattened, and it has been suggested that they are related to the ancestors of the flatworms, the phylum Platyhelminthes. But the majority of zoologists believe that the appearance of homology is misleading. Haeckel believed that a primitive hydrozoan was the ancestor of the bilateral phyla, basing his opinion as usual upon the Biogenetic Law. The evidence is insufficient, and it seems at least as probable that flatworms were first derived from the same planula-like stock which gave rise to the coelenterates. But in the flatworms, the middle cell layer (mesoderm) became more highly developed, with organized muscle layers, reproductive system, and an excretory system. Yet no coelom, or body cavity, developed within the mesoderm, as it does in most higher groups. The nervous system is formed from the ectoderm, although it is imbedded in the mesoderm. It is not a diffuse nerve net as in the coelenterates, but rather it is somewhat centralized, being organized about cerebral ganglia at the head end and longitudinal cords. There are organized sense organs, including eyes. Food is still distributed to the various parts of the body by the branches of the digestive tract, which is now called an intestine rather than a gastrovascular cavity. Thus the flatworms are much advanced beyond the tissue grade of construction which characterizes the radiate phyla, for definite organ systems are present, principally in the mesoderm.

Hadzi has urged a radically different phylogeny. He believes that multinuclear ciliates became acoelous flatworms by formation of cell membranes. In evidence, he points out that both are ciliated; hermaphroditism of flatworms he finds homologous with conjugating of ciliates; trichocysts of ciliates he believes are represented by sagitocysts, rod-like inclusions in some of the epidermal cells of acoelous flatworms. He believes that the Anthozoa were derived from flatworms by adoption of the sedentary life, and that the other coelenterates were then derived from anthozoans. The higher Metazoa were also derived from flatworms, and are grouped into only four phyla, a procedure which unites some highly diverse groups.

At this point it may be well to recall the admonition of Hyman* that "the exact steps in the evolution of the various grades of invertebrate structure are not and presumably never can be known. Statements about them are inferred from anatomical and embrylogical evidence and in no case should be regarded as established facts." Although the phylum Platyhelminthes, and especially the most primitive class of this phylum, the Turbellaria, is commonly treated as the stem group from which the higher phyla arose, this is by no means established. Closely related to the Platyhellminthes, and on the same general level of organization, is the phylum Nemertinea, a small group of marine flatworms. This group is less well known than the former because it

* By permission from Hyman, L. H., "The Invertebrates," Vol. I, McGraw-Hill Book Co., Inc., 1940.

is predominantly marine, and because it is a difficult group to study. But the Nemertinea show some characters which qualify them for special consideration as potential forerunners of the higher invertebrates. For the first time, there is an anus present, so that the digestive system is said to be complete. Also, there is a simple blood circulatory system, and the blood contains hemoglobin. Particularly important according to Kofoid is the fact that the nervous system is based upon cerebral ganglia and eight longitudinal nerve cords, two dorsal, two ventral, and two on each side. He has pointed out that this lends itself to the formation of the principal invertebrate nervous systems by the development of the ventral cords and the suppression of the others; and to the formation of the chordate nervous system by the development of the dorsal cords and the suppression of the others. But the fact the nemerteans capture their food by means of an extensible proboscis, a mechanism found in no other group, suggests that they are a terminal groups.

Yet another possibility is that a third group, an unknown phylum of very primitive flatworms derived from the primitive planula by the development of the mesoderm may have been ancestral to both the Platyhelminthes and the Nemertinea. Whatever their origin, once formed, the platyhelminths diverged along three main lines of descent. The first of these is represented by the class Turbellaria, comprising the free-living flatworms, of which *Planaria* is the best known example. The other two classes have become greatly modified for parasitism, and most of them have developed complicated life cycles to permit their transfer from one host to another. These are the class Trematoda, the flukes, which have retained all of the organ systems of the Turbellaria, and are generally internal parasites of vertebrates; and the class Cestoda, the tapeworms, which; have undergone an extreme degenerative evolution. They are all intestinal parasites of vertebrates.

On the same general level of organization as the above discussed acoelomate groups, there are a number of minor phyla, of very uncertain relationships, which are characterized by the possession of a pseudocoel. A coelom is, by definition, a body cavity formed within mesoderm. A pseudocoel, on the other hand, is a remnant of the cavity of the blastula, and it may be partially filled by large, vacuolated cells. The taxonomy and relationships of these pseudocoelomate organisms are very uncertain. It seems best here simply to acknowledge that the exist, and that they are among the most primitive of bilaterally symmetrical animals, without attempting to unravel their relationships. These are the Rotifera, a phylum of microscopic animals which are known to every student of elementary biology as fascinating contaminants of most protozoan cultures; the phylum Gastrotricha, which are small, worm-like animals of wide distribution; the phylum Kinorhyncha, comprising a group of minute, marine, wormlike animals; the phylum Nematoda, a very important group, the so-called round worms, which includes many little-known free-living worms as well as some of the best known parasites of both plants and animals; the phylum Nematomorpha, or horsehair worms, consists of long, slender worms which are parasitic during larval life, but are generally free-living when mature; and the phylum Priapuloidea, comprising a few species of small, externally segmented marine worms. These six phyla are sometimes grouped together as classes of a single phylum, Aschelminthes. The phylum Acanthocephala, or spiny-headed worms, are generally small worms the young of which are parasitic on invertebrates, while the adults are parasiti on vertebrates, including man. A final pseudocoelomate phylum is altogether different from these worm-like phyla. This is the phylum Entoprocta, a group of colonial animals, often forming encrusting colonies. Formerly regarded as a class of the phylum Bryozoa, it is now clear that they differ fundamentally from the Bryozoa. The intestine is U-shaped, and both the mouth and the anus are included within a ring of ciliated tentacles, the lophophore, which sets up a current for plankton feeding. This gives them a rather hydroid-like appearance superficially.

THE PROTOSTOMOUS PHYLA

The remaining major phyla of the Animal Kingdom can be arranged in two divergin lines of descent, largely on the basis of embryological criteria. One line culminates in the Annelida, Arthropoda, and Mollusca, while the other culminates in the Echinodermata and the Chordata. Certain minor phyla can be associated with one or the other of these main lines with varying degrees of satisfaction. Haeckel believed that the echinoderm-chordate lien was derived from the Turbellaria, while Kofoid believed that both lines were derived from the Nemertinea. It is at least as likely that both were derived from the unknown, primitive, acoelous flatworm from which both the Playhelminthes and the Nemertinea were probably derived.

The cleavage divisions of the annelid-arthropod-molluscan line are both spiral and determinate. In spiral cleavage, spindles are at right angles to those of the preeding division, so that the cells of each layer alternate with the next layer like bricks in a building. This is not true of the large, yolky eggs of arthropods, yet their position in this series is secure, for their origin from annelids is clear (see below). Determinate cleavage proceeds according to a set pattern, with the part of the body to be formed from each blastomere fixed from the start. Destruction of a blastomere results in a deficient larva. One can designate which blastomere will form ventral surface, which the gut, and so on. The mesoderm is formed from stem cells which multiply to form a pair of ventral bands, growing forward from the posterior end of the larva. The coelom is formed by splitting of these bands, hence these phyla are said to be *schizocoelous*. The blastopore becomes the mouth of the adult, and hence this whole series of phyla is called Protostomia.

Generally, though by no means always, development leads to a trochophore larva. This is a more or less spherical larva with an *apical tuft* of cilia dorsally, and a prototroch, or girdle of cilia around the equator, by which it can swim weakly. There is a digestive system consisting of a mouth, a short forget, an enlarged stomach, a short hindgut, and an anus, a complete digestive system. Some mesoderm is present, and some mesodermal organs, such as a kidney, are formed. This type of larva is characteristic of the annelids and molluscs, but unique larval stages have been developed by the arthropods.

Other characters are also held in common by the members of these phyla, but those mentioned are sufficient to indicate the probability of relationship among them. The actual relationships are, however, largely a matter of conjecture. Origin from some primitive, acoelous flatworm is probable, yet there is little basis for a decision as to which of the possible groups of flatworms was the ancestor, beyond the fact that spiral cleavage is also found in the Nemertinea and in the polyclad Turbellaria. Because of the widespread occurrence of the trochophore larva, it is generally supposed that an independent, trochophore-type animal was an ancestor intermediate between the flatworms and the present-day phyla of the annelid-arthropod-molluscan series. The only evidence for this is the interpretation of the embryological evidence according to the Biogenetic Law, and it has already been pointed out how insecure this is. Yet, although it may be true, direct evidence is not likely to be obtained, for such an organism would be very poor material for fossilization, and it would have to be sought in pre-Cambrian rocks, for the major invertebrate phyla were all present in the Cambrian.

Phylum Mollusca

Whatever the source, the phylum Mollusca was early established as a group divergent from the others. This is a great phylum, the second largest in the Animal King do, having around 80,000 species encompassing a great variety of forms, and inhabiting almost every type of marine, freshwater, and terrestrial habitat. All of the extant classes of Mollusca were present already in the Cambrian, and literally hundreds of species are known. The classes were just as distinct then as now, and so paleontology is of no help in deciding what the relationship within the phylum may be.

In addition to the general characters of all members of the protostome line, the phylum Mollusca is distinguished by many structural characters. In all, the body is divided into four regions: a muscular foot, a head, a visceral hump, and a mantle which generally secretes a calcareous shell. Basically, they are bilaterally symmetrical, but this is obscured in the adults of some classes. The coelom is very much reduced. Mollusca are traditionally described as unsegmented, yet in 1957 *Neopilina,* a new species from the deep waters of the Gulf of Mexico, was described. This animal has several segmentally arranged organ systems, and its affinities are with a class, Monopalcophoa, which had been thought to be extinct since the Devonian. It indicates that the primitive molluscs may have been segmented. On the basis of anatomical evidence, the class Amphineura is regarded as very primitive, but not necessarily as the progenitor of the more specialized classes. This class comprises the chitons or sea cradles, animals which are familiar parts of the fauna of rocky sea coasts, but there are no freshwater species. The chitons have a broad, flat, elongated foot, over which lies the visceral hump and at the anterior end of which is the head. The mantel coverts the visceral hump and typically secretes a series of eight calcareous plates, the valves, which may be beautifully ornamented.

The class Gasteropoda, which includes the snails, slugs, limpets, and their less well-known allies, has developed a much enlarged visceral hump. In most gasteropods, this hump has grown asymmetrically, with the familiar coiling as a result. Corrrelated with this, there has been a torsion of the visceral hump through 180º, so that structures which were originally posterior have moved to an anterior position, and vice versa. The class Scaphopoda, or tooth shells, is a small group which is very much specialized for digging. All species are marine, and there are only a few of them. The class Lamellibranchiata includes the bivalves, the familiar clams, mussels, and their allies. They are compressed bilaterally, and enclosed within a two-valved shell, hinged dorsally, and secreted by the two lobes of the mantle. The giils are immensely enlarged to form broad ciliated tracts which produce a current for feeding upon plankton and detritus. The head is very much reduced. The foot is generally wedgeshaped, unlike the broad, flat foot of the more primitive molluscs, but it can be protruded between the valves to serve a locomotor function. In some species, it is specialized for burrowing.

The class Cephalopoda includes the octopi, the squids, the chambered *Nautilus*, and its extinct allies. They are the most complex of mollusca, comparing favourably with the most advanced insects and vertebrates in degree of complexity. The visceral hump, like that of the gasteropods, is much enlarged, but it is still symmetrical. The mouth is surrounded by a ring of tentacles for seizing food. The head and foot are fused so completely that there is no agreement among specialists as to which these tentacles are derived from. There is a siphon by which water may be forcibly ejected from the mantle cavity. The coelom is better developed than in most molluscs. The nervous system is highly centralized, and efficient eyes, superficially quite similar to those of vertebrates, are present. The more primitive members of the class have a chambered shell of which the animal always lives in the most recently secreted chamber. Each chamber corresponds to a stage in the most recently secreted chamber. Each chamber corresponds to a stage in the growth of the animal, like the successive moults of an arthropod. The shelled cephalopods are today represented only by a single genus, *Nautilus,* but this is the last remnant of a once dominant group in the oceans of the world. The nautiloids appear in the fossil record in the late Cambrian, and rapidly assumed a dominant position in the marine fauna. They reached their peak of development in the Silurian, and then gradually declined. During the Devonian, they gave rise to another suborder, the Ammonoidea, with which they competed unsuccessfully for a long period of the earth's history. The ammonites were the dominant marine molluscs throughout most of the Mesozoic Era, yet they dwindled and became extinct near the end of the Cretaceous. Meanwhile the nautiloids survived in small numbers, and are known from a single genus at the beginnig of the Cenozoic Era. No longer being in competition with the ammonites, the nautiloids underwent a rapid evolution at the beginning of the Cenozoic, for seven new genera appear during the Paleocene epoch. But only one of these, *Nautilus,* has survived to the present time. Meantime, forms with the shell much reduced and internal, the octopi and squids, have become the principal cephalopods.

Phylum Anneliba

The second main branch of the protostome line is the annelid-arthropod branch. The phylum Annelida is best known by the common earthworms, but it includes a wide variety of worms arranged in four classes. All annelids are segmented, that is, the functional and structural units of the body are repeated serially along the length of the body, and the segments may be marked externally. A well-developed coelom separates the digestive tract from the muscular body wall. There is an advanced nervous system based upon a pair of cerebral ganglia and a pair of ventral cords on which ganglia are located in each segment of the body. Whenever there is a larval stage, it is a trochophore. Most annelids have a thin cuticle, and the typical species have segmentally arranged chitinous bristles called chaetae.

The phylum Annelida is best typified by its most primitive class, the Polychaeta. These take their name from their numerous chaetae, which arise from limb-like lobes, the parapodia, on each body segment. There is typically a distinct head which may have appendages. The sexes are separate, and fertilization is external. There is a trochophore larva. The polychaetes are adapted to a wide range of habitats, including pelagic and bottom-dwelling, surface-crawling and burrowing, and permanent tubed welling, and they show very striking morphological adaptations. In numbers of species, they undoubtedly exceed the remaining classes combined. Yet they are less well-known because they are almost exclusively marine.

Better known, but far less typical of the phylum, is the class Oligochaeta, including the earthworms and many small freshwater annelids. The class appears to have been formed from the polychaetes by a process of reduction and simplification. The head region is much reduced and never including appendages. The chaetae are reduced both in size and in numbers, and they are no longer set on parapodia. Oligochaetes are all hermaphroditic. Development is direct, thee being no larval stages. The embryos develop in a cocoon, an adaptation to terrestrial life. The class is much more uniform than is the Polychaeta.

The class Archiannelia is a small group of extremely simplified marine annelids. These are all small worms which generally lack parapodia and chaetae. They are ciliated, like the young of polychaetes. Their nerve cords retain the primitive connection with the epidermis. The coelom is only slightly developed. The Archiannelida were named with the intention of inferring that this was the most primitive class of the phylum and the probable progenitor of the other classes. But embryological and morphological studies have led to the conclusion that this class is not truly primitive, but rather has been derived from the Polychaeta by extreme simplification.

The class Hirudinea, or leeches, appear to have been derived from oligochaete ancestors, for they share many characters with that class. But all leeches are ectoparasites, and they are extensively modified for that mode of life.

Phylum Arthropoda

The protostome series is climaxed by the great phylum Arthoropoda, the most successful of all groups of animals if numbers of species be an indication of success, for about three fourths of all species of animals are arthropods. This is also the most varied of phyla, for there are arthropods adapted to every imaginable habitat from the abyssal depths of the ocean (crabs and pycnogonids) to aerial heights (insects). Perhaps the key to arthropod evolution is their early development of a thick, chitinous cuticle. This made necessary the joints—thin places in the cuticle—from which the phylum takes its name. In order to move the hard pieces of the cuticle, the continuous muscular wall, inherited from plychaete ancestors, became broken up into specialized muscles, attached to ingrowths of the cuticle. The limbs, being jointed, are far more adaptable than the parapodia of the polychaetes, and they have become specialized for many functions, including sensation, feeding, and locomotion. This thick cuticle undoubtedly minimizes water loss by evaporation, and it may be that this is a major factor in their successful invasion of the land, for no other invertebrate phylum has so large a portion of terrestrial species. The concentration of the nervous system and the sensory organs in the head region (cephalization), begun in the polychaetes, is carried much further in the arthropods. The coelom is much reduced, and has become largely replaced by a hemococle. Segmentation is prominent externally, but less so internally.

As might be expected with so large a phylum as the Arthropoda, there is no general agreement on the; taxonomic rank to be accorded to the major divisions of the phylum. For present purposes, five major subphyla will be recognized, and no attempt will be made to treat their component classes because of the vastness of the groups. These are the Trilobita, the Crustacea, the Myriapoda, the Insecta, and the Arachnida. These well-defined groups, with the exception of the Trilobita, are universally familiar, and so need not be defined here.

The trilobites are of especial interest because they are the most ancient arthropods known, and because their structure was very generalized and could conceivably have given rise to the other major groups of arthropods. The trilobites were dominant from the Cambrian through the Silurian. They then declined until the Permian, from which period only a single species is known. With this, the group became extinct. The trilobite body was enclosed in a chitnious skeleton, and was divided into three longitudinal lobes by two longitudinal furrows, hence the name of the group. The body consisted of a head and a segmented trunk. The segments were generally movable, but a variable number of posterior segments were joined together to form a rigid unit, the pygidium. There was a single pair of antennae. The remaining appendages were all simple, undifferentiated, biramous appendages. None were specialized as mouth parts, but all had gnathobases, basal processes which could be used for biting. All species were marine. A typical trilobite is illustrated in Fig. 9.4. Nothing is known of the internal structures.

Fossil remains of transitional organisms from trilobites to the other major arthropod types are entirely lacking and it may be that the several groups arose independently. Indeed, this would avoid some preplexing inconsistencies. Comparative morphology indicates that the crustaceans, myriapods, and insects from one line of descent, while the arachnids must have diverged very early. In the Crustacea, this transformation has involved the division of the body into a highly fused cephalothorax and an abdomen in which the original segmentation is retained. The appendages have become greatly diversified, while still retaining the biramous plan. Almost all crustaceans are aquatic, but there are some terrestrial species, such as the sowbugs. Like the trilobites, the crustaceans are represented in early Cambrian rocks, and it is not improbable that they arose in pre-Cambrian times.

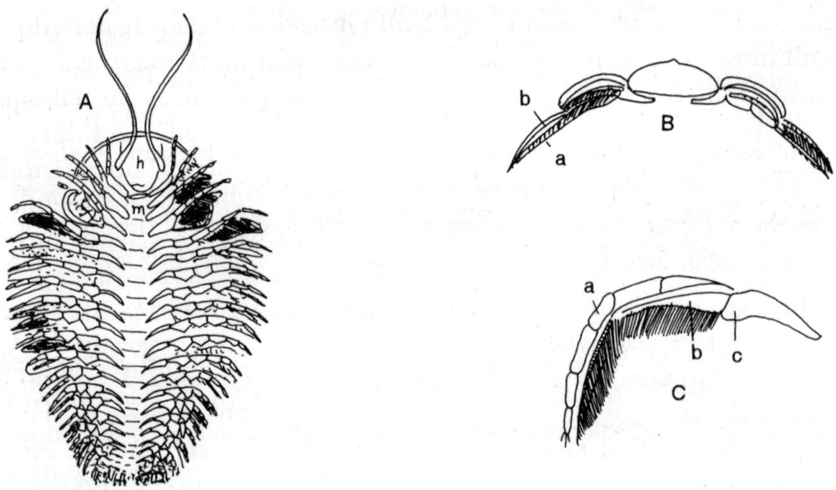

Fig. 9.4: A trilobite, *Triarthrus becki*. *A*, Ventral view. *B*, section through a thoracic segment. C, a thoracic leg. *a*, endopodite; *b*, *exopodite*; *c*, *protopodite* with gnathobase. (After Beecher, from Borradaile and Potts, "The Invertebrata.")

The Myriapoda, including the millipedes, centipedes, and their allies, may have been derived from a crustacean progenitor by the reduction of the exoskeleton and the loss of the gills. The latter were replaced by a system of tracheae, small tubes which carry air to the tissues of the body, thus permitting direct respiration. While not well represented in the fossil record, yet they are known from rocks as old as the Devonian. Insects probably also arose in Devonian times or earlier, but good fossils are first seen in rocks of Pennsylvania age. They may have been derived from myriapods, or they may have come directly from crustacean progenitors. Whichever ancestry they may have had, the tracheal system is always present and is undoubtedly one of the important adaptations which has permitted successful invasion of the land. The body became divided into three well-defined regions, head, throax, and abdomen. The appendages of the; head segments are all specialized either for sensation or for eating. The thorax bears three pairs of walking legs. Dorsally, it bears two pairs of wings, but one or both pairs may be lacking. The abdominal appendages are all suppressed. The immense variety of insects which has been 660,000, or more than the combined numbers of species of all other living groups. And the numbers of individuals of many of these species are truly immense.

The subphylum Arachnida includes a highly varied array of organisms which have developed along lines quite divergent from the series just discussed. It comprises the horseshoe crabs, scorpions, spiders, harvest men, ticks, mites, and their allies. It is difficult to derive them from the trilobites, but it may be possible. There is a tendency toward suppression of segmentation, with only two body regions, the anterior prosoma and the posterior opisthosoma being marked off. There are six pairs of appendages, of which four pairs are walking legs. Unlike all other arthropods except the trilobites, none of the appendages are specialized as jaws. Gnathobases of the anterior limbs serve this function.

The class Pycnogonida, the sea spiders, is a small and little-known group which is usually placed with the Arachnida, simply because of their superficial resemblance to spiders. Yet their morphology is utterly different. Hedgpeth has studied the group thoroughly, and has concluded that they are undoubtedly arthropods, but that they are so widely divergent that they can not reasonably be grouped with any of the other arthropod types. Even less justification can be found for the common practice of making the Tardigrada a class of archnids. This is a small and little-known group of minute, freshwater organisms which shows some relationship to the Arthropoda, and has usually been treated as a class of the Arachnida. But specialists in the field feel that it should be regarded as an independent phylum of uncertain relationship to the Arthropoda.

The Onychophora—A Unique Evolutionary Link

It was mentioned above that the derivation of the Arthropoda from the Annelida is more certain than is the derivation of any other phylum. This depends upon the Onychophora, a group of about eighty species all of which are assigned to a single genus, *Peripatus* (Fig. 9.5). This group of animals shows a peculiar mixture of annelid and arthropod characters. Among the annelid characters may be mentioned the general appearance of the organisms, for they look much like polychaetes in which the parapodia do not bear chaetae. The cuticle is thin like that of annelids, and the muscles of the body wall are continuous. The excretory organs of both annelids and onychophorans are mesodermal tubules segmentally arranged (coelomoducts), while those of the Arthropoda are usually entodermal or ectodermal in origin. The reproductive ducts of the Onychophora are ciliated, but cilia are unknown among the Arthropoda. The eyes of the annelids and onychophorans are simple, whereas those of the arthropods are compound. On the other hand, in contrast to the annelids, the Onychophora and the Arthropoda have jaws derived from appendages. The coelom in each is much reduced and largely replaced by a hemocoel, while the coelom of the annelids is highly developed. The circulatory system of the Onychophora also resembles that of the Arthropoda rather than that of the Annelida. Finally, the respiratory system of the Onychophora consists of a set of tracheae, a characteristic known nowhere else but in the Arthropoda.

Because of this strange mixture of characters, the taxonomic position of the Onychophora has always been a much vexed question. They were orginally treated as a class of the Annelida. But because of the presence of a hemocoel and especially because of the tracheal system, they are now generally treated as a class of the Arthropoda. But these are not the only possibilities. They are sometimes treated as an independent phylum intermediate between the Annelida and the Arthropoda. And Light has urged that, as they are not properly separable from either of the major phyla, the entire annelid-onychophoran-arthropod series ought to be recognized as one great phylum under the name Articulata. With no other series of phyla are such considerations possible and hence the statement with which this discussion began, that the origin of the Arthropoda from the Anneida is more certain than the origin of any other phylum. It is an unfortunate fact that the Onychophora are scantily known in the fossil record.

Fig. 9.5: *Peripatus.* (Courtesy Ward's Natural Science Establishment.)

Minor Protostomes

A few minor phyla also show the general characters of the protostome line. Two of these, the Sipunculoidea and the Echiuroidea, are worm-like burrowers of the tide flats. They are generally treated as minor annelids, an arrangement which is more convenient than accurate. Yet they probably are more closely related to the Annelida than to any other phylum. The three remaining phyla, the Bryozoa, the Phoronida, and the Brachiopoda, are more difficult to lace. Like the Entoprocta, they feed by means of a lophophor, but unlike the phylum they have a coelom. The Bryozoa are small, colonial animals which superficially resemble the Entoprocta, but actually differ from them fundamentally. The Phoronida include only two genera. The animals are elongate, wormlike creatures which dwell in tubes on tide flats. When the tide is in, the lophophore projects into the water to feed on plankton and detritus. The phylum Brachiopoda, or lamp shells, bears a superficial resemblance to the molluscs because of their bivalve shells, but these are dorsal and ventral rather than right and left. Internally, they do nto suggest the molluscs at all. They have a very prominent lophpore. These marine animals are present in the earliest Cambrian deposits, and perhaps the greatest interest in the group drives from the fact that a single genus, *Lingula,* has persisted from the Ordovician to the present time, a span of 400,000,000 years. It may be the oldest genus inexistence. These three phyla were formerly groped together as a single phylum, the Molluscoidea. Yet they have little in common with one another except the lophophor, and there is little indication that any of them have a close relationship to the Mollusca. Hence it seem best to treat them as independent phyla of protostomes, of uncertain relationships to the larger phyla.

THE DEUTEROSTOMOUS PHYLA

The other major branch of coelomate animals is the Deuterostomia, and it comprises only five phyla: Chaetognatha, Pogonophora, Echinodermata, Hemichordata, and Chordata. This series contrasts with the Protostomia in that group of embryological characters which was used above to characterize the latter. The cleavage divisions are not spiral, and neither are they determinate. The mesoderm is not formed from stem cells, but rather from outpocketings of the entoderm of the gut. This simultaneously establishes the coelom, which is said to be *enterocoelous* (meaning simply that the cavity is established from the gut). The original blastopore becomes the anus, and a new mouth is formed in this series of phyla, hence the name Deuterostomia.

Development does not lead to a uniform larval type. The Chaetognatha have a unique larval type. The Echinodermata have several types of larvae, but all of them first pass through a *Dipleurula* stage to which especial theoretical significance is attached. A typical gastrula is formed, then the entoderm and ectoderm first fuse at one end, then break through to form a mouth. The blastopore is now the anus. The digestive tube now buds off an anterior vesicle which first divides into two lateral compartments, and finally into three segments on each side. These are the coelomic pouches. The cilia which covered the blastula and gastrula evenly now become concentrated in a series of bands arranged around the concave ventral surface of the animal. This is the Dipleurula larva. Because of its universality among the echinoderms, it is generally believed that the echinoderms must be descended from a Dipleurula-like ancestor which; was bilaterally symmetrical and free-swimming.

The Hemochordata have a type of larva, the Tornaria, which is quite similar to the Dipleurula, and even more similar to the Bipinnaria larva of starfishes. When first discovered, the Tornaria was described as a larval starfish, and only much later was the error discovered. This resemblance of the larvae is one of the major arguments for a relationship of the Hemichordata to the Echinodermata. Larvae are not of general occurrence among the Chordata, but the tunicates and amphibians have tadpole larvae, while the cyclostomes have a unique larva, the Ammocoetes.

Phylum Chaetognatha

The Chaetognatha are a small and uniform group of marine worms, the arrow worms, which show little evidence of relationship to any larger group of animals. They are included here because of conformity to the general deuterostome characters, yet they show no more specific affinity to any of the remaining deuterostome phyla, and it seems probable that the phylum branched off very soon after the formation of the deuterostome line. They show a superficial resemblance to Amphioxus, but this is undoubtedly misleading. Aside from a few doubtful specimens, the phylum is not represented in the fossil record.

Phylum Pogonophora

The Pogonophora is a phylum of deep-sea worms, only recently discovered and still but little known. The body consists of small protosoma and mesosoma, and a very elongate metsoma. The protosoma may bear one or more tentacles. It has an unpaired coelom, drained by a pair of nephridial ducts. In the other body segments, the coelom is paired. A nervous mass and ring in the protosoma give rise to a paired dorsal nerve cord. The circulatory system consists of two longitudinal vessels. Musculature is made up of subepidermal longitudinal fibers. There is no digestive system whatever. Larvae are unknown. What little is known of these animals seems to ally them with the Deuterostomia.

Phylum Echinodermata

The Echinodermata are best known by the starfishes, and these perhaps typify the phylum well. All echinoderms have secondarily established radial symmetry in the adult, after beginning life as a free swimming, bilaterally symmetrical larva. The radial symmetry of the adults is generally based upon a pentamerous (five radial segments) plant, or upon a plan derived from such. The development of radial symmetry may have been related to the change from a pelagic to a sessile mode of life by primitive echinoderms, for radial symmetry is a general characteristic of sessile organisms. The fossil record of the echinoderms is one of the best, going clear back to early Cambrian times, and excellent phylogenie scan be constructed within each of the five extant and two extinct classes. But the record does not throw light; upon the origin of the phylum, nor upon its possible relations to other phyla. The questions depend, at present, entirely upon the embryological evidence, with all of its limitations.

Phylum Hemichordata and the Origin of the Chordata

The Hemichordata are a small phylum of worm-like marine animals which have been extensively studied because of their supposed relationship to the Chordata. In Haeckel's phylogeny of the vertebrates, the hemichordates were given as the next stage after the primitive flatwrorm. They were originally classed as a subphylum of the Chordata, because they show the three basic diagnostic characters of the Chordata: a dorsal nerve tube, a pharynx modified for respiration, and a notochord. Yet each of these is equivocal. The dorsal nerve tube is confined to the collar region, and the main nervous system is a ventral nerve cord like that of many invertebrates. The pharynx is pierced by numerous gill slits, yet it appears that these function primarily as exit pores for the feeding current rather than as respiratory organs. Yet some respiration probably does occur in the gills. But feeding many have been the primary function in the gills of true chordates also. Finally, the notochord is a small structure developed as an outgrowth from the digestive tract into the proboscis. It has also been homologized with the pituitary gland, and so its taxonomic value is doubtful. It

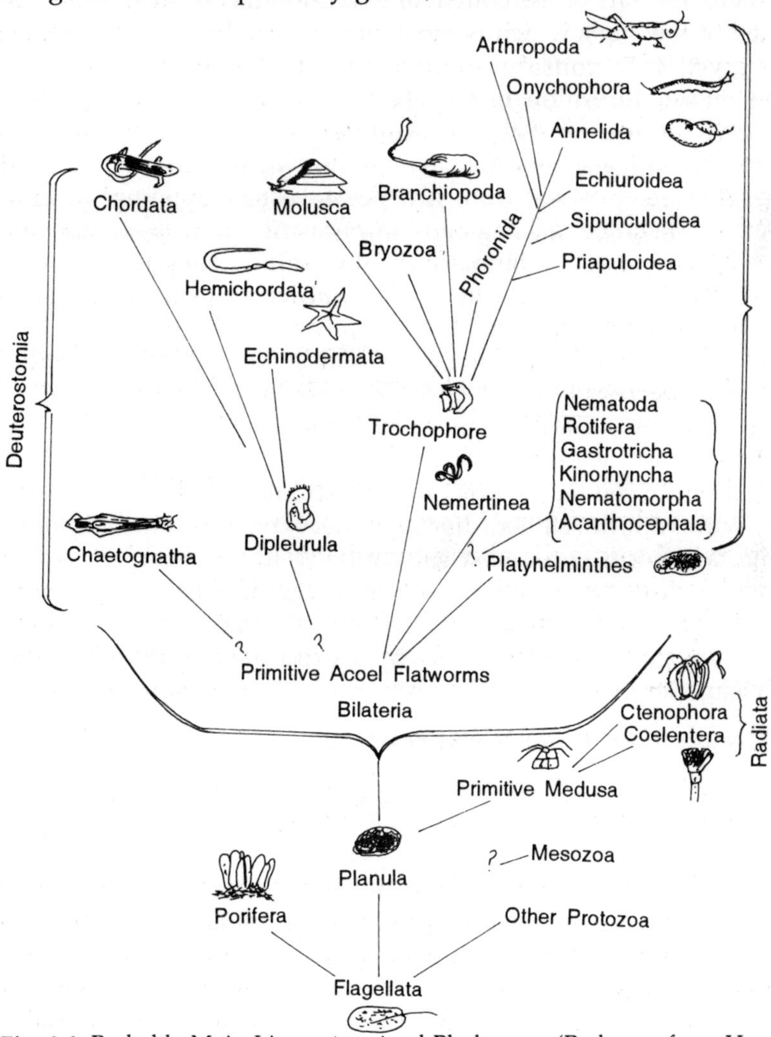

Fig. 9.6: Probable Main Lines ot Animal Phylogeny. (Redrawn from Hyman.)

seems probable that these worms are related to the chordates, but how is not clear. Their relationship to the echinoderms, as indicated by a comparison of the Tornaria and Bipinnaria larvae, has already been discussed.

As the phylum Chordata, including as it does the vertebrates and man, will be discussed in the next chapter, it need nto be taken up here beyond the obvious statement that this phylum forms the climax not only of the obscure (from the viewpoint of relationship) Deuterostomia, but also, perhaps, of the entire Animal Kingdom. Figure summarizes one view-point on animal phylogeny.

SOME EVOLUTIONARY GENERALIZATIONS

Now that the greater part of the course of evolution has been sketched, and before taking up that part of the story which is most closely related to those who study it, it may be well to give some consideration to broad tendencies and principles of evolution. At the outset, evolution is not always "upward" or "progressive." Many examples are known of the evolution of simpler or more degenerate types from originally complex types. Thus the fungi may have evolved from algae by the loss of chlorophyll. Grasses have evolved from lily-like ancestors by simplification of parts, especially of flowers. Mistletoe, an angiosperm parasitic upon trees, has undoubtedly evolved from free-living ancestors. Similarly in the Animal Kingdom, many examples of retrogressive or degenerative evolution are known. The development of sexual reproduction and its great evolutionary importance was emphasized. The Rotifera were undoubtedly evolved from bisexual ancestors, and some species are still bisexual. Nonetheless there are species in which the males are unknown. These still reproduce sexually, for the ova develop parthenogenetically, but the major advantages of sexual reproduction are lost. The development of the parasitic habit almost always involves degenerative evolution. In the tapeworms, this is extreme. Although derived from free-living flatworms, with well-developed digestive, nervous, reproductive, muscular, and other systems, the tapeworm is reduced substantially to an absorptive sac containing gonads. But, while such degenerative evolution is characteristic of parasites, it is by no means confined to them. As already pointed out, the class Archiannelida was most probably derived from polychaete ancestors by a process of simplification and loss of parts. Thus it is clear that evolution can be retrogressive as well as progressive.

Origin of New Groups from Primitive Ancestors.

A second very important generalization relates to the form of the tree of life. It is more properly a shrub than a tree, for new groups do not arise from the most advanced and specialized members of their parent groups, but from the primitive, unspecialized ones. Thus the primitive flagellates have given rise to many additional plant and animal groups, but the more specialized protozoan and algal groups are generally terminal. Again, if indeed the Hemichordata and Chordata were derived from echinoderms, it seems certain that the more advanced phyla must have arisen from the ancestral Dipleurula in a very primitive stage before radial symmetry developed. One more qualification relative to the shape of the phylogenetic tree is necessary for plants. Because of the phenomenon of allopolyploidy—to be discussed in detail below—hybridization of related species may result in new species. Hence branches may form, them fuse again, so that a network results. This phenomenon is of great importance for plants, less so for animals.

Rates of Evolution

The rates of evolution have not always been uniform. In general, periods of major geological change have been periods of rapid evolution, while periods of geological uniformity have been periods of slow evolution. This might be expected *a priori,* for the selective forces themselves should be in a state of flux during geologically unstable periods, while they should be quite stable during geologically uniform periods. Thus in a period of uniformity, organisms generally should tend to become well adapted to their environments. In such a situation, almost any change would be disadvantageous, and would tend to be eliminated by natural selection. But in periods of geological change, only those organisms which also changed could possibly have a selective advantage. Thus there was a rapid burst of evolution in the Siluria and Devonian when plants and animals were colonizing the land. Not only were vast new habitats thus opened up, but this was a time of mountain-building and of great changes in sea level. In the Mississippian, physical conditions were again quit stable, and evolution was slow. Many such alternations of periods of rapid and slow evolution have occurred. It appears that we are at present in the midst of a period of rapid evolution.

Two special situation favour rapid evolution. The first of these is adaptation of a group to a new mode of life, as invasion of land by aquatic plants and animals. During the transition, the organisms must be rather ill-adapted, hence strong selection pressure favors rapid change, and only those that can respond will leave descendants. This will be greatly facilitated if, in the earlier environment, characteristics developed which are *preadaptive* to the later habitat. Thus many Devonian fishes developed lungs, which were adaptive to life in stagnant, drying ponds. This preadaptation late speeded the adaptation to life on land. Simpson has called this "quantum evolution" as it involves sudden change from one"adaptive orbit" to another. He regards the term as unsatisfactory, because quantum events in physics are very small, while quantum evolution is on a large scale. Perhaps *macroadaptive* evolution would be a better term.

Second, when a group has invaved a new major mode of life, many variations of habitat are open to it with a minimum of competition, and hence rapid diversification is favored until these empty ecological niches are largely filled. This is the adaptive radiation which was discussed above.

Entirely apart from changes in the general rate of evolution, the rates in different lines of descent have not been equal. It some, change has been extremely slow for long periods of time. The braciopod *Lingula,* for example, has been substantially unchanged since the Ordovician. In this connection, it should be pointed out that the warm, shallow seas which it inhabits constitute perhaps the most stable environment on earth. On the other hand, the mammals have evolved very rapidly during the Cenozoic Era.

Trends in Size

A very common trend in evolution, sometimes called Cope's Law, is one toward increasing size of individuals. The original studies of the phenomenon were made upon vertebrates, but comparable studies have shown the same tendency in many groups of invertebrates and plants. A review of the paleontology of almost any group shows that its largest representatives are not its earliest ones, though not necessarily its latest ones either. Newell has pointed out that species now living are the largest known representatives of the vertebrates, crustaceans, echinoderms, pelecypods, gastropods, cephalopods, and annelids. Yet the tendency toward size increase has been by no means universal. As already mentioned, the rise of herbs and shrubs is a recent thing, and they have been derived from trees and other large plants. Hooijer has pointed out that progressive size decrease has been characteristic of many vertebrate groups during the Quaternary period, which is now in progress.

Complexity and Efficiently

It is obvious that the general progress of evolution has involved the development of new organ systems and increasing complexity. Yet development of increased efficiency often involves reduction in number and complexity of parts. Vestigial organs in general could perhaps be considered in this light, but it is equally true of actively functional structures. Thus the teeth of fishes are very numerous and usually indefinitely replaceable. They are less numerous in amphibians and reptiles, still less so in mammals, where they reach their maximum degree of specialization and efficiency. Much the same thing is true of the ververtebrae, and of the bones of the skull. This tendency can also be exemplified by plants, for example by the reduction of numbers of stamens in specialized plants.

Increased efficiency is also often obtained by fusion of originally separate parts. Thus the sacrum of mammals is formed by the fusion of three to five originally separate vertebrae, thus making a much stronger attachment of the hind limb to the vertebral column than would otherwise be possible. Another good example is afforded by the pectoralis muscles which, in tetrapods, arise near the midline of the chest region and insert upon the humerus. Muscle slips from many adjacent body segments join to form these muscles. Among plants, the corollas of flowers such as the cucurbits or petunia are formed by the fusion of originally separate petals.

Dollo's Law

Many times during the long history of life, advanced organisms have returned to ancestral habitats and modes of life. This gives selective value to adaptations similar to those of the ancestral species, and raises the question whether evolution might be reversible. Study of such cases shows that always a gross similarity between ancestral and descended structures is achieved without any genuine reversal at all. Thus many reptiles and mammals have reverted to an aquatic mode of life. They have assumed a generally stream-lined, fish-like form, and the limbs have become shortened, webbed, and fin-like. Yet the skeleton of such flippers is always distinctly that of the class to which the animal belongs rather than that of a fish fin. Similarly, many angiosperms have returned to the water and assumed alga-like appearances, but their morphology is still that of flowering, vascular plants. The evidence indicates that major evolutionary steps, once taken, are never reversed. This is known as Dollo's Law. It even might be expected *a priori,* for major evolutionary steps are compounded of many smaller steps, each preserved by natural selection. That such a sequence, occurring by chance once, should by chance be exactly reversed would be a most extraordinary thing. If not impossible, it is at least most improbable for whole organisms. Attempts to apply Dollo's Law to individual characters have failed, for these are, indeed, reversible by mutation.

The Significance of Extinction

Some closing remarks should be made upon the subject of extinction, for this has been the fate of most species since the origin of life. Extinction may have completely different significance in different instances. The dinosaurs were a highly specialized line of Mesozoic reptiles which dominated the earth for a long time. But when conditions became unsuited to their survival, they became extinct and left no descendants. They were succeeded by other unrelated forms. The cynodont reptiles also became extinct, but they were succeeded by their own descendants, the adaptively superior mammals. Thus extinction may mark the end of a line of descent or it may be the accompaniment of the origin of new and superior types. It ought to be added that the major adaptive types—phyla and classes—very rarely become extinct.

These, then, are some of the conclusions which may be drawn from a study of the course of evolution. The causative factors which have led to these results will be taken up in parts III and IV.

REFERENCES

Borradaile, L. A., and F. A. Potts, 1958. "The Invertebrata," 3rd Ed., Macmillan Co., New York, N.Y. An authoritative and comprehensive text on the invertebrates.

Hadzi, J., 1953. "An attempt to reconstruct the system of Animal Classification," *Systematic zool.,* 2, 145–154. A radical revision, yet to be adequately assessed by other zoologists.

Hardy, A. C., 1953. "On the Origin of the Metazoa," *Quart. J. Microscop. Sci.,* 94, 441–443. A brief but stimulating paper urging the origin of the Metazoa from plants.

Hyman, Libbie H., 1940. "The Invertebrates," McGraw-Hill Book Co., New York, N.Y. Five volumes of this series have been published, and at least three more are planned. As far as it goes, it is the best, most comprehensive, and most profound treatment of the invertebrates in the English language. (Cleveland, Kofoid.)

Ivanov, A. V. 1955. "The Main Features of the Organization of the Pogonophora," *Systematic Zool.*, 4, 170–178. New phyla are rarely described, but this is such a description.

Marcus, E., 1958. "On the Evolution of Animal Phyla," *Quart. Rev. Biol.*, 33, 24–58. A scholarly presentation of a phylogeny similar to the present one in general, but differing in important respects.

Simpson, G. G., 1953. "The Major Features of Evolution," Columbia University Press, New York, N.Y. The viewpoint of a genetically minded paleontologist. This book includes much material applicable to the last part of this chapter. (Cope, Dollo.)

Tiegs, O. W., and S. M. Manton, 1958. "The Evolution of the Arthropoda," *Biol. Rev.*, 33, 255–337 A penetrating review for mature students.

Chapter—10

Evolution of Land Plants

It is almost a point of definition for the algae that they are aquatic plants, although some of them have invaded moist habitats on land. The useful fossil record begins with the Cambrian period of the Paleozoic Era, over 500,000,000 years ago, and the Cambrian record, so far as plants are concerned, consists entirely of a wide variety of algae and bacteria. In fact, the early Paleozoic is often referred to as the Age of Algae and Invertebrates. But fossils from the Silurian period, beginning about 360,000,000 years ago, include primitive land plants, and it is probable that their first appearance was in the preceding period, the Ordovician. When these first colonists left the waters to invade the more difficult but more varied habitats on land, the algae remained the dominant members of the earth's flora. But soon (geologically speaking) the land dwellers surpassed their aquatic progenitors. One of the crucial problems which had to be solved before plants could invade the land was the protection of the zygote against drying. In all land plants, this is accomplished, with important differences in the details, by the retention of the zygote and the developing embryo within the sex organs of the maternal plant. For this reason, the land plants are known collectively as the subkingdom Embryophyta. This subkingdom includes only two phyla, the Bryophyta and the Tracheophyta, but the latter is greatly varied.

THE BRYOPHYTES

The phylum Bryophyta is a relatively small group comprising the mosses, the liverworts, and the hornwohese. These are the amphibians of the plant world, for they have met only minimum requirements of adaptation to the terrestrial environment. They are restricted to wet habitats, and all of them require water for reproduction, at least as a film over the surface of the plant in which sperm can swim. The bryophytes share with other land plants certain adaptations which permit them to utilize terrestrial habitats. As already mentioned, the embryos, which are always multicullular, are retained within the female sex organs, thus protecting them from drying. The plants are always oogamous, that is, the egg is a large, nonmotile cell which must be sought by the sperm. The sex organs of both sexes include a jacket layer of sterile, protective cells. All of the aerial parts of the plant are covered by a Waxy cuticle which protects the plant against drying. Finally, alternation of generations is well developed, with the sporophyte begin substantially a parasite upon the gametophyte in the bryophytes, for the former is always attached to the latter, and it contains inconsequential amounts of chlorophyll, if any.

Most of these characters, other than relention of the embryo in the maternal sex organs, are already known in the algae, although no alga has all of them in combination. Other bryophyte characters tie these plants in with the algae more closely. Like the algae, the bryophyte body is a *thallus,* a rather simple cell mass which is not differentiated into roots, stems, and leaves. Structures which resemble these parts are present, but the same may be said of many of the algae. The gametophyte is larger than the sporophyte, as in most algae, and photosynthesis is largely restricted to the gametophyte.

Life Cycle of a Moss

A typical moss may serve as an example (Fig. 10.1). 28 a spore (the agent of dispersal) falls on suitable ground, it germinates to form a network of threads not unlike a filamentous alga and is called the protonema. The protonema sends fine, root-like rhizoids into the substrate for attachment, and also for procurement of water and salts. Buds on the protonema develop into shoots bearing leaf-like structures. At the tips of these shoots, the sex orans, antheridia in the male or archegonia in the female, develop. In some species, the sexes are separate (dioecious), while in other species both male and female shoots may occur on the same plant (monoecious). In either case, the sperm are released and swim to the archegonia, where fertilization occurs. The embryo now develops within the archegonium, and remains attached there by a foot when it grows to maturity. In addition to the foot, the mature sporophyte consists of a stalk and a terminal capsule which contains the spore mother cells. Here the meiotic divisions occur, and the spores are released to start the cycle over again. The gametophyte, consisting of the haploid protenema, rhizoids, shoots, and leaf like structures is much larger than the sporophyte, and is independent because of the possession of chlorophyll. The sporophyte, consisting of foot, stalk, and capsule, is smaller and dependent on the gametophyte because it is lacking in chlorophyll, or nearly so. In neither the gametophyte nor the sporphyte are there any specialized vascular (conducting) tissues, although a limited amount of conduction of foods and water does occur. Nor are there any fibrous tissues. Lack of these elements (vascular and fibrous tissues) is probably the most important single factor which has prevented the development of large plants in this phylum.

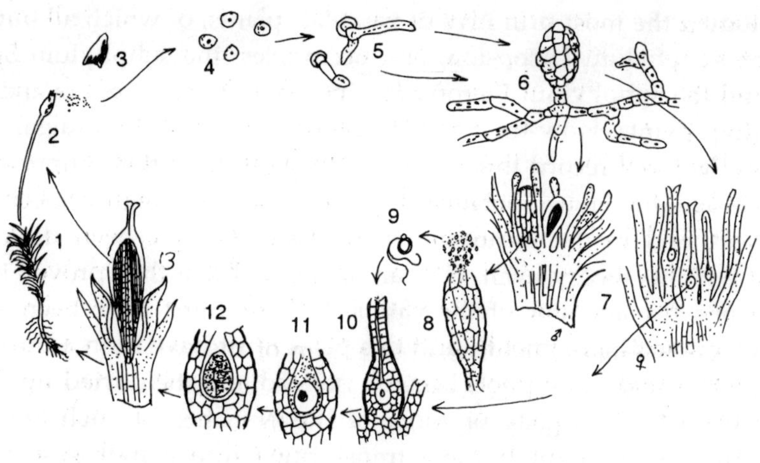

Fig. 10.1: The Life Cycle of a Moss. 1. Mature gametophyte with a sporophyte (2) growing upon it and discharging its spores. 3. The cap of the spore capsule. 4. Ripe spores. 5. Germinating spores. 6. Young gametophyte with protonema. 7. Portions of male and female gametophytes showing ahtheridia and archegonia. 8. Antheridium discharging sperm (9). 10. Archegonium with egg. 11, 12, and 13. Developmental stages of the sporophyte. (From Pauli, "The World of Life," Houghton Mifflin Co., 1949.)

Phylogeny of the Bryophytes

In view of the simplicity of the bryophytes, and in view of the many characters which the share with the green alagae, it is highly probable that the bryophytes were derived from an ancestor among the green algae, most probably a tide-flat dweller for which terrestrial adaptations would already be advantageous. One group of bryophytes, the hornworts, suggests relationship to higher plants. While the gametophyte is a small, simple thallus, the sporophyte is larger and quite differentiated, much the dominant phase of the hornwort. It is well supplied with chlorophyl, but still depends upon the gametophyte for absorption of water and minerals from the substrate. The capusule has a central axis of elongate cells which suggests vascular tissue. The epidermis has stomata like those month leves of vascular plants. With such an array of traits suggesting those of vascular plants, it is difficult to avoid the inference that modern hornworts may be but little changed from an ancestral stock by which bryohytes gave rise to vascular plants. Nonetheless, this inference appears to be misleading, for vascular plants appear in the fossil record in the Silurian, and the bryophytes appear more than 100,000,000 years later. Obviously, if these data are correct, bryophytes cannot be ancestal to tracheophytes. And so the weight of opinion among botanists now favors direct origin of the vascular plants, the phylum Tracheophyta, from a chlorophytan ancestor.

ORIGIN OF THE VASCULAR PLANTS

The phylum Tracheophyta is an ancient and much varied group, which includes the dominant plants of today. It is subdivided into four subphyla, the subphylum

Psilopsida, including the most primitive of vascular plants, of which all but two genera are extinct; the subphylum Lycopsida, or club mosses; the subphylum Sphenopsida, or horsetails; and the subphylum Pteropsida, including ferns, confiers and their allies, and the flowering plants. How the vascular plants arose from their algal ancestors is not known, and the fossil record throws very little light upon this important question. Nearly sixty ears ago, the French botanist Lignier published a highly speculative theory upon this subject, and in the meantime such fossils as have been discovered have been consistent with his theory, even if it can not be said that they proved it. According to Lignier's theory, the ancestor of the vascular plants must have been a green alga characterized by branching filaments, and this plant must have been a tide-flat dweller. As the land was elevated, tide pools become isolated and then dried up. Much of the flora became extinct. But if one or more of the branches of such an alga should penetrate the ground, it might become transformed into a root system capable of supplying the plant with water and minerals. Some of the branchings might then straighten out, leaving a main stem or trunk with branches. Because the entire plant is no longer immersed in water, a conducting system is now necessary, and only those plants which develop one can survive. Thus the stems become thickened, and the ends flatten out to form specialized organs for photosynthesis, the leaves. Now the conducting system must operate in both directions, carrying water and salts up from the roots and carrying organic compounds down from the leaves. Finally, only those plants which developed a cuticle over the aerial parts could escape extinction by drying. While critical proof of this theory is still lacking, evidence accumulated in the past sixty years is consistent with it, and no more probable theory has yet been proposed.

Subphylum Psilopsida

The subphylum Psilopsida includes two orders, and perhaps the sharpest distinction between them is simply one of time: the Psilophytales are known only from rocks of Silurian and Devonian age, while the Psilotales are represented by two living genera, *Psilotum and Tamesipterus.* A series of fossils connecting these two orders is lacking, and it has been suggested by some botanists that they are actually unrelated. But the similarities of the plants of these two orders are so close that it seems very improbable that they could be coincidental. The first discovered member of the group, *Psilophyton,* was described in 1858 by Sir William Dawson on the basis of fossils from the Devonian of Canada. The plant looked much like a small, rootless, leafless shrub (Fig. 10.2). It did not fit into the botanical classifications then in use, and Dawson's discovery was ignored or disparaged. But in 1917 Kidston and Lang discovered three genera of similar plants in a silicified bog of Devonian age at Rhynie, Scotland. The preservation of the Rhynie fossils was so good that the cell walls can be seen in thin sections. The fossils are complete and abundant. In the meantime, at least three more genera of psilophytes have been discovered, so that thee is now a wide variety of material available for study of this very primitive group which may be the connecting link between the green algae and the vascular plants.

Psilophyton was a small plant growing to a height of about three feet. It consisted almost entirely of a green, branched stem. True leaves were absent, but ere suggested by numerous small spine-like projections of the superficial layers of the plant. The green stem was the main photosynthetic organ. Sporgania were located at the tips of some of the branches. such shoots arose not from true roots, but from rhizomes, which were nothing but subterranean stems bearing rhizoids for absorption of water and salts. There were vascular bundles in the stems. Thus *Psilophyton* turns out to be remarkably similar to the hypothetical primitive land colonist of Lignier.

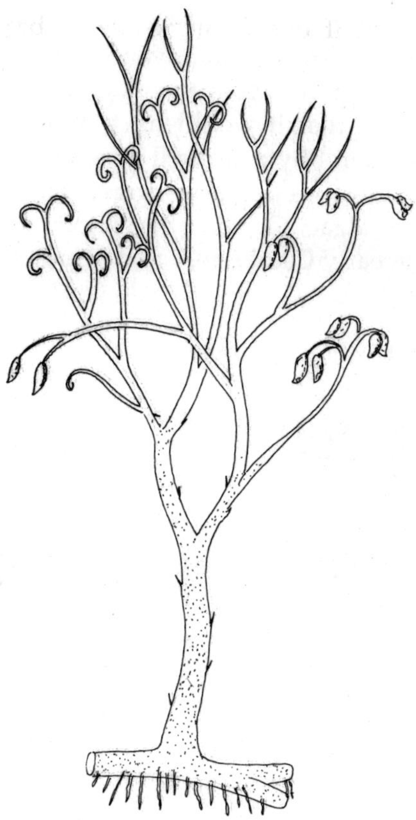

Fig. 10.2: *Psilophyton.* (After Dawson, from Fuller and Tipp, "College Botany," 2nd Ed., Henry Holt & Co., Inc., 1954.)

Current opinion among botanists favors the Psilophytales as ancestral to the remaining three subphyla of vascular plants, not only because the Psilophytales are obviously extremely ancient and primitive vascular plants, but also because different genera show tendencies toward specialization in the direction of each of these subphyla. Thus some psilophytes, such as *Psilophyton* and *Asteroxylon*, had small scale-like leaves which were developed as projections of the superficial layers of the stem. The Lycopsida (club mosses) are characterized by similar scale-like leaves. Another genus, *Hyenia,* had whorled branches, and may have been ancestral to the Sphenopsida (horsetails). Finally, some genera, such as *Pseudosporochnus,* had the branch tips flattened out, and it is generally believed that this led to the formation of broad leaves, as in many of the Pteropsida.

Thus the psilophytes appear to occupy a key position in the evolution of the higher plants. They are the most primitive vascular plants known—one might almost say they are the most primitive vascular plants imaginable—and it is highly probable that they arose from the green algae. The psilophytes appear to have given rise to the three higher subphyla independently, that is, each of the higher subphyla has arisen from a different psilophytan ancestor, so that these represent parallel developments rather than a sequence from primitive to advanced. It has also been suggest that the bryophytes may have been derived from the psilophytes, but this does not seem very probable in view of the fact that it would have required a reversal of the tendency toward dominance of the sporophyte, as well as a supression of the vascular bundles. Current opinion favors the independent origin of both the bryophytes and the psilophytes from the green algae. On the other hand, spores resembling those of land plants are known from the early Cambrian, and it may well be that even the Devonian

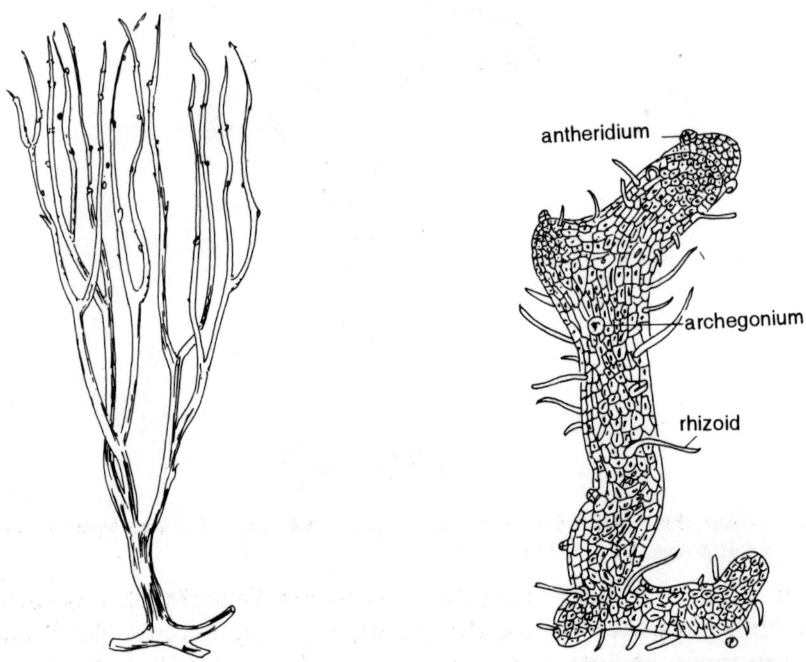

Fig. 10.3: *Psilotum Sporophyte* (left) and Gametophyte (right) (After Lawson, from Fuller and Tipp, "College Botany," 2nd Ed., Henry Holt & Co., Inc., 1954.)

psilophytes are late remnants of much more ancient stages in the origin of the higher plants.

The living Psilotales (Fig. 10.3), *Psilotum and Tmesipterus,* are quite similar to *Psilophyton*. In nature, they are confined to tropical and subtropical regions, but they can be grown in greenhouses anywhere. The major morphological difference between the living and extinct species is that the sporangia are located in the axils of the "leaves" of the living plants, whereas they were located at the tips of branches in the extinct species. Another point of considerable interest relates to the alternation of generations. Only the sporophytes of the psilophytes have been preserved in the fossil record. But the gametophytes of the modern genera are known. The gametophyte looks very much like a fragment of rhizome, bearing numerous archegonia and antheridia which are rather similar to those of bryophytes. These may give an indication of the character of the gametophyte of the extinct psilophytes, but there is no assurance that they do. In any event, the Psilotales are regarded as the little-changed descendants of the ancient psilophytes.

Subphylum Lycopsida

The Lycopsida, comprising the club moses, appear to have been derived from psilophytes in the Devonian period. All of the living lycopsids (there are only four genera) are small plants, generally less than a foot high, but this has not always been true, for during the Carbonferous period (Mississippian and Pennsylvanian combined) there were giant club mosses as tall as 135 feet. These were the dominant plants of the time, and their fossil remains are a major part of the coal beds. But these giant club mosses became extinct in the Permian period, perhaps as a result of inability to adapt to the severe Permian climate, for this was a time of extensive glaciation.

The club mosses (Fig. 10.4) are advanced over the psilophytes in many respects. The differentiation into root, stem, and leaves is complete. Their leaves, which are small and spirally arranged, are supplied with vascular tissue. The sporangia are enclosed in specialized leaves called sporophylls, clusters of which occur at the tips of branches. Such clusters are called cones or strobili. The life cycle of *Lycopodium,* one of the living genera, is quite simple, and perhaps not very different from that of the psilophytes. The spores are all alike, and are scattered by the wind. Those which fall on favorable ground germinat to form small, subterranean, thallus-like gametophytes. These are monoecous. Sperm swim to the archegonia to fertilize the eggs. The zygote then produces a young sporophyte which is at firs a parasite upon the gametophyte. But a root is soon formed and the gametophyte decays. In another living gens, *Selaginela,* the life cycle is modified in a fashion quite suggestive of the seed plants. There are two types of spores, megaspores which are contained in sporophylls in the lower part of the storbilus, and microspores which are contained in the upper sporopylls. While still contained within the sporangia, these megaspores and microspores germinate to produce female and male gametophytes respectively. The gametophytes are thus parasites upon the sporophyte, although the female gametophyte may contain some chlorophyll. When the microspores are released, they fall down onto the lower sporophylls. When wet by rain or dew the microspore wall splits and sperm are released. These swim to the archegonia of the female and fertilize the eggs while the female gametophyte is still contained in the spore wall, and hence in the parental sporophyte. The embryo is thus produced in the female gametophyte while the latter is still contained in the sporophyte, a condition very similar to that of the seed plants.

One might expect that a group with so many progressive characters would have produced further, more progressive descendants, but this has not been the case. The Lycopsia are a terminal group. Soon after its origin, the group produced the dominant plants of the coal swamps, but these became extinct before the end of the Paleozoic Era, and only four genera of small, insignificant lycopods have survived to the present. These are often referred to as living fossils because their closest relatives are long-extinct plants.

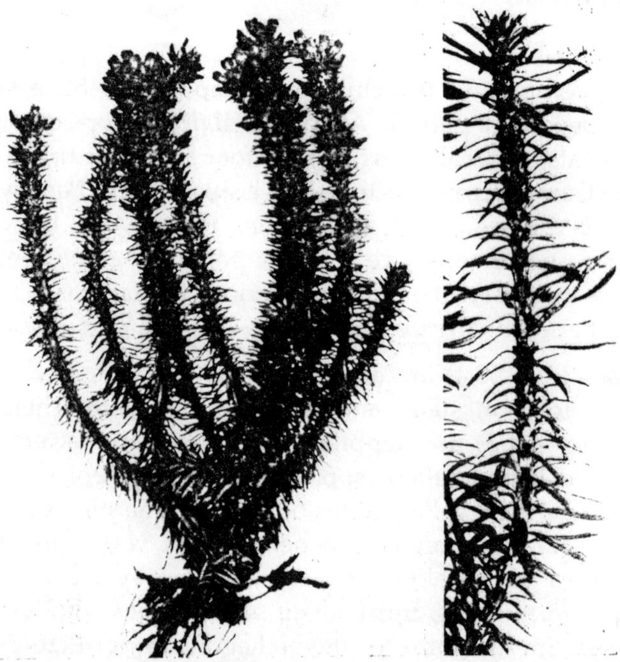

Fig. 10.4: A Club Moss, *Lycopodium*, including an enlarged view of a single stem. (From Weatherwax, "Plant Biology," 2nd Ed., W.B. Saunders Co., 1947.)

Subphylum Sphenopsida

The Sphenopsida, the horsetails, have played a smaller role in the history of the plant kingdom than the Lycopsida. Like the latter, they probably arose in the Devonian from psilophyte ancestors by the differentiation of the body into definite root, stem, and leaves. The stem of the horsetails has a pith region which is commonly hollow.

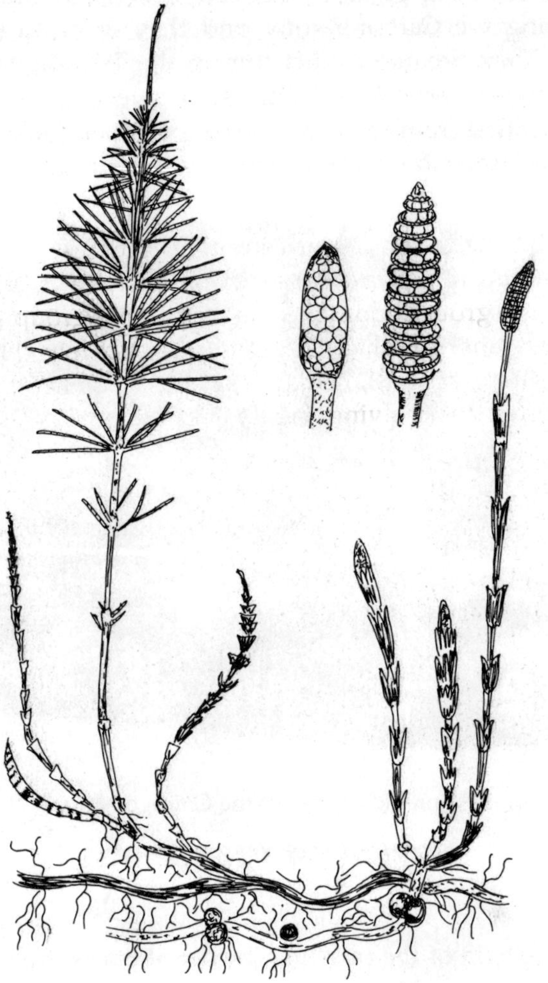

Fig. 10.5: A Horsetail, *Equisetum*. (From Weatherwax, "Plant Biology," 2nd Ed., W. B. Saunders Co., 1947.)

The stem grows by means of concentrations of meristem at definite nodes, which can be easily disjointed, to the delight of many a child (Fig. 10.5). The leaves are arranged as whorls, and they are relatively small. Nonetheless, the weight of opinion favors their origin by flattering of the tips of branches rather than as emergencies from the stem. The leaves are not adequate for the photosynthetic needs of the plant, and the entire stem has retained this function.

The Origin of the horsetails from psilophytes is strongly indicated not only by the similarity of shoots and leaves but by the fact that *Hyenia* is an excellent intermediate between the two groups and might with equal justification be assigned to either. From the Hyenia stock there were developed herbaceous species, comparable to the horsetails of today, and gigantic, tree-like species as much as forty feet tall. These flourished during the Carboniferous, and they from an important part of the coal beds (Fig. 10.6). They became extinct during the Triassic, the first period of the Mesozoic Era. The modern horsetails comprise a single genus, and it is uncertain whether they are descended from the herbaceous species of Paleozoic times, or whether they have been derived from the horsetail trees by reduction.

Fig. 10.6: Carboniferous Flora. (Courtesy of the Chicago Natural History Museum.)

SUBPHYLUM PTEROPSIDA

Ferns

The last of the subphyla of the phylum Tracheophyta is the Pteropsida, comprising ferns, confiers, and flowering plants. Like club mosses and horsetails, the ferns—class Filicineae—appear to have developed from psilophyte ancestors of Devonian age. Some of the psilophytes, for example *Pseudosporochnus* and Protopteridium, show a tendency toward leaf formation bMy flattening of whole branches, and it is probable that the ferns were derived from such psilophytes. Like all tracheophytes except the psilophytes, the plant is differentiated into true roots, stem, and leaves. The stem is generally subterranean, and so is a rhizome, and small roots arise all along its length. The leaves are large fronds, and these are the only aerial parts of the plant. The sporangia occur in clusters called sori on the undersurfaces of the leaves. The spores, when released, germinate on moist ground to form monoecious gametophytes, which are always small, thallus-like plants. The sperm of one plant swim to the archegonia of another, and there fertilization takes place. A new sporophyte then develops from the zygote.

Although the ferns most familiar in temperate latitudes are all of moderate size, in the tropics thee are tree ferns in which the stem forms an erect trunk as high as seventy-five feet. The leaves from a palm-like cluster at the top. The most primitive order of ferns, the Coenopteridales, developed great forests of fern trees during the Carboniferous period. These, together with the giant club mosses and horsetails (erroneously called fern allies) formed the dominant vegetation of the times, and their remains have come down to the present as coal beds. Like the the psilophytes, these primitive ferns bore the sporangia at the tips of the branches. Like so many other ancient groups, these giant ferns became extinct during the Permian, but not before giving rise to the three modern orders of ferns.

Gymnosperms

The second class of the subphylum Pteropsida is the class Gymnospermae, the dominant members of which are the conifers, but their allies include the extinct seed ferns, the cycads or sago palms, the maidenhair trees, and the very aberrant order Gnetales. The class may have arisen in the Pennsylvanian period from some group of ferns via the seed ferns. The seed ferns were long classified with the typical ferns, but accumulating data have shown that their affinities are close to the gymnosperms. The basic datum is the fact that they did reproduce by seeds, an innovation of the greatest importance. The seed ferns formed an important part of the Carboniferous forests, but they later declined in importance and finally became extinct during the Jurassic period, but no without leaving more advanced gymnosperms as descendants. The most successful of these is the order Coniferales, including all of the well-known evergreens. These reached their greatest development in the Mesozoic Era, and have been declining since. Three other orders have survived to the present time, however. These are the Cycadales, or sago palms, with nine living genera; the Ginkgoales, represented only by a single species, *Ginkgo biloba,* which has survived only because it has for centuries been cultivated in Chinese temple gardens; and the Gnetales, including three genera of very aberrant plants, of which the most important is perhaps *Ephedra* from which the drug ephedrine is obtained.

Unlike the low tracheophytes, gymnosperms are generally large, woody trees, although some are shrubs. Most are evergreens. But their most important characters are concerned with the reproductive cycle. At the outset, the gymnosperms produce two types of spores, megaspores which develop into female gametophytes, and microspores or pollen which develop into male gametophytes. In spite of their names, the two types may be of equal size, or the microspores may actually be the larger. The cones in which the spores are formed consist of specialized sporophylls spirally arranged about a central axis. They are thus modified leaf clusters. The microspore divide while still within the spore wall to form the male gametophyte. The pollen is shed in great numbers while in the four-cell stage, and is carried by the wind, sometimes for great distances. Some of the pollen will reach female cone and become stuck in a sticky fluid exuded by the ovules, complex structures which include the female gametophyte. The pollen produces a tube-like growth which enters the ovule. Down this pollen tube pass the two sperm nuclei, one of which fertilizes the egg. The developing embryo (sporophyte) is pushed into the mass of the female gametophyte, which is now called the endosperm and which serves a nutritive material for the embryo. This endosperm is, in turn, surrounded by a seed coat, which is actually a part of the parent sporophyte. The seed is shed, and if it falls on favorable ground the seedling may develop into a mature sporophyte.

Some significant aspects of this reproductive cycle should be indicated. For the first time in the phylogenetic series, reproduction is independent of water. The pollen is carried to the female cones by wind, and the sperm—mere nuclei rather than flagellate cells—are carried to the egg by the protoplasmic pollen tube. Both the male and female gametophytes are reduced to minute structures consisting of only a few cells, and thus the relative predominance of the sporophyte in the life cycle has become great. The gametophytes lack chlorophyll and are completely dependent upon the sporophyte. Finally, the seed, a new structure in the phylogenetic series, consists of an embryo (sporophyte) contained within the endosperm (gametophyte) which is in turn contained within a seed coat which is sporophytic tissue of the parental generation. Thus the appearance is very much as though the embryo sporophyte were produced directly by the parent sporophyte, with the gametophytes being simply organs of the parent sporophyte. Only comparison to lower plants reveals the true situation.

The Flowering Plants

The third and final class of the subphylum Pteropsida is the class Angiospermae, the flowering plants. This is much the dominant class of the world's flora today, including on the order of 10,000 genera and 195,000 species. The variety of the angiosperms is unlimited, ranging from great trees to grass; generally they are land plants, but they have become adapted to almost every available habitat, including marine; and while they are typically free-living green plants, not a few have become parasitic and some are saprophytic.

The angiosperms share the major characters of the gymnosperms, being differentiated into true roots, stems, and leaves, and reproducing by means of true seeds. They have a highly developed vascular system. Fertilization is by means of polen which are independent of water. The sporophyte is much the dominant generation, the gametophyte being minute and completely dependent. In addition, the sporangia of the angiosperms are within flowers, which are modified cones surrounded by modified and often highly ornamental leaves. The microspores develop into pollen (male gametophytes) with only three nuclei, one tube nucleus and two sperm nuclei. The macrospores develop into female gametophytes with only eight nuclei, including the egg nucleus and two polar nuclei which fuse to form the fusion nucleus. The macrospores are completely enveloped within modified sporphylls, the carpels, which become the fruit when mature. Pollination may be by wind as in the

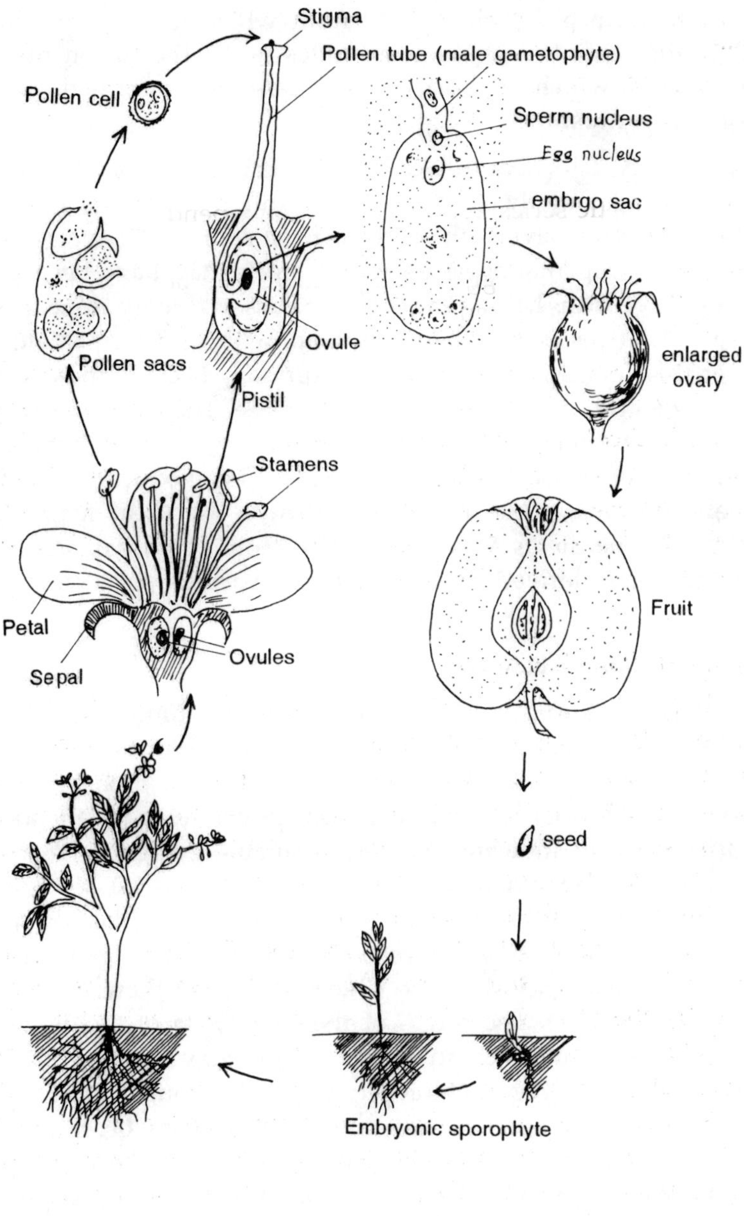

Fig. 10.7: The Life Cycle of a Flowering Plant, (From Young, Hylander, and Stebbins, "The Human Organism and the World of Life," Harper Brothers, 1938.)

gymnosperms, but it is commonly effected by insects, occasionally by birds, and rarely by water (in the case of aquatic plants). A unique characteristic of the angiosperms is the phenomenon of double fertilization. As usual, a single sperm nucleus unites with the egg nucleus to form a diploid zygote from which the embryo develops. But in addition to this, the other sperm nucleus unites with the fusion nucleus to form a triploid (3 n) cell from which the bulk of the endosperm is formed. The life cycle of a typical angiosperm is illustrated in Fig. 10.7.

About a century ago, Darwin called the origin of the flowering plants an "abominable mystery," and there the problem still rests. While botanists can agree that the angiosperms must have been derived from some primitive gymnosperm stock, there is nothing but spepculation as to which stock may have been the correct one. The fossil record is of little help. The angiosperms suddenly appear in considerable numbers in the Cretaceous period (late Mesozoic) with no intermediate plants in older rocks. Whatever their origin, the angiosperms quickly became the dominant plants of the world, and they appear to be stil! on the increase. The class Angiospermae include two subclasses, the Dicotyledoneae and the Monocotyledoneae, which are separated mainly on the basis of embryological characters. The monocots include the grasses, lilies, palms, and orchids, while the dicots include the great majority of flowering plants. Anatomically the dicots appear to be the more primitive, and so it is probable that the monocots were derived by reduction of various parts from some primitive dicot.

The Main Lines of Plant Evolution

The main lines of plant evolution may now be summarized with the aid of Fig. 10.8. Whatever the origin of viruses and the bacteria may have been, the seven phyla of algae must have arisen from bacterial ancestors, probably from among the autotrophic bacteria. Whether the different algal phyla arose independently or from a common stock is scarcely indicated by any available evidence. The Cyanophyta or blue-green algae are the earliest plants found in the fossil record; they are extraordinarily primitive in their lack of a distinct nucleus or plastids; and they contain chlorophyll. In view of these facts, it is plausible that the Cyanophyta may be ancestral to some other algal phyla, including the Chlorophyta, yet there is very little evidence that this is actually the case. The origin of the two phyla of molds, Myxomycophyta and Eumycophyta, is also much in doubt. They may have come from bacterial ancestors, or form algae by loss of chlorophyll, or even from Protozoa. However much vexed the problem of the origins of these lower groups may be, it appears to be quite clear that the land plants arose from the green algae, with the two phyla, Bryophyta and Tracheopyta arising independently. The bryophytes have become differentiated into three major groups, the mosses, liverworts, and hornworts, but have not produced any more progressive types of plants. The primitive tracheophytes (psilophytes), on the other hand, having first succeeded in colonizing the land, quickly gave rise to three more subphyla, the Lycopsida, the Spenopsia, and the Pteropsida. The first two were dominant groups for a time, but were reduced to minor groups about the end of the Paleozoic Era. The pteropsid line was at first represented only by the ferns, but these gave rise to the gymnosperms which in turn appear to have given rise to the angiosperms. The more primitive subclass of angiosperms, the dicots, has given rise to the more specialized monocots.

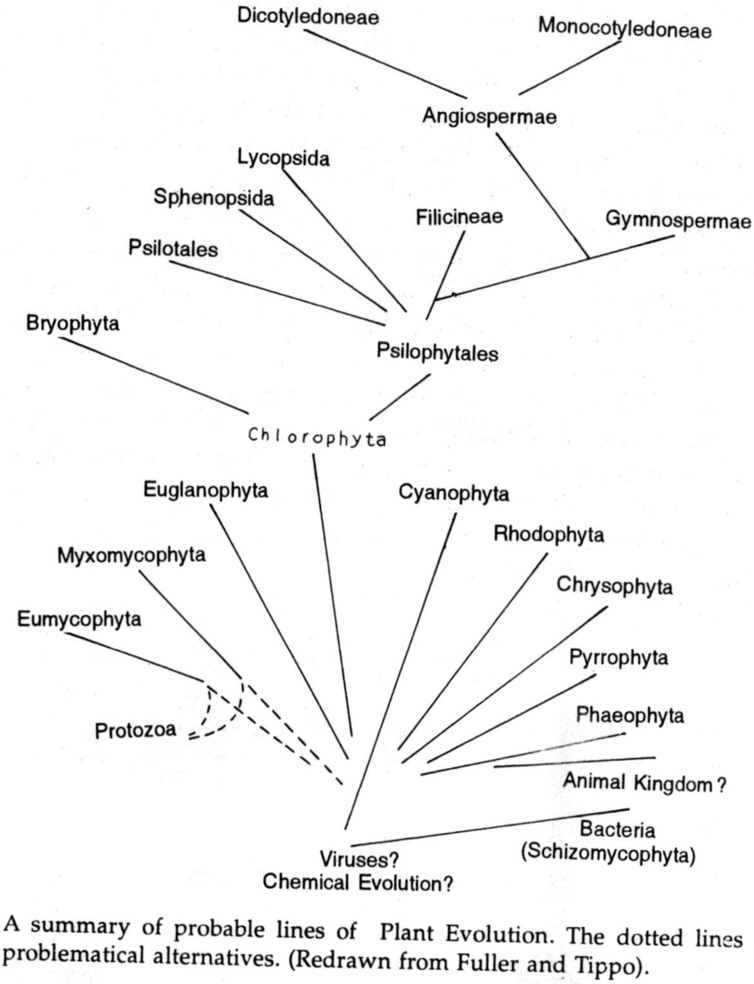

Fig. 10.8: A summary of probable lines of Plant Evolution. The dotted lines present highly problematical alternatives. (Redrawn from Fuller and Tippo).

Some general trends of evolution in the Plant Kingdom may be noted. Among the most primitive plants, increasing complexity of the nuclear apparatus, culminating with the typical cell structure as found in all plants above the blue-green algae, is a major trend. Among the algae, development of the colonial habit and multicellularity is one major trend. A second is the evolution of sexual reproduction, together with the alternation of generations. Finally, a transition from aquatic to terrestrial habitats occurred, but little specific information is available regarding the steps in this process.

This required the development of vascular and strengthening tissues, the development of which is one of the most characteristic features of the land plants. These tissues are, however, found only in the sporophyte, and so the sporophytic generation has increased in size and functional importance relative to the gametophytic generation until finally, in the angiosperms, the gametophytic generation is recognizable as distinct from the sporophyte only by comparison to the lower plants. This trend is illustrated in Fig. 10.9. In a general way, thee has been a tendency to increase in size. This is obviously so among the algae, where the most primitive species are all unicellular and the more advanced ones may be (but need not be) multicellular. While the most primitive land plants were smaller than large algae, all of the classes of tracheophytes have produced species much larger than the largest of the algae. Nonetheless, it is far from axiomatic that evolution is necessarily accompanied by progressive size increase. The most successful members of many groups are characterized by small size. The club mosses and horsetails, for example, were once represented by great trees, but only the small members of the group have survived to the present. Among the angiosperms, the trees seem to have been primitive, and from these the shrubs and grasses have been derived. This has been accompanied by the development of the annual habit, presumably as an adaptation to prevent extinction by winter-killing, for a dormant seed may easily survive severe weather which would kill a mature plant. Thus development of small size, rapid growth and maturity, and the annual habit comprise an adaptation to arctic and subarctic conditions, and such plants appear to be on the increase in temperate lands as well.

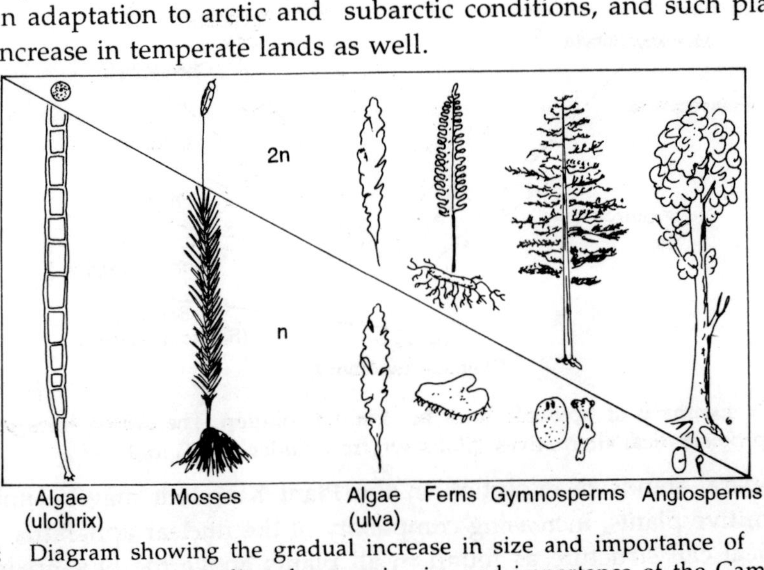

Fig. 10.9: Diagram showing the gradual increase in size and importance of the Sporophyte, and the corresponding decrease in size and importance of the Gametophyte. (From Fuller and Tippo, "College Botany," 2nd Ed., Henry Holt & Co., Inc., 1954.)

REFERENCES

Andrews, H. N., 1947. "Ancient Plants and the World They Lived In," Comstock Publishing Co., New York, N.Y. A readable summary of paleobotany.

Axelrod, D. I., 1959. "Evolution of the Psilophyte Paleoflora," *Evolution,* 13, 264–275. Evidence is adduced that the higher plants arose much earlier than generally supposed, and from algal ancestors.

Bold, H. C., 1957. "Morphology of Plants," Harper Brothers, New York., N.Y., A penetrating treatment, going far beyond the minimum presented in this chapter.

Fuller, H. J., and O. Tip, 1954. "College Botany," 2nd Ed., Henry Holt & Co., Inc., New line New York, N. Y. An excellent text, and the source of the classification used in this chapter (Lignier, Kidston, and Lang).

Thomas, H. H., 1936. "Paleobotany and the Origin of the Angiosperms," *Botan. Rev.,* 2, 37–418. Documentation for the statement that the origin of the angiosperms is still "an abominable mystery."

Chapter—11

Evolution of the Pentadactyl Limb

Ever since Richard Owen's demonstration that the limbs of tetrapod vertebrates can be understood as transformations of one another, the whole set of variations defining the typical form known as the pentadactyl limb, the phylogeny and the ontogeny of this basic vertebrate structure have been matters of considerable interest and dispute. Darwinians, reifying the abstract concept of a typical form, sough a common ancestral tetrapod limb structure in fossil amphibians such as *Eryops,* and antecedents in crossopterygian relatives such as *Sauripterus,* the limbs of later vertebrates being regarded as adaptive modifications of the ancestal form. Embryologists have looked for evidence of a basic pattern of cartilage condensation in the limb buds of developing amphibians, birds, and mammals prior to secondary modifications such as loss or fusion of carpal and tarsal elements. They were following the exception, derived from the theory of recapitulation, that the ancestral form may appear transiently in the early ontogenesis of the limb. However, no consensus on an archetypal limb form has emerged.

This should not surprise us: the quest was initiated by an error of conceptualisation, the generalised form of which unfortunately runs through most of contemporary biology. This was Darwin's suggestion that the concept of the pentadactyl limb should not be treated as an abstract generalised form, as it was defined by Owen, but as an actual, materialised form which had (and perhaps still has) a historical existence in an ancestor common to the line of terrestrial vertebrate descent. This is explanation of form by historical contingency, not by morphogenetic law, the latter being the goal of the rational morphologists, such as Geoffroy St Hilaire, Cuvier, and Owen. The contemporary descendant of explanation by historical continuity and contigency is to be found in the concept that embryogenesis is a result of the operation of a genetic programme which is a historical record of adaptive success in a particular phylogenetic sequence. In this paper, we will not pursue this line of reasoning, but return instead to the concept that organismic morphology, and in particular vertebrate limbs, are to be understood in terms of general principles of spatial organisation, together with particular constraints (including the effects of specific gene products) which we develop in terms of the classical concept of the morphogenetic field.

Our objective is to demonstrate the simple proposition that tetrapod limb morphogenesis may be understood in terms of some basic generative principles capable of producing a great variety of limb forms which are all transformable one into the other under modifications of the limb-generating process. None of these forms is special in any way, none intrinsically more 'primitive' than any other, so that none can be selected as an ancestral form. Indeed, one of our purposes is to show that the study of principles or laws of biological form must be clearly separated from historical considerations, since the search for the invariants which define homology (equivalence of structure under a transformation which leaves certain spatial relations unchanged or invariant) depends solely on form and not on ancestry. Historical conservation of forms in particular lineages may also occur, but this observation does not help one to understand problems of biological organisation, such as the nature of the generative principles underlying the transformation equivalence of tetrapod limbs to which we now turn.

THE GENERALISED TETRAPOD LIMB FORM

It is convenient, with a generalised tetrapod limb form, to start by simply defining the essentials of its structure. For this we use a pattern similar to those found in texts on comparative anatomy such as Smith's *Evolution of chordate structure* (1960), though we emphasise again that, whereas he interprets this as a primitive ancestal limb form, the so-called 'cheiropterygium,' we use the form shown in Fig. 11.1 to illustrate the general structure to be considered and the relationships of their components to the different parts of the tetrapod limb. The Figure is form a contour plot produced by a computer programme based upon a model of limb morphogenesis which is discussed briefly below and in detail in the Appendix (pp. 94-8). Each of the enclosed elements is a contour of a hill with a central maximum of a field function $u(x)$, as shown in a three-dimensional perspective in Fig. 11.2, and is taken to represent a domain where cartilage condensation will subsequently occur in a developing limb. Figure 11.1 thus describes an early stage in the formation of a 'pentadactyl' limb with (i) the usual number of elements in the stylopodium and the zeugopodium, where there is the greatest uniformity of structure in tetrapods, and (ii) an autopodium of 12 carpal or tarsals, 5 metacarpals or metatarsals, and a total of 13 phalanges arranged in 5 digits whose phalange formula is 23332. Since our objective is not to model the detailed development of any particular tetrapod limb but to explore the generative principles common to this type of limb and thus to gain some insight into the concept of homology from a developmental perspective, we have concentrated on the proximo-distal sequence of limb elements and ignored the secondary modifications which occur in most limbs with respect to the antero-posterior axis. Thus all the limb patterns shown have antero-posterior symmetry. Models incorporating these spatial details can quite readily be developed, but we have deliberately avoided these asymmetries in the present analysis in order to focus on the notion of transformational equivalence on which homology depends.

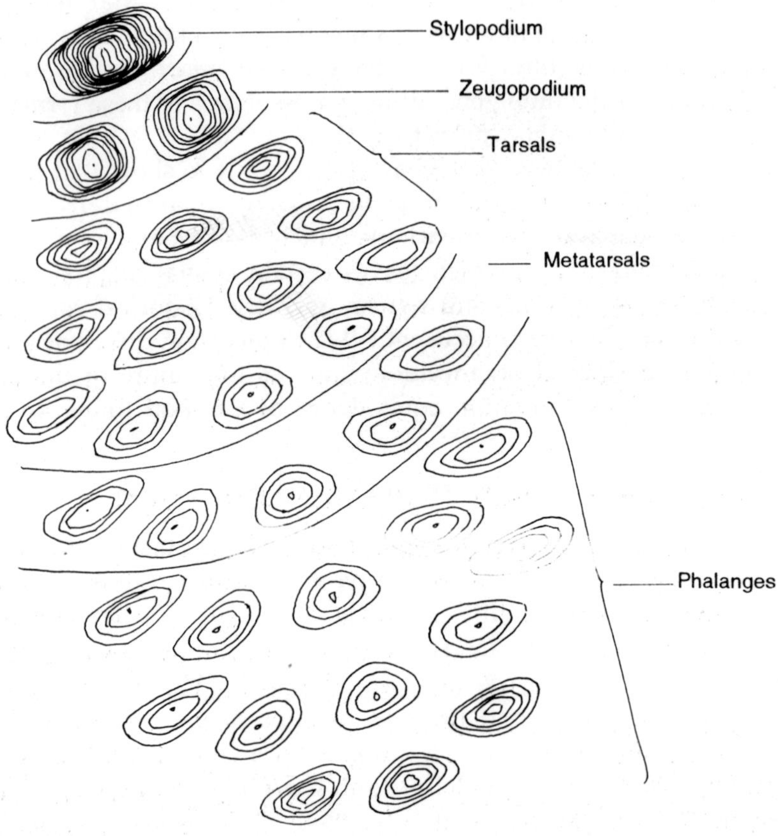

Fig. 11.1: Contour plots of the sequence of field solutions generated by the model of pattern formation for a hind limb with five digits. Each closed curve represents a particular level on a hill which describes a part of the field solution at some proximo-distal position. Thus the five metatarsal elements correspond to a single field solution with five peaks of the function which is selected by a particular 'energy' level at this position in the proximo-distal axis.

The principles used to model the morphogenetic field for limb formation in this paper correspond closely to those used in a previous publication dealing with a field description of the embryonic cleavage process (Goodwin & Trainor, 1980). We assume the existence of an order parameter, $u(x)$, which defines the state of the limb field at point x in terms of some relevant biological variable which describes that aspect of the limb-forming process which is the immediate causal antecedent of the emergent order resulting in a specific pattern of chondrogenic sites. This order parameter is assumed to obey a 'smoothness' principle of the general type used in describing the spatial variation of fields which is familiar to biologists from such examples as diffusion-like processes (e.g. Lawrence's (1970) 'sand model') or the intercalation process used in the clock model of limb regeneration (French *et al.*, 1976). This together with

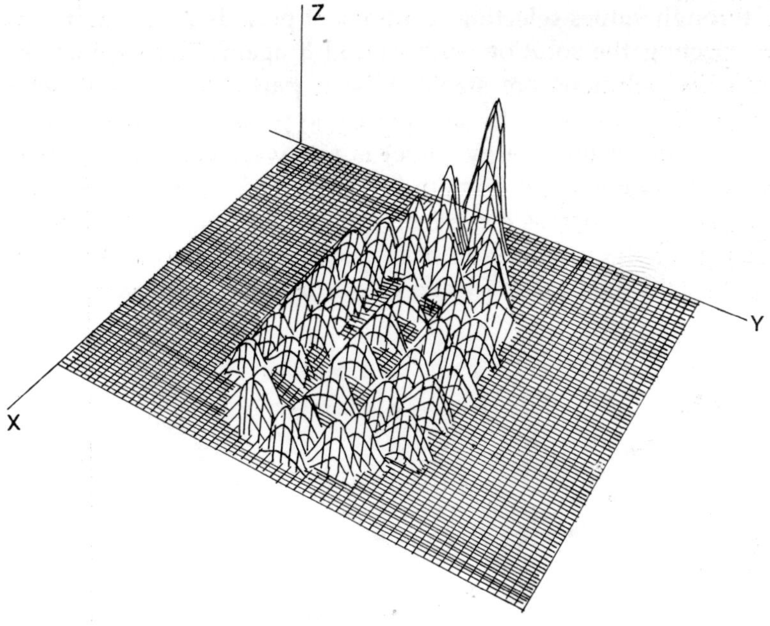

Fig. 11.2: A three-dimensional view of the pattern shown in Fig. 11.1 (but lacking the three terminal phalanges). Each hill represents a presumptive centre of cartilage formation in the developing limb bud. For variables, see Appendix p. 97.

an assumption that the field value are everywhere bounded (limited in size) leads to a field equation for *u(x)* who solutions depend upon the geometry of the field and the values of the function on the boundaries of this domain, as well as on the specific value of a quantity generally analogous to an energy within the domain. In our model, a semicircular 'domain of competence' is assumed to propagate proximo-distally, starting at a particular stage in limb bud development, and to lay down a sequence of field solutions as shown in Fig. 11.1, each antero-posterior array of elements being generated by a particular field solution within the domain. This assumption is similar to that used by Ede (1976) in his model of limb morphogenesis. Thus, the presumptive centre for the humerus or femur is laid down first be a solution that produces a single maximum of *u(x)* within the first hemispherical domain of competence. Then a solution with two maima is established in a more distal domain generated by the advancing wave of competence; then a solution with three maxima, then four, then a series of four solutions with five maxima, and finally a solution with three maxima are laid down in proximo-distal order. The particular sequence thus obtained is, in this model, dependent upon a proximo-distal gradient in the limb-bud which may be compared to 'energy' levels stabilising or selecting particular solutions of the field equation. Thus to get the solutions generated in Fig. 11.1, the gradient is assumed to decrease progressively from a value which selects the solution with one period, giving the stylopodium, through values selecting solutions of periods 2, 3, 4, and 5, and finally rising to a value selecting the solution with period 3 again. The gardient is assumed to be continuous, since solutions are stable only at particular 'energy' levels in this type of model; i.e. we are dealing with an eigenvalue problem. Repetition of the same solution, such as 5, occurs if the energy level is relatively constant so that solutions are set up as soon as a large enough domain is established by the advancing wave. To get the absence of a solution such as 4 between the distal 5 and 3 solutions, we assume a rather rapid change in the gradient so that a solution of period 4 has insufficient space or time to get established as the wave of competence propagaes distally.

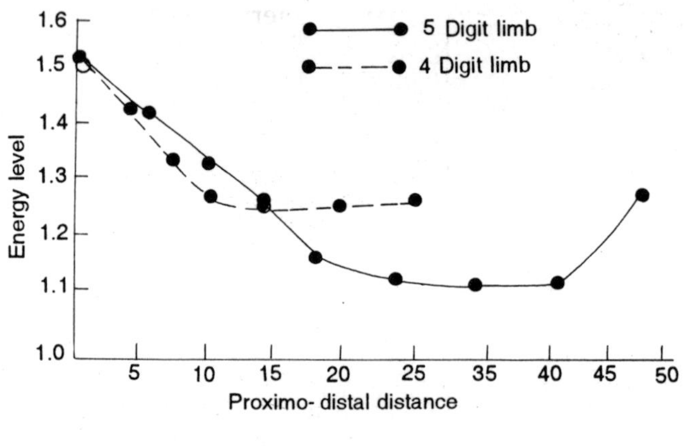

Fig. 11.3: The proximo-distal 'energy' profile which selects or stabilise a particular series of solutions of the field equation in successively more distal domains of competence in the limb bud. The solid curve gives rise to the sequence shown in Fig. 11.1, each solid circle on the curve defining an 'energy' level which corresponds to a particular eigensolution of the field equation, with a specific number of peaks across the antero-posterior axis. The dashed curve generates the pattern shown in Fig. 11.4, the curve flattening out at an 'energy' level which gives eigensolutions of period 4. The pattern generated is not dependent on limb and size.

The sequence in Fig. 11.1 corresponds to the 'energy' profile shown in Fig. 11.3, which falls through a series of values generating the increasing series of elements, then levels off to give a sequence of four 5-unit solutions, and finally rises to give a terminal 3-unit solution. For comparison, a profile of a four-digit limb is included, which will be considered later.

Evidently it is the shape and the height of the proximo-distal 'energy' gradient which determine the sequence of elements which is laid down in a limb in this particular model, so that changes in the details of the gradient determine changes in pattern. This is only one of the possible ways of generating patterns by field solutions. A somewhat different mechanism is proposed by Wilby & Ede (1976) and by Newman & Frisch (1979), who used diffusion-reaction equations for their field and selected different solutions by means of changes in an effective diffusion constant. The similarity of the models is that both use solutions of the same field operator (the Laplacian) to give a prepattern defining where the skeletal elements will later appear in the limb. We could say of the present model that the proximo-distal wave of competence, together with the treatment of the process as an eigenvalue problem, thus giving quantised solutions across the anteroposterior axis, is what transforms a one-dimensional energy profile into a pattern which is periodic in two dimensions. (The third dimension of the limb is ignored in this treatment, since we are concerned only with the bone sequence. We assume that the periodicity of the solutions in the third dimension (dorso-ventral) is always one, making the elements in Fig. 11.1 all solids. Any treatment of muscle and cartilage pattern would of course have to consider three-dimensional field solutions, with boundary conditions partly determined by the cartilage condensation patterns.)

A final point about the model is that no attempt is made to describe the modifications following initial chondrogenesis, such as the differential growth of elements in different parts of the limb or fusions of basipodial elements. Thus our concern is solely with the variety of forms which is possible in the primary condensation pattern, and how this may be generated.

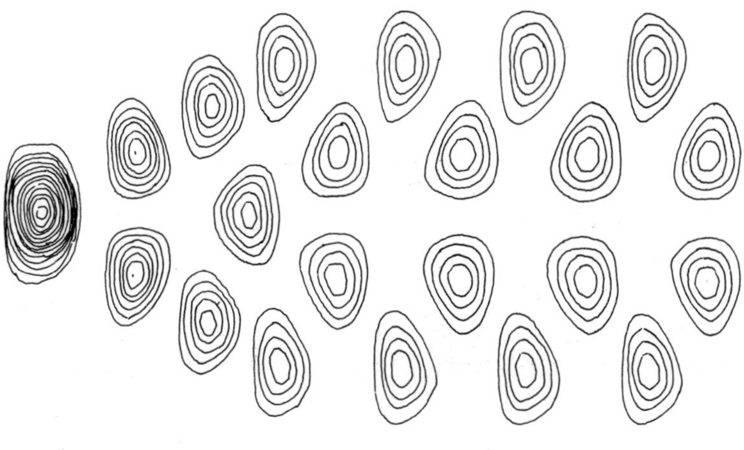

Y:	1.500							
R:	3.83	3.83	5.14	6.38	7.59	7.59	7.59	7.59
F:	0.70	0.71	0.74	0.78	0.81	0.82	0.82	0.82
A:	0.00	1.00	4.00	7.00	10.00	15.00	20.00	25.00
N:	1	1	2	3	4	4	4	4

Fig. 11.4: The limb pattern generated by the 'energy' profile of the dotted curve in fig. 11.3. *Y* is an arbitrary scale factor. *R, F,* and *A* are as described in the Appendix, while *N* describes the Bessel function *Jm* which has been selected by the energy value in accordance with equation (A_9) The first two Bessel functions corresponding to N = I at position A = 0 and I.0 fuse to give the first chondrogenic site, after which the functions are separated.

INVARIANCE AND VARIATION

A simple modification of the form in Fig. 11.1 involves a reduction in the total number of element as a result of both a shorter proximodistal series and a reduced periodicity across the antero-posterior axis, as shown in Fig. 11.4. This pattern correlates with an altered 'energy' profile, as shown in Fig. 11.3 (dashed curve). The level generating field solutions of period 5 is not reached, so the limb is one with only four digits. This is the type of limb found in urodeles such as the axolotl (*Ambystoma;* forelimb) or the salamander, *Necturus,* which is shown in Fig. 11.5 in a drawing taken from Kent *Comparative anatomy of the vertebrates* (1969). This drawing shows alternative ways in which a five-digit limb could undergo conversion to one with four digits, by 'loss' of a digit either preaxially or postaxially. However, our generative model suggests a much simpler and more plausible solution: in a transformation from one limb from to another, there is no conservation of elements so that there can be no question of one digit being lost and others conserved. Biologists have been misled by this type of question ever since Darwin adopted a view of evolution which stressed the adaptive detail of organismic morphology (i.e. the particular 'content' of organisms) rather than principles of form, and problems of inheritance rather than those of organisation. Great

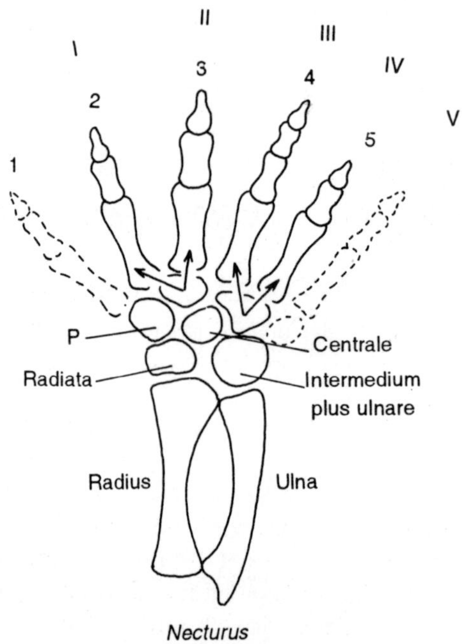

Fig. 11.5: Hypothetical alternatives in the evolution of a four-digit limb from a five-digit 'common ancestor', according to the Darwinian concepts of homology and inheritance. The original pentadactyl limb has digits I–5, or I–V, and either digit I or digit V is then selected against to give a four-digit limb, 2–5, or I-IV. P = pisiform. (After Kent, 1969.)

attention has been paid to the exact number of phalangs per digit, their relative length, and other features which could identify their individuality and so allow them to be recognise as units of inheritance when they turn up in *Erypos,* in the salamander, in the wing of the domestic hen, or in some other organism with an even greater 'loss' of digits. Thus the thee digits of the hen's wing are identified as digits 2, 3, and 4 of the ancestral five-digit limb, the two of the antelope are labelled 3 and 4, and the single toe remaining in the horse is identified as digit 3. Just how is such identification possible, especially when so much variation can occur in the details of the bone pattern within a single species, or even in an individual with multiple regenerate: a question to which we shall turn in a mmoment.

Fig. 11.6 shows examples of the general categories of limb reduction to 3, 2 and 1 terminal digit, without any effort to reproduce the detailed pattern of any particular species. What emerges very clearly at this level of abstraction is just the point made by the pre-Darwinian rational morphologists: biological patterns must be seen and understood as wholes which undergo transformation, so that there is no identity of elements in different patterns. It is wrong even to suggest that the most proximal element of the limb is the 'same' in all tetrapods, even though the element in this position in nearly always single. What is the same is the generating principle of the total pattern, so that there are invariants or universal which are recognised as common to the whole set of forms. These forms are thus equivalent under transformation (an idea embodied in Geoffroy St. Hilaire's Principle of Connections' and later in Owen's concept of homology). Each element is a component of a total pattern, but has no inherited individuality in a genealogical sequence.

EMBRYOLOGICAL TRANSFORMATION

The conclusion reached above suggests that it may be possible to perturb limb morphogenesis in a single vertebrate species in such a way that the basic skeletal patterns of other vertebrates are produced (not, of course, the secondary modifications). This is because, if the only universal constraint is the generative mechanism of the presumptive chondrogenic pattern, then perturbations to the initial and the boundary conditions of the limb field should result in perturbations in the pattern even against a fixed genetic background, since in a field description genes cease to play the role of specific causes of form (cf. Webster & Goodwin, 1982). They contribute to the particular conditions which stabilise or select a specific field solution from the possible set, or a specific sequence of solution which generate the field morphology, but their influence can be over-ridden by other particular conditions. Because of the absence of inherited individuality in the units of the limb, we should not expect to be able to identify elements in the perturbed limbs, but only general categories of tetrapod pattern. It is even possible that patterns will be observed which are not found in any tetrapod, since the expectation is that the set of possible forms which the generating mechanism can produce will be greater than the set which has been realised in evolution, i.e. the potential set is greater than the historically realised set, and individual species embody this potential in their fields, independently of genotype.

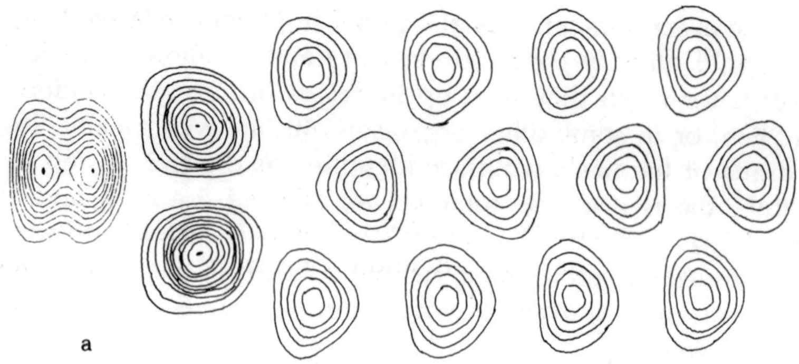

a

Y:	1.200							
R:	3.83	3.83	5.14	5.14	6.38	6.38	6.38	6.38
F:	0.70	0.71	0.75	0.76	0.79	0.80	0.80	0.80
A:	0.00	1.50	4.00	5.00	8.00	12.00	16.00	20.00
N:	1	1	2	2	3	3	3	3

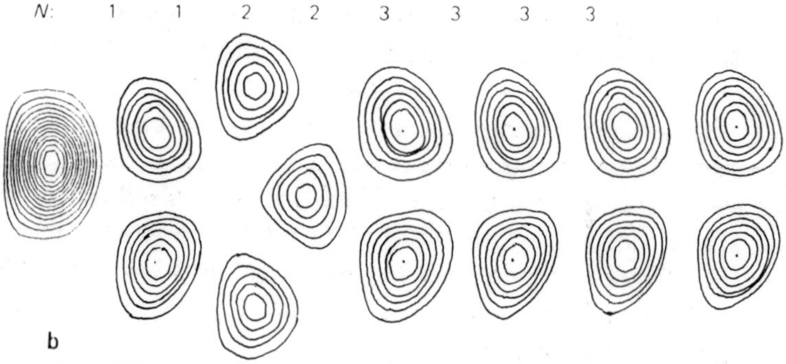

b

Y:	1.200							
R:	3.83	3.83	5.14	6.38	5.14	5.14	5.14	5.14
F:	0.80	0.81	0.84	0.85	0.88	0.85	0.84	0.83
A:	0.00	1.00	4.00	7.50	12.50	16.50	20.50	24.50
N:	1	1	2	3	2	2	2	2

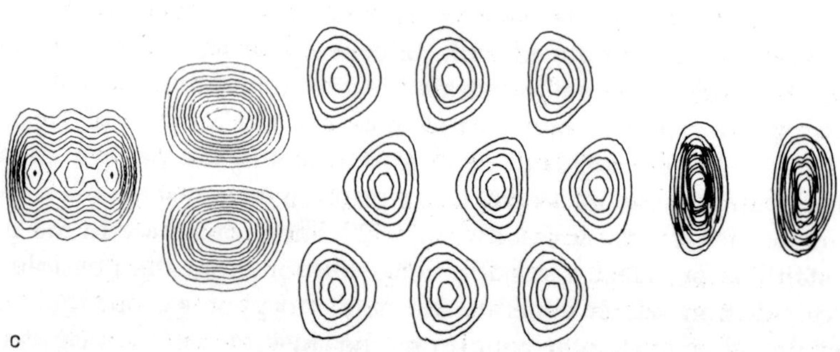

c

Y:	1.300										
R:	3.83	3.83	3.83	5.14	5.14	5.14	6.38	6.38	6.38	3.83	3.83
F:	0.70	0.71	0.71	0.75	0.76	0.77	0.81	0.82	0.81	0.72	0.71
A:	0.00	1.50	3.00	6.00	7.00	8.00	11.00	15.00	19.00	25.00	29.00
N:	1	1	1	2	2	2	3	3	3	1	1

Fig. 11.6: Limb patterns generated by modifications in the proximo-distal `energy' profile of the limb bud, corresponding to limbs with (*a*) three, (*b*) two, and (*c*) one digit. *Y,R,F,A* and *N* are as described n the legend to Fig. 11.4 and in the Appendix.

The investigation of limb regeneration in urodeles provides an opportunity for this type of observation, and we reproduce in Fig. 11.7 some patterns which have resulted from work with the axolotl, *Ambystoma mexicanum,* by M. Maden. A more complete study of this material is in progress, but we report here some observations of relevance to the present investigation. The sample shown in Fig. 11.7 provides evidence supporting the general principles developed above, illustrating how plastic is the limb field in the patterns which can be generated and reinforcing the proposition that the field is the generative unit in development and evolution.

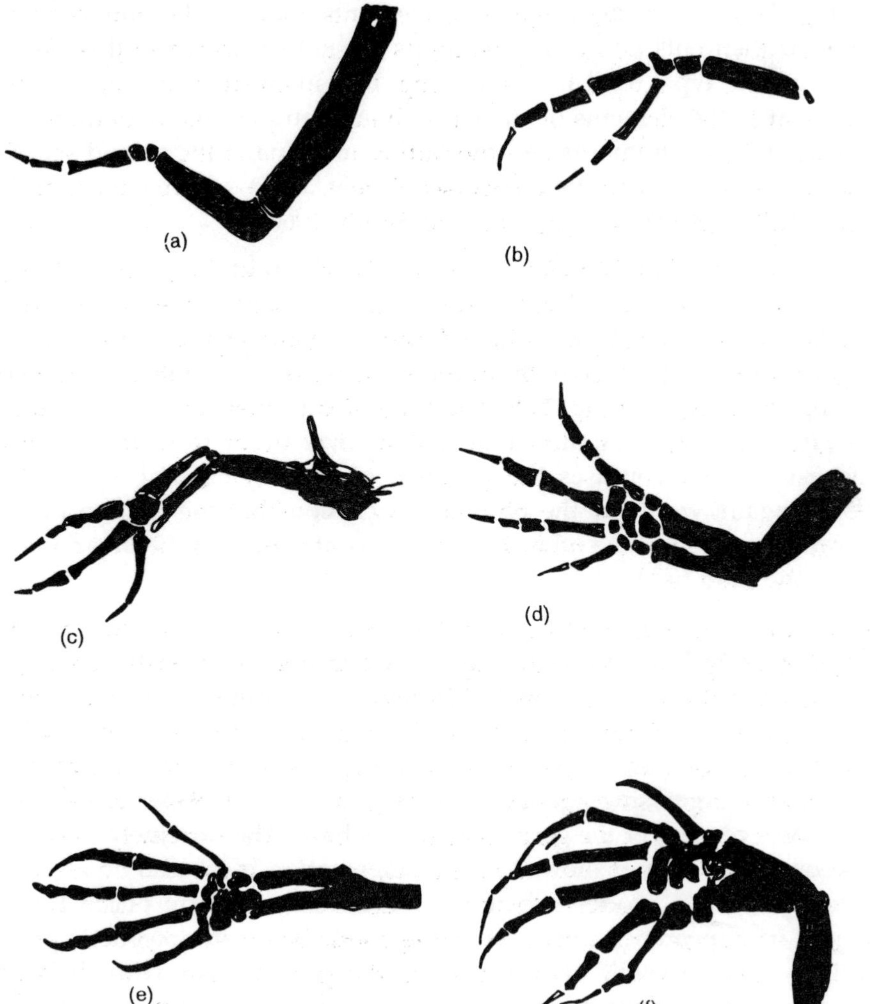

Fig. 11.7: Cartilage patterns observed in supernumerary limbs produced after 180° ipsilateral rottions of blastemas in the axolotl, *Ambystoma mexicanum* (*a*) one, (*b*), (*c*), three and (*f*) six-digits together with normal forelimb ((*d*) four-digit) and hindlimb ((*e*) five-digit) patterns. (Courtesy of M. Maden.)

TRANSFORMATIONS INVOLVING EXPANSION

The limb pattern of Fig. 11.7*f* shows an expansion of elements across the proximo-distal axis, rising to six phalanges. Such a limb is usually interpreted to result from a fusion of two four-digit limbs, with the 'loss' of a digit in the process. However, there is no need to regard this type of pattern in this way if we are using a field description of limb morphogenesis, since it is perfectly possible for the generating process to involve 'energy' levels giving a periodicity of six (or more) across the antero-posterior axis. An example of such a limb is to be found in the fossil marine reptile *Ichthyosaurus*, shown in Fig. 11.8, where the number of elements spaning the antero-posterior axis rises to six and then falls off again. The digits are not separated in this case, since the limb was of paddle type, used for swimming. It is interesting to note that no attempt is made to identify the elements of such a limb in relation to the hypothetical 'common ancestor', except for the humerus and the radius and ulna, if the second set of elements in proximo-distal order constitute a pari (see Romer, 1956)—radius, ulna, and pisiform if these constitute a triplet (see Fig. 11.8 and Smith, 1960).

A computer simulation of this type of limb is shown in Fig. 11.9. In this particular example, the most proximal element arises from a 'fusion' of the maxima of functions with periods 1 and 2, giving a Y-shaped chondrogenic centre. Thus the number of elements goes from one to three without the presence of a separated solution of period 2, duplicating the pattern of Fig. 11.8. The type of variation seen in the bone sequence of *Ichthyosaurus* and the possibility that fusions may occur from the earliest stage of chondrogenesis further confuses the problem of identifying limb elements between species. Once again, we draw the obvious conclusion that the pattern of Fig. 11.8 is simply a transformation of the other limb forms conforming to the invariant generative principles of the limb field.

This conclusion is also reinforced by the absence of any transient archetypal limb pattern during early limb morphogenesis, as has been demonstrated so clearly by Hinchliffe & Griffiths (this volume). However, the notion that there is inherited continuity of elements of pattern such as digits, as in Darwin's concept of continuity of descent from a common ancestor, or that there is a specific, isolatable material acting as a central organising agency of biological form (Webster & Goodwin, 1982), as in Weismann's concept of the germ plasm, dies hard. The extremely valuable studies of Nieuwkoop (this volume) show that the latter notion must also be abandoned as a general principle, sin in urodeles there is no continuity of germ plasm for generation to generation, its appearance during embryogenesis begin an epigenetic phenomenon. It seems likely that this will turn out to be the general case, with different species occupying different positions on the regulative-mosaic spectrum, so that the classical instances of apparent germ plasm continuity are resolvable in terms of early mosaicism in oogenesis, an earlier regulative phase nevertheless occurring. These observations all lead in the direction of an abandonment of the rather simplistic notions of the inheritance of sufficient casual elements of morphogenesis in an entity with the properties of a central organising agency, now called the 'genetic programme', and its replacement by an exact theory of morphogenesis in terms of universal generative principles expressed by field constraints, together with the particular influences operating during ontogenesis (including those arising internally from the genes) which result in an organism or specific form.

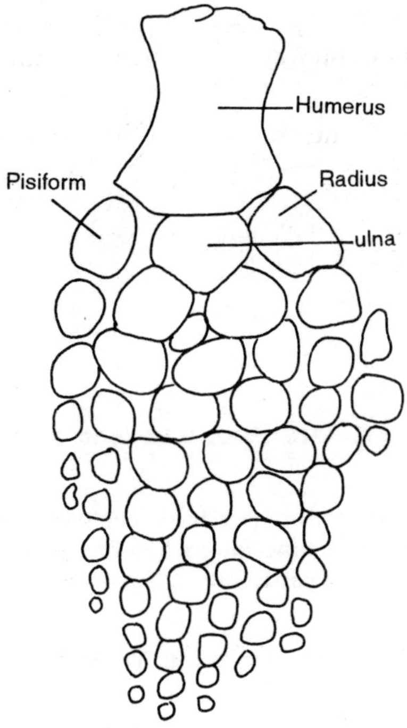

Fig. 11.8: Bone pattern in the limb of a fossil reptile, *Ichthyosaurus*. (After Smith, 1960.)

DISCUSSION

We have made obvious the transformational equivalence of different tetrapod limb forms by concentrating on what we regard as the essential generative principles responsible for the patterns of chondrogenic sites. In so doing, we have ignored what we regard as the secondary influences in the limb field responsible for the systematic difference of bone structure observed across the antero-posterior axis, and the factors responsible for the differential growth of the chondrogenic sites in the proximo-distal axis. That these are considered as secondary, not primary, is one feature that distinguishes our treatment of limb morphogenesis sharply from a model such as that of Wolpert *et al*. (1975), in which the limb elements are derived from primary axial gradients by a process of 'interpretation' and there is no prepattern of elements in a field isomorphic with the manifest limb pattern. A major difference between these views is in the number of parameters involved in generating the pattern and in

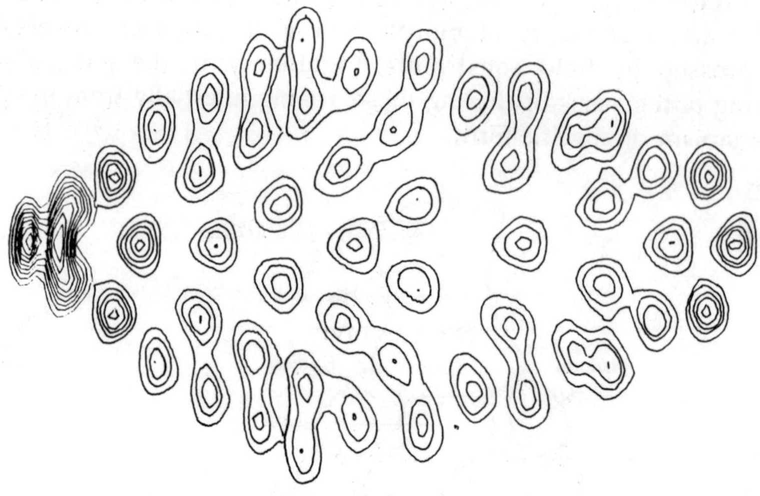

Y:	1.400											
R:	3.83	3.83	5.14	6.38	8.77	9.94	11.09	12.23	11.09	9.94	8.77	6.38
F:	0.80	0.81	0.83	0.86	0.90	0.94	0.97	1.00	1.00	0.95	0.92	0.85
A:	0.00	2.00	2.50	5.50	8.50	12.00	15.50	18.50	26.50	33.50	39.00	45.50
N:	1	1	2	3	5	6	7	8	7	6	5	3

Fig. 11.9: Contour plot of a pattern with higher harmonic solutions (eigensolutions) than that with a periodicity of 5, simulating the pattern of Fig. 8 and showing partial contour fusions which could result in variability of bone pattern. For variables, see Fig. 4 and Appendix.

changing from one pattern to another. Applying Wolpert's chick limb model to the pentadactyl limb, and using the threshold mechanism of Lewis *et al.* (1977), about 24 parameters are needed to generate 5 discrete elements across the antero-posterior axis (a minimum of 4 parameters for each threshold value, of which there are 5, and the morphogen gradient mechanism, involving another 4 parameters or more). Each digit 'gained' or 'lost' involves changes in several parameters, for the threshold points must be properly positioned relative to the morphogen gradient to produce discrete, separate chrondrogenic sites. The proximo-distal pattern requires more threshold points in another monotonic gradient involving some 35 parameters or more. Clearly, genetic programmes of this type are expensive on particulars because there are no generative universal; and since, in such models, parameters are specified by genes, we would expect a large number of mutants affecting the details of limb morphology, for which there is little evidence. A polydactylous mutant with six digits requires either the introduction of a new threshold (four parameters from one gene) or a change in the gradient such that a 'hidden' threshold is revealed, in which case it might be concluded that the 'common ancestor' had six digits.

On the whole, the evidence seems strongly to favour a more robust pattern-forming mechanism involving fewer parameters, whose basic characteristic is the generation of spatially ordered arrays (whole patterns) of repeating elements which are primarily the same but differ by secondary variations superimposed on the primary pattern generator. This takes us back to the prepattern fields of Stern (1954) and Turing (1952), and away from positional information and interpretation, whose specific objective was to replace prepatterns by monotonic gradients and thresholds wherever there is evidence that the units produced are different from one another (difference in this context being called non-equivalence, so that equivalence comes to have the extremely restrictive connotation of identity rather than its more general meaning of similarity under some transformation, which is how we use it in relation to the concept of homology). The model proposed in this paper, involving a field theory which generates a morphogenetic prepattern of a type different from those based upon reaction and diffusion but with some fundamental characteristics in common, is economical on parameters, since the basic pattern varies in response to the proximo-distal 'energy' profile. This profile can be described by curves of three parameters, only one of which needs to be varied to generate the range of patterns shown in Fig. 1, 4, and 6. Genes are assumed to be involved in specifying the details of these curves, but the rest of the pattern-forming mechanism is assumed to be universal to the tetrapods. We must then assume a secondary mechanism (say another three-parameter gradient) across the antero-posterior axis to introduce variations in the intensities of the order parameter, u, leading to asymmetries in the units across this axis.

Tetrapod limbs, as a general category of vertebrate structure, are characterised in such an analysis by a high-level constraint relating to the generative particulars which distinguish limbs from fins and other appendages, presumably involving genes implicated in processes such as limb bud formation and chondrogenesis. However, as well as these particulars, the basis of evolutionary change, there will continue to be universal mechanisms involved which, in our model, are the morphogenetic fields themselves, with solutions characterised as eigensolutions in our formulation. Thus, at every level of a generative analysis, there is a combination of universal and particular factors at work.

The view of biological form advanced here is based upon the general scientific principle that diversity of structure is logically intelligible only if there is some unity or invariance which limits not only observed diversity (which is the sample that history happens to have given us), but possible or potential diversity. In the case of the typical vertebrate form which has come to be known as the pentadactyl limb (somewhat unfortunately, since there is nothing special about five digits), this invariance is to be found in the generative principles which are involved in primary limb morphogenesis. Our particular model involves constraints deriving primarily from the mechanism whereby a continuous 'energy' profile selects a series of eigen-function solutions of a field arising in a distally propagating domain of competence in the developing limb bud. In consequence, limb forms are defined by a series of discrete chondrogenic sites whose antero-posterior arrays differ in number of elements by a small integer (usually one) as one moves distally. If limbs commonly occurred with elements increasing distally in a sequence such as 1, 2, 4, 8, 16, 32, . . . then different generative rules would be involved. Such a sequence describes the number of cells in holoblastic cleavage, which has been shown to be describable by a field treatment similar to that used here for vertebrate limb morphogenesis, but involving different selection (generative) rules on the field solutions (Goodwin & Grainor, 1980). Thus, we see that the general field approach to biological form gives a hierarchy of constraints within a global theory whereby different forms can be distinguished from one another while still united by general principles of spatial organisation as embodied in the field treatment. In this way both the unity and the diversity of biological forms may be grasped, and a generative basis provided for a rational taxonomy, as envisaged by the pre-Darwinian rational morphologists.

Another constraint arising from our field description of limb morphogenesis is very similar to that which Holder (this volume) has deduced in relation to the polar coordinate model of limb formation. His prediction is that limbs will not be found whose primary patterns have external digits larger than internal ones. Anomalies such as the lobster-claw syndrome are excluded because they arise from a secondary process which effectively destroys part of the primary pattern. A similar prediction can be made for the pattern-generating mechanism which we describe in this paper. This is that tetrapod limbs will not be fond with more phalanges in the external digits than in the internal ones. The reason for this is that there cannot be an eigensolution which has blank spaces in the middle of an antero-posterior array. The converse can certainly occur, as illustrated in Fig. 11.1, where the interior three digits each have one more phalanges than the external digits, because of drop in the 'energy' profile distally. With antero-posterior asymmetry, there could be a very long external digit on one side of the limb, as in certain species of whale, but not on both.

The approach to morphogenesis described in this paper has the further consequence that the historical sequence of forms emerging during evolution is logically secondary to an understanding of the generative principles defining the potential set of these forms and their transformations; so that Darwinian taxonomy, which is genealogy, is secondary to rational taxonomy, and both are logically derivative from ontogenetic principles. The external selection principle on historically realised forms which is added by the functional or stability considerations embodied in the concept of fitness, while of obvious importance in many instances, is supplementary to those which operate within organisms in the generative process itself, and certainly can never be used as a causal substitute for the latter. Within the rationalist approach to biological morphology being proposed, the interesting possibility emerges that a stability constraint (such as 'fitness') on the set of possible morphological transformations may embody a general principle of directed or time-dependent transformation of the type which in physics gives such directional stability criteria as the second law of thermodynamics, imposing temporal asymmetry on the time-reversal symmetries of mechanics (Goodwin, 1982). It is evident that a much more highly elaborated theory of form and stability than that currently available is required before this type of theory can emerge, although it remains a part of the research programme which follows from the principles underlying the treatment of biological form and organisation advanced in this paper.

Grateful acknowledgement is made to Charles Trainor for substantial assistance with the computation involved in this work, to John Totafurno for the computer graphics of Fig. 2 and To Malcolm Maden for the photographs of limbs shown in Fig. 11.7. We are also indebted to the National Sciences and Engineering Research Council of Canada for financial support.

REFERENCES

Ede, D. A. (1976). Cell interactions in vertebrate limb development. In *The cell surface in animal embryogenesis and development*, eds. G. Poste and G. L. Nicolson, p. 495–543. Elsevier: Amsterdam.

French, V., Bryant, P. J. & Bryant, S. V. (1976). Pattern regulation in epimorphic fields, *Science* 193, 696–81.

Goodwin, B. C. (1982). Development and evolution. *J. theor, Biol.* 97, 43–55.

Goodwin, B. C. & Trainor, L. E. H. (1980). A field description of the cleavage process in embryogenesis. *J. theor. Biol.* 85, 757–70.

Kent, G. C. (1969). *Comparative anatomy of the vertebrates.* C. V. Mosby Co.: St Louis.

Lawrence, P. A. (1970). Polarity and patterns in the postembryonic development of insects. *Adv. Insect Physiol.* 7, 197–265.

Lewis, J. H., Slack. J. M. W. & Wolpert, L. (1977). Thresholds in development. *J. theor, Biol.* 65, 579–90.

Newman, S. A. & Frisch, ,H. L. (1979). Dynamics of skeletal pattern formation in the developing chick limb. *Science* 205, 662–8.

Romer, A. S. (1956). *The vertebrate body.* W. B. Saunders Co.: Philadelphia.

Smith, H. M. (1956). *Evolution of chordate structure.* Holt, Rinchart and Winston: New York.

Stern, C. (1954). Two or three bristles. *Am. Sci.* 42, 213–21.

Turing, A. M. (1952). The chemical basis of morphogenesis. *Phil. Trans. R. Soc. B* 237, 37–72.

Webster, G. C. & Goodwin, B. C. (1982). The origin of species: a structuralist approach. *J. Soc. biol. Struct.* 5, 15–47.

Wilby, Of. K. & Ede, D. A. (1976). Computer simulation of vertebrate limb development. In *Automata, languages, development*, eds. B. Lindenmeyer and G. Rozenberg, pp. 143–61. North-Holland: Amsterdam.

Wolpert, L., Lewis, J. & Summerbell, D. (1975). Morphogenesis of the vertebrate limb. In *Cell patterning*, Ciba Foundation Sympsoium No. 29, pp. 96–119. Associated Scientific Publishers: Amsterdam.

APPENDIX: A GENERATIVE MODEL FOR THE PENTADACTYL LIMB

We assume that the process of chrondrogenesis is 'staged' by a set of contingent factors which pick out, at each stage, a particular or characteristic solution from the set of all possible solutions of the morphogenetic field equations. The contingent factors (which may be related at the subcellular or cellular levels to metaboic activity, ultra-structural detail, membrane properties, etc.) define at the field level the domain of activity, the values of the field on the domain boundary, and the overall field 'energy'.

In our model, we regard the primary morphogenetic field $u(x,y)$ to be an 'order parameter', whose value at the point (x, y) in the limb bud describes the degree of order there; a high degree of order favours chondrogenesis.

Figure 11.2 in the text gives a three-dimensional perspective of the ordering field; peaks correspond to centre of chondrogenic activity. As described in the text, we concentrate on the proximo-distal aspect of the developing field with a two-fold simplication: first we ignore dorso-ventral distinctions and treat the morphogenetic field as a two-dimensional field in the plane containing the proximo-distal and antero-posterior axes secondly, we ignore antero-posterior asymmetrie and consider only field solutions which are intensity symmetric about the proximodistal axis.

As any particular stage in chondrogenic development we assume that the level of activity is quadratic in both the field amplitude and its gradient, an assumption which has numerous successful analogues in physical theory. We further assume that an optimality principle operates which requires investigation of the field energy (Hamiltonian)

$$E=\int H r dr\theta=\int\frac{1}{2}\left[\left(\frac{\partial\mu}{\partial r}\right)+\frac{1}{r^{3}}\left(\frac{\partial\mu}{\partial\theta}\right)^{2}+\sigma\mu^{2}\right]r dr d\theta, \qquad \text{(AI)}$$

subject to a normalisation of the field intensity

$$\int u^{2}\, r dr d\theta=1 \qquad \text{(A2)}$$

In these equations, σ is analogous to an elastic constant whose value establishes the relative importance of field strength as opposed to field smoothness (gradient). We use polar coordinates with $\theta = 0$ along the proximo-distal axis and $r = 0$ at the proximal centre of activity. The field domain is discussed below.

The equivalent minimisation problem is

$$\delta\int L\, dr d\theta=0,$$

where

$$L=r\left(H-\frac{\alpha u^{2}}{2}\right)$$

and α is a Lagrange parameter. Defining $\beta = \sigma - \alpha$, and using the notation

$$u_{r}=\left(\frac{\partial u}{\partial r}\right),\ u_{\theta}=\left(\frac{\partial u}{\partial\theta}\right),$$

the Euler–Lagrange equation

$$\frac{\partial}{\partial r}\left(\frac{\partial L}{\partial u_r}\right)+\frac{\partial}{\partial \theta}\left(\frac{\partial L}{\partial u_\theta}\right)-\frac{\partial L}{\partial u}=0$$

becomes

$$r^2\frac{\partial^2 u}{\partial r^2}+r\frac{\partial u}{\partial r}+\frac{\partial^2 u}{\partial \theta^2}-\beta r^2 u=0. \tag{A3}$$

Separation of variables then yields the equations

$$\frac{d^2 g}{d\theta^2}+\lambda g=0, \tag{A4}$$

$$\frac{d^2 f}{d\theta^2}+\frac{Idf}{rdr}-\left(\beta+\frac{\lambda}{r^2}\right)f=0, \tag{A5}$$

where λ is a separation constant and

$$u(r,\theta)=f(r)g(\theta). \tag{A6}$$

If we take the domain of a field activity at any stage to be a semi-circular region of radius R with base perpendicular to the proximo-distal axis, as shown in Fig. 11.10, and require $u = 0$ on the boundaries, the following solutions are obtained:

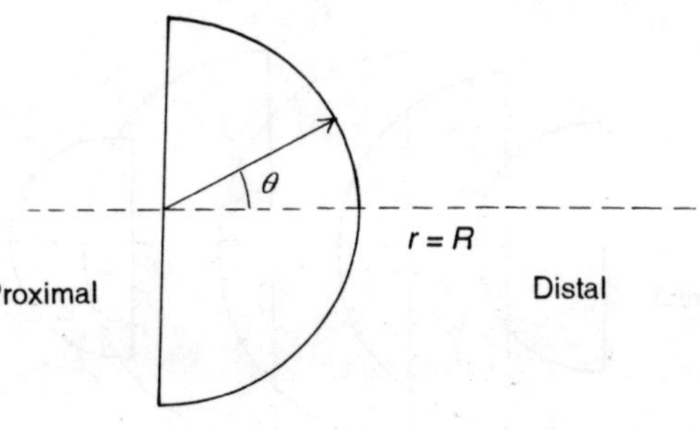

Fig. 11.10: Semi-circular domain of field activity or of competence in the developing limb bud, within which a field solution is set up whose form is determined by the boundary values on the domain and the energy level within it.

$$g=g0\,{}^{\sin m\theta}_{\cos m\theta} \text{ where } m=0,\pm1,\ \pm2,\ldots,$$

and the λ values which are possible are restricted by the eigenvalue condition

$$\lambda=m^2\geq 0.$$

Thus for $m = 0$, a null solution occurs,

for $m = \pm1$, g = cos θ,

for $m = \pm2$, g = ±sin 2θ,

for $m = \pm 3$, $g = \cos 3\theta$

for $m = \pm 4$, $g = \pm \sin 4\theta$.

(We have taken $g_0 = I$ for convenience.)

The radial equation the becomes

$$\frac{d^2 f}{dr^2} + \frac{Idf}{rdr} - \left(\beta + \frac{m^2}{r^2}\right) f = 0, \tag{A7}$$

with solutions indexed by $|m| = 1, 2, 3, \ldots$, according to the possible values of m^2.

Writing $\beta = -k^2$ and defining $\varrho = kr$, this equation becomes Bessel's equation

$$\frac{d^2 f}{d\varrho^2} + \frac{I\, df}{\varrho\, d\varrho} + \left(I - \frac{m^2}{\varrho}\right) f = 0 \tag{A8}$$

with eigensolutions $f \longleftrightarrow J_m(\varrho) = J_m(kr), m = 1, 2, 3, \ldots$ The corresponding $u\ (r,\theta)$ solutions are then (up to a normalisation constant)

$$u_1 = J_1(kr) \cos \theta,$$

$$u_2 = J_2(kr) \sin 2\theta,$$

$$u_3 = J_3(kr) \cos 3\theta,$$

$$u_4 = J_4(kr) \sin 4\theta \text{ etc.}$$

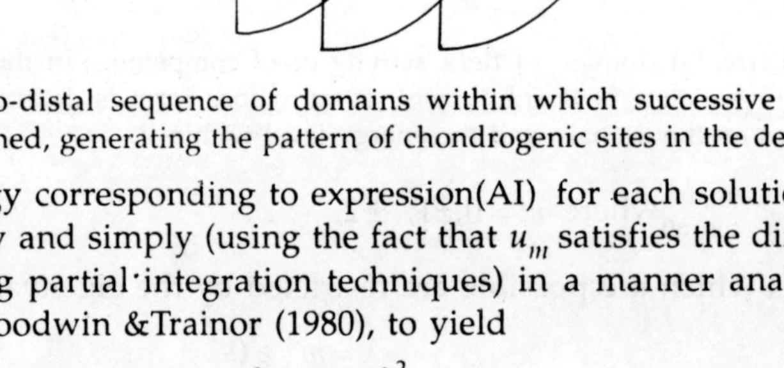

Fig. 11.11: Proximo-distal sequence of domains within which successive field solutions are established, generating the pattern of chondrogenic sites in the developing limb.

The total energy corresponding to expression(AI) for each solution u_m can then be calculated explicitly and simply (using the fact that u_m satisfies the differential equation (A_3) and employing partial integration techniques) in a manner analogous to that in the Appendix of Goodwin &Trainor (1980), to yield

$$E_m = \frac{\sigma - \beta_m}{2} = \frac{\sigma + k_m^2}{2} \tag{A9}$$

The computer simulations used in this paper, and referred to in the main text, assume that successive stages of chondrogenesis take place in hemispherical domains of activity which shift distally and which may expand or contract depending upon contingent conditions (see Fig. 11.11). Although the order field $u(r,\theta)$ obeys a linear equation, the $u(r,\theta)$ solutions at successive stages develops at different times, so that earlier (more proximal) fields have generated fixed or determined presumptive chondrogenic patterns before later (more distal) fields develop. The result is that, rather than amplitudes, it is the intensities, which are proportional to u^2, that are additive.

At each state, which; Bessel solution will be selected depends on the metabolic energy available at that stage, since equation (A9) must be satisfied. The overall pattern obtained depends then on which succession of Bessel solutions is selected and in what size domains (R value), as well as on the domain locations along the proximo-distal axes. For example, Fig. 11.1 is a contour plot and Fig. 11.2 is a simulated landscape of u^2 *intensity,* corresponding to the legend values in Fig. 11.1: the main value (Y) indicates which Bessel function f_m has been selected according to equation (A9) by the (assumed) energy profile given by the solid lines in Fig. 11.3: the R values are the radial positions of the first zeros of the Bessel function while the F values are the reciprocals of the scale factors k_n, in the Bessel function, which are themselves determined according to (A9); finally the A values are the distal shifts in the origin of coordinates for successive stages, corresponding to a distal progression in the stages of early cartilage formation.

The process can be visualised as follows: the basic geometry together with minimisation condition leads to eigensolutions of Bessel's equation. Stable solutions develop only when the metabolising condition satisfy (A9). At each stage of chondrogenesi a particular solution occurs and a pattern is laid down. As the metabolising level falls (or rises), successively higher (lower) solutions of Bessel's equation are selected and more (less) complicated patterns develop. The boundary conditions $u(r, \theta) = 0$, together with energy optimality, select out the first zero of the appropriate Bessel function to coincide with the domain of chondrogenic activity. From this perspective, successive stages of chondrogenic activity are defined by the successive time frames in which the metabolic activity reaches a level which satisfies some particular eigien solution of Bessel's equation.

The model illustrates the importance of organising principles in developmental biology. While genetic factors play an important contingent role, for example in the determination of the energy profiles in Fig. 11.3, the process of patterning reflects properties of the morphogenetic field and is common to early limb development in many diverse species.